CAMBRIDGE LIBRARY COLLECTION

Books of enduring scholarly value

Botany and Horticulture

Until the nineteenth century, the investigation of natural phenomena, plants and animals was considered either the preserve of elite scholars or a pastime for the leisured upper classes. As increasing academic rigour and systematisation was brought to the study of 'natural history', its subdisciplines were adopted into university curricula, and learned societies (such as the Royal Horticultural Society, founded in 1804) were established to support research in these areas. A related development was strong enthusiasm for exotic garden plants, which resulted in plant collecting expeditions to every corner of the globe, sometimes with tragic consequences. This series includes accounts of some of those expeditions, detailed reference works on the flora of different regions, and practical advice for amateur and professional gardeners.

A Selection of the Correspondence of Linnaeus, and Other Naturalists

After the death of the younger Carl Linnaeus in 1783, the entirety of the Linnean collections, including the letters received by the elder Linnaeus from naturalists all over Europe, was purchased by the English botanist James Edward Smith (1759–1828), later co-founder and first president of the Linnean Society of London. In 1821, Smith published this two-volume selection of the letters exchanged by Linnaeus *père et fils* and many of the leading figures in the study of natural history, revealing some of the close ties of shared knowledge and affection that bound the European scientific community at that time. Where necessary, Smith translates the letters into English, with the exception of those written in French, which are presented in the original. The varied correspondents of Linnaeus senior, whose letters appear in Volume 2, include the botanists Johann Dillenius and Bernard de Jussieu, and the philosopher Jean-Jacques Rousseau.

Cambridge University Press has long been a pioneer in the reissuing of out-of-print titles from its own backlist, producing digital reprints of books that are still sought after by scholars and students but could not be reprinted economically using traditional technology. The Cambridge Library Collection extends this activity to a wider range of books which are still of importance to researchers and professionals, either for the source material they contain, or as landmarks in the history of their academic discipline.

Drawing from the world-renowned collections in the Cambridge University Library and other partner libraries, and guided by the advice of experts in each subject area, Cambridge University Press is using state-of-the-art scanning machines in its own Printing House to capture the content of each book selected for inclusion. The files are processed to give a consistently clear, crisp image, and the books finished to the high quality standard for which the Press is recognised around the world. The latest print-on-demand technology ensures that the books will remain available indefinitely, and that orders for single or multiple copies can quickly be supplied.

The Cambridge Library Collection brings back to life books of enduring scholarly value (including out-of-copyright works originally issued by other publishers) across a wide range of disciplines in the humanities and social sciences and in science and technology.

A Selection of the Correspondence of Linnaeus, and Other Naturalists

From the Original Manuscripts

VOLUME 2

EDITED BY
JAMES EDWARD SMITH

CAMBRIDGE
UNIVERSITY PRESS

University Printing House, Cambridge, CB2 8BS, United Kingdom

Published in the United States of America by Cambridge University Press, New York

Cambridge University Press is part of the University of Cambridge.
It furthers the University's mission by disseminating knowledge in the pursuit of education, learning and research at the highest international levels of excellence.

www.cambridge.org
Information on this title: www.cambridge.org/9781108069717

This edition first published 1821
This digitally printed version 2014

ISBN 978-1-108-06971-7 Paperback

A SELECTION OF THE CORRESPONDENCE OF LINNÆUS, AND OTHER NATURALISTS,

FROM THE

Original Manuscripts.

By Sir JAMES EDWARD SMITH,
M.D. F.R.S. &c. &c.
PRESIDENT OF THE LINNÆAN SOCIETY.

IN TWO VOLUMES.

VOL. II.

London:
PRINTED FOR LONGMAN, HURST, REES, ORME, AND BROWN,
PATERNOSTER ROW.

1821.

LONDON: Printed by JOHN NICHOLS and SON, 25, Parliament Street.

CONTENTS OF VOL. II.

Biographical Memoir

OF

DANIEL CHARLES SOLANDER,

LL. D. F. R. S.;

AND

His Letters,

FOLLOWED BY THOSE OF

THE REV. DR. STEPH. HALES AND OTHERS,

TO MR. ELLIS.

DANIEL CHARLES SOLANDER, LL. D. F. R. S. Under Librarian of the British Museum, was a native of Sweden, and a favourite pupil of Linnæus, as appears by several of the letters in our first volume (see pp. 134, 136, &c.) We learn also, from the same source, that he came to England in 1759, being consigned by his great preceptor, with peculiar earnestness, to the care of Mr. Ellis. He was universally esteemed here, for his polite and agreeable manners, as well as his great knowledge in most departments of Natural History. Being engaged by the illustrious Banks to accompany him in his voyage round the world, with Capt. Cook, he was ever after the companion and friend of that distinguished patron of science, and was domesticated under his roof, as his secretary and librarian. To Dr. Solander was allotted the technical description of all the acquisitions of that voyage, especially of the plants destined to appear in a magnificent work, the plates of which have long been engraved, but the manuscripts of Solander remain unpublished in the Banksian Library. The intended work made no progress after his death. The fears of Linnæus respecting the fruits of this celebrated voyage, as expressed in his letter to Ellis (see vol. I. p. 267—270), have proved almost prophetic; not indeed from the undertaking of another expedition, as was once proposed, but from the interruption caused by other avocations, the dissipation of London society, to which so agreeable a companion was always acceptable, and the indolence induced by a sedentary and luxurious life, suddenly terminated by a stroke of apoplexy in 1782, at the age of 46. Indications of these habits appear in Dr. Solander's growing neglect of epistolary correspondence, of which Linnæus complains in some of his last letters to Ellis; and which was, even much earlier, experienced by Solander's mother (see vol. I. p. 222); several of whose letters to her son were found *unopened* after his death!

It ought nevertheless to be remembered, that if the talents and liberality of this eminent man were not so directly useful, in the way which might have been expected, any more than those of his great patron Sir Joseph Banks, whose loss, never perhaps to be compensated, we have now to deplore, they have, like his, been otherwise pre-eminently beneficial. They have proved the example and the spur of all that has been done for natural science, during half a century, in Britain; perhaps in every quarter of the world. It was Solander who reduced our garden plants to order, and laid the foundation of the *Hortus Kewensis* of his friend Aiton. His instructions made every body correct and systematic, and introduced Linnæan learning and precision, in spite of opposition, which sometimes assumed considerable authority (see vol. I. pp. 35, 36). No one ever came so near his great teacher in the specific discrimination of plants. In generic distinctions he was prone rather to combine than to separate; in which also he followed his master's example. Natural orders or affinities seem never to have entered into his contemplation. In nomenclature and terminology he was always classical and correct, never yielding to the fashions or corruptions of the day. Conchology eminently engaged his attention, and he laboured successfully, with Ellis, in that study; as also in the investigation of the more obscure tribes of marine vegetables as well as animals.— Dr. Solander soon became familiar with the English language. The first and second letters only, of the following collection, have been found to require any material correction.

A letter from Mr. Stanesby Alchorne to Mr. Ellis is here prefixed, as illustrative of what follows. This gentleman was Assay-master in the Mint, and a good English Botanist. He died in 1799.

A few miscellaneous letters, following those of Dr. Solander and the Rev. Dr. Hales, close the Correspondence of Mr. Ellis in this volume.

Correspondence.

MR. ALCHORNE TO MR. ELLIS.

DEAR SIR, Tower, Nov. 27, 1761.

Dr. Solander tells me you have desired some account of Hill's new work; but not having seen it himself, the Doctor has asked me to send you my opinion of it; which I shall do with pleasure, having first begged that you will not place too great dependence on what I may say; for I have had but a cursory view of it, and what I am now to write is merely from memory.

I think Dr. Hill's introduction does as good as tell us, that the best recommendation he can give of the book is, that himself had very little hand in it; and he intimates that it was executed under the care, and at the charge, of a certain Nobleman *. But however such noble person might contribute towards defraying the expence of it, I would not willingly have so mean an opinion of his judgment as to suppose he could compare the plates with nature and approve them. My present circumstances will not permit me to enter so critically into this subject as I could wish; but I must tell you, in general, that the author, after finding great fault with every vegetable system extant, promises *hereafter* to give a much better one than has yet appeared.

* The Earl of Bute.

Meantime he gives a sort of system (under the name of an *artificial index)* dividing herbs into forty-three classes, by which he pretends any plant may be easily known. And in this order he proposes to figure and describe every known vegetable *(not from single specimens, but from whole beds of them, where their characters may be justly ascertained)*; and this first book contains, I think, above 80 plates, comprehending only the *Flores compositi radiati*, about one fourth of Linnæus's *Syngenesia*; so God knows when the work can be complete. I am afraid the present specimen will gain him no great applause or encouragement, for which I am very sorry. Really 'tis pity that so pompous a work, which, properly executed, might have been extremely useful, should, for want of care, be thus rendered of no value. Whoever had the drawing of the plates I know not; but certainly they are more like patterns for ladies to work by, than figures to inform a botanist. Some of the flowers indeed are indifferently well done; but the leaves are mostly so contorted as to be quite unintelligible. I remember two figures, among others, which I showed a friend of mine conversant in this study, and he agreed with me 'twas impossible to know by them the plants whose names they bore; viz. *Jacobæa folio non laciniato, & Verbesina folio tripartito diviso*; both of Ray's *Synopsis*. In short, whatever pains may be taken to recommend this work, I think, Sir, you would have no difficulty to fix upon the author of it at first sight, though his name were not in the title-page; for really it is worthy only of the great Dr. Hill.

I am, Sir, your most obedient, humble servant,

S. Alchorne.

DANIEL CHARLES SOLANDER, LL. D. TO MR. ELLIS.

[English. Orthography corrected]

Sir, London, Nov. 29, 1761.

By Mr. Alchorne's letter you may see, Sir, that I have been a good deal mistaken about Dr. Hill's new work. It contains not, as I thought, all the compound and aggregate flowers, but only the radiated flowers. I never met with any person who had it, or could show it me; therefore I desired Mr. Alchorne to give me his idea, because he has had an opportunity of looking it over; and he has been so obliging as to give me this letter to send you.

This summer, in August, they have discovered a new marble in Gottland, in Sweden. I have seen two tables of it, sent as a present to a merchant here in London. It is of a very odd colour, almost like tallow cut into small square pieces, and mixed with light-green soap. In some places there are spots of dark-red jasper, and dark-green porphyry. Besides this, they have lately discovered a black marble in the same island, said to be very fine and heavy, almost as heavy as the black one we had before; but I have not seen a specimen.

Just now, Sunday, betwixt 11 and 12 o'clock in the forenoon, it snows, so that when I look out of my windows I can easily imagine that I am in the Northern part of Sweden, for I feel it pretty cold too.

Dr. Ribe and Mr. Varney present their humble respects.

I am always, with the utmost regard and friendship, Sir, your most obedient, humble servant,

DANIEL C. SOLANDER.

FROM THE SAME TO THE SAME.

[Slightly corrected.]

DEAR SIR, London, March 5, 1762.

Last night I was at the Royal Society. It was a long meeting, but very few things of consequence. One Rev. Dr. Foster had sent two letters; in one he will prove, against Mr. Collinson, that swallows really, during winter, immerse themselves in water. He says he has observed them to assemble in large flocks in autumn, fly very high, quite out of sight, and then come down again, sit a while upon reeds or willows, and then plunge into the water. It was into one of his own ponds; but he forgot to search for them in the water afterwards; and this autumn they came not to that place. In the second letter he says it is the *Forficula auricularia* of Linnæus (Earwig), which makes the labyrinth-like furrows under the bark, upon old trees, because he has found several of these insects in such holes. He likewise mentions having found there a kind of white maggots, which he took for *larvæ* of these *Forficulæ*. But that he is mistaken in these observations is very certain. The white maggots are the artificers of these labyrinths, and are the caterpillars either of *Dermestes typographus*, or *D. micographus* of Linnæus. The *Forficulæ* have only gone into such holes for shelter, as they do into all the cavities they meet with. In the first letter likewise mention is made of frogs in winter, during a hard frost, being found frozen, apparently dead, being hard and brittle like flint (glass?), so that they break with a blow. But if taken into a warm room, they come to life again. I never heard this observation before, and wonder from whence he has taken it. I have desired Mr. Collinson to write to him, and enquire.

There was also read an account of some antiquities found in the Isle of Wight, by Dr. Stukeley; and an account of the transit of Venus. An improvement in digesters was shewn to the Society; and a couple of new members proposed. This is all that came under consideration.

The cold weather still continues, and is very disagreeable. I never thought it could be so cold in England; and I am surprised to see, by the newspapers, how many persons have perished by the inclemency of the late storm of snow.

Mr. White tells me he intends to publish his figures in a large work, of which he has hopes that the King will be at the expence. He is now in the country, but will return in six weeks, when he means to write to Dr. Linnæus, about some alterations he wishes to have made in the *Systema Naturæ*. Mr. White will publish his figures in a systematical form; and he wishes his work should agree with that of Dr. Linnæus. But I am afraid the latter will, on some points, not agree with him in opinion, especially as to classification. For instance: Mr. White means to reduce the *Cete* to fishes, or else to make a new class on purpose for them. He proposes, moreover, to refer swallows to the *Accipitres* (because, he says, their gape is like a hawk, and they catch insects as hawks do other birds or animals, which, in my opinion, is not a sufficient character); or else he would make a new order for them, not enduring to see them ranged with sparrows. There are many other things, relating to system, in which I think Dr. Linnæus cannot coincide with him.

I have the pleasure always to be, &c.

MY DEAR SIR, Plymouth, Aug. 25, 1768, 2 o'clock.

It is to you, Sir, I chiefly owe my favourable reception in this country. It was you that first introduced me to such of your friends as have afterwards made England so agreeable to me. I assure you it is not without reluctance I now leave it, notwithstanding it is only for a time, and in the best of company, and in hope of gratifying my most favourite desires. I should be void of all gratitude if I did not always recollect your friendship and goodness.

Pray present my best compliments and warmest good wishes to Miss Ellis *; I hope at my return to be able to make some additions to her collection. She shall have something from every place, to put her in mind of her father's friend, and what a straggler he has been.

When you see Dr. Fothergill give him my respects, and tell him that we here in Plymouth met with a friend of his, Mr. Cookworthy, as worthy a man as can be, full of knowledge, and very communicative: we are very much obliged to him for his civilities.

We have been detained longer than we wished by continually blowing westerly winds; but just now the wind is turned fair, and we have got the signal for repairing on board.

I am, and shall always be, dear Sir, your humble servant and sincere friend, DAN. CH. SOLANDER.

P. S. Mr. Banks gives his compliments. Give mine to Mrs. Butler and all friends.

I left a paper with Mr. Fabricius that I had promised Messieurs Davis and Reymers; I hope they have got it. I was at last so hurried that I had not time myself to translate and to send it, but he promised to do it.

If you see any of the Museum people, tell them I wish them all well.

* Afterwards Mrs. Watt.

Endeavour, off the Bay of Funchal, Sept. 18, 1768.

My dear Sir,

We just now, at going out of Madeira, met with an opportunity of sending a letter, but must write in a hurry, as the ship we are to send it with is under sail. We have seen a great many fine *Mollusca;* got drawings and descriptions of them; but as no ship this year is expected to sail from Madeira to England, we resolved to send what we have by Lisbon.

Pray be so good as to send the inclosed to Mr. Lindegren, with my compliments, and beg him to forward it. You shall have a large packet by way of Lisbon.

Mr. Banks desires his compliments to you; mine to all friends, particularly Mr. and Mrs. Webb, Miss Wilson *, all at the Museum, &c. &c. &c.

I am, for ever, your most truly and sincere friend,

Dan. Solander.

New Burlington Street, Oct 8, 1771.

Dr. Solander presents his and Mr. Banks's compliments, with many thanks for the loan of Plumier's Plants.

Dr. Solander has sent inclosed some of the *Gumbo* or Sassafras Leaves, as they are called in Florida, where they are used in many different ways, chiefly with fish and rice. The ordinary way that people of fashion use them is in beef-broth, with onions,

* I have scrawled three or four lines to them.

spinach, or calliloo, and Cayenne pepper; one tea-spoonful to a pint. It must stand 10 or 12 hours in cold water; then pour it backward and forward, for a minute, into different vessels; after this, mix it intimately with a pint of the liquor, and pour the whole into the pot, where it must not boil above a quarter of an hour.

DEAR SIR, London, Oct. 9, 1773.

Many thanks for the last. At my return to town I met with Mr. Aiton, who told me that he had procured Mr. Taylor ripe fruit of the Coffee; and that if he had known you yourself had wanted any, he could have procured you some from Gunnersbury, where they now have pulled all the ripe fruit; but he thinks that there are still some unripe ones on the tree: if they come to perfection, he will send you them.

The French have not been so successful in their visits to the New Zealanders as we were. Monsieur Marion was sent out from Old France, a little after our return, to visit New Holland and New Zealand. In January 1772 he visited the southern parts of New Holland, but did not like the inhabitants; he therefore proceeded to New Zealand, where, after he had passed through the Channel between Cape Maria, Van Diemen, and the Island of Three Kings, he at last anchored in a bay which we call Motuaro, or Bay of Islands, within six miles of our anchoring-place. There, I suppose, he ventured on shore, without previous alliance, and the consequence was, that he himself and 25 of his men were killed, and in all probability afforded the inhabitants a good meal. This discouraged the rest so much, that they made the best of their way to Manilla, from whence this intelligence was sent by one of the surviving officers.

My best compliments to the ladies, and to Mr. and Mrs. Jackson, Mr. Deponthieu, &c.

I am, for ever, dear Sir, your most obedient servant and sincere friend, DAN. SOLANDER.

DEAR SIR, London, Friday night, July 22, 1774.

Nothing but the arrival of an Otaheite friend could have made me so forgetful, in regard to you and many more of my English friends. But as I am one of the three that he can converse with, I have been obliged to give him almost all my time, in hopes that my other friends will forgive me. However, I have waited upon Dr. Fothergill, to let him know that I would, according to your desire, deliver back to him all the Corals which are already figured; but he chose rather that they should remain in your chambers till his return. He sets out next Monday or Tuesday. I have delivered to Mr. Fabricius the walnut twig.

Now I will give you a short account of the voyage performed by Capt. Furneaux in the Adventure, who left England in the summer of 1772, with

Capt. Cook in the Resolution. They kept company to the Cape of Good Hope; and in the beginning of December left that place, steering south, in search of Cape Circumcision, which they did not find: they therefore proceeded as far south as they could go; but in latitude 67° 10′ south, the ice prevented their proceeding further. They afterwards steered a little northward, and so east, shaping their course towards New Zealand, where Charlotte Sound was to be their rendezvous. Soon after they had left the ice, near the supposed place of Cape Circumcision, the two ships by accident parted company. Capt. Cook all along steering a more southerly course, chiefly in latitude 60°, he at last anchored in Dusky Harbour, on the southernmost part of New Zealand. Capt. Furneaux, keeping two or three degrees more to the northward, at last made the south part of New Holland, where he anchored in Bay Frederick-Henry. He staid there but a few days, and then proceeded to the place of rendezvous in Charlotte Sound, where he arrived two months before Capt. Cook came up to him from the southern district of New Zealand. From thence the two ships kept company to Otaheite, where they were remarkably well and friendly received. As their stay in New Zealand was longer than first proposed, they could stay but a short time in Otaheite, not above eight or ten days. A most cruel war had caused a great scarcity, and many of our friends were killed in a battle, where several thousands were slain by the people of Little Otaheite, who, for the sake of securing the title of king to their chief, had ventured a second time to attack the people of the Great Otaheite, and both times proved victorious. The two ships afterwards visited Huaheine, where *Omai* the Indian, who is now residing in Mr. Banks's house, came on board Capt. Furneaux's ship. He is a native of Ulaietea, and was at Otaheite when we were there. Oridi, a native of Bola Bola, at the same time embarked in Capt. Cook's ship. After a week's stay at Huaheine, they went to Ulaietea, where they staid five or six days; and afterwards, in their way to the south, fell in with the islands of Middleburg and Amsterdam, two very delightful spots, well described by Tasman. They staid but three days in those islands, and then proceeded again towards Charlotte Sound in New Zealand, to take in wood and water. The Resolution, which is a better sailing ship than the Adventure, got into the harbour on the 29th of November last, when the latter was blown off the coast, and obliged to run to the northward, and anchor in Tolaga, where the inhabitants rejoiced to see the friends of Tubaya and all of us. After a short stay there, Capt. Furneaux made the best of his way towards Charlotte Sound, but did not get in till the latter end of December, when he learnt, from a letter left in a bottle, that Capt. Cook had left the place four days before his arrival. The inhabitants seemed now, as well as before, well pleased with their guests; but still a very dismal catastrophe made him repent his coming thither. One day he had sent out a boat, with an acting-lieutenant (Mr. Rowe), a midshipman, the ship's steward, and seven common men, to gather wild celery, in Grass Cove, a bay which we frequented every other day in our month's stay in the harbour. This place was not six miles from the anchoring-place; and as the boat did not return the same evening, the Captain, the following morning, sent another boat to look out for them, when they soon learnt the fate of their comrades. They had been killed, cut up into proper joints, roasted, and partly eaten; several baskets were found full of roasted pieces, some still hot, &c. &c. Capt. Furneaux, not having got any orders or rendezvous appointed by Capt. Cook, resolved to do as much as possible towards executing the tenor of the general orders, which was, to look out for land in the southern seas. He therefore sailed directly south from New Zealand, till he came into latitude 55°; and afterwards, between that and 60°, continued his course eastward, chiefly in sight of ice, looking out for St. Sebastian's Land, and again for Cape Circumcision; but arrived, the 18th of March last, at the Cape of Good Hope, without having seen an inch of new land. From thence he arrived in England the end of last week.

Obs. Notwithstanding he has not discovered any new lands, he has still made a glorious voyage; he has sailed round the globe, in a higher southern latitude than any ship before. He has proved that there is no large southern continent, and that the French pretended discoveries are small islands, instead of continents; or perhaps, as my friend Omai calls ice, things that the sun drives away, or causes to vanish. — I have been so long writing this letter, that Omai is almost asleep. He will come out to Hampstead, and visit you, as soon as he has had the small pox. To-morrow we are to go out with him to Baron Dimsdale, who is appointed to perform the operation of inoculation; which I with all my heart wish may be attended with success. He is a well-behaved, intelligent man.

I am, for ever, your most obedient servant and sincere friend, DAN. SOLANDER.

DEAR SIR, New Burlington Street, Aug. 10, 1774.

I am very much obliged to you for your last letter and the bottle from Greg, which I received just as I was setting out for Hertford to attend my friend Omai during his confinement for the small pox. I left him yesterday, when he was declared to be out of all danger. The small pox was come out, and seemed to be of a mild sort. I promised him to go down again next Friday, so that my short stay in town will not permit me to visit you till my return next week, which will be on Tuesday.

Capt. Furneaux has brought home a few seeds, which we divided among Aiton, Gordon, Lee, Forsyth, and Malcolm. A few have also been given to gentlemen. Mr. Topham Beauclerk has a greater number from Mr. Bayley than the Captain had. I hope he will take good care of them, as his seeds seemed to be better preserved than the Captain's.

I am, with great regard, my dear Sir, your most humble servant and faithful friend,

DAN. SOLANDER.

MY DEAR SIR, London, Oct. 13, 1774.

I have read through Dr. Garden's Account of the Electrical Eel, and think it a paper well worth publishing; I have also shewn it to Sir John Pringle, who is of the same opinion. If the weather is not very bad, I purpose to wait upon you on Sunday next, and will then bring back this letter, and talk over what you else have wrote about.

I am, for ever, my dear Sir, your most sincere friend, DAN. SOLANDER.

MY DEAR SIR, London, Thursday, Oct. 27, 1774.

Do not imagine that any thing shall alter my principles in regard to the friendship I owe to you as one of my first and best friends; you shall always find me a true friend and well-wisher. I am sorry if you should ever find reason to think otherwise of me. The reason why I so seldom can dispose of myself in the forenoon, is my attendance at the Museum; so I have only Saturday and Sunday forenoons to myself, and then often such things call me away that cannot be foreseen or suffer any delay. That was the case last Saturday, when I was obliged to go into the City to get from the Custom-house a bottle-case with animals from the Mosquito Shore, preserved in spirits. Mr. Deponthieu was at Mr. Banks's the same day, when I shewed him a Chocolate fruit from thence, dried, of a different variety from what is cultivated in the West India Islands. It is spotted, and therefore by the Spaniards called the Tiger Chocolate. It is said to yield the best Chocolate; and is now introduced on the Mosquito Shore.

Be assured I shall not omit to arrange and write the proper names on the Corals at your chambers. You shall soon find it done.

Mr. Banks and Omai returned from the country yesterday; but Omai goes down with Lord Sandwich again on Saturday next.

I have re-examined the Chocolate flowers, and will give you a copy of my description.

Mr. Aiton has sent me two plants from the seeds Mr. Bruce brought from Abyssinia; both are new, one a species of the *Ajuga*, the other a *Coreopsis*.

I am, with the utmost sincerity, my dear Sir, your humble servant and faithful friend, DAN. SOLANDER.

MY DEAR SIR, London, Nov. 7, 1774.

To-day, when I leave the Museum, I shall go into the City, and make a further enquiry after the box and bottle which Mr. Hodgson has sent you in the Sarah and Elizabeth, Capt. Foote. What Mr.

Banks had from the Mosquito Shore came by a Capt. Miller.

The man who shewed the Electrical Eels in Carolina, as described by Dr. Garden, is now arrived; but, unluckily, all the five eels died during the voyage, or at least before he came up to London. One was alive when he landed at Falmouth, where several persons felt the electrical shocks; but that one died on the passage from thence to London. However, he has benefited by Dr. Garden's advice, to put them into spirits in case they should die. He has brought us four complete specimens, well preserved; for which we propose to raise by subscription, or some other method, a sum of money, to enable the man to go out again. Mr. John Hunter danced a jig when he saw them, they are so complete and well preserved.

This week I certainly will settle and mark your Corals; it vexes me very much I have not been able to do it long ago.

We shall have a drawing made of one of the Electrical Eels; and John Hunter has promised an anatomical description, to accompany Dr. Garden's account, when presented to the Royal Society.

My best compliments to your friends about you, particularly Mr. Scott.

I am, with great regard, my dear Sir, your sincere friend and humble servant, DAN. SOLANDER.

My dear Sir, Dec. 21, 1774.

I hope you have received safe the coloured prints I sent you the day before yesterday.

I have taken an exact copy of the drawing I now return you; so from that I can at any time make out what may be necessary for the plate of the *Theobroma.* Whenever I see the drawing, which is to be engraved, I shall accordingly shape and proportion the parts of fructification. Do not you think that the inclosed are too large?

Mr. Banks has bought Miller's *Herbarium,* and we have been busy these two weeks in getting it home and into some order. As there are a great many of Houstoun's plants from Vera Cruz, &c. I think it a valuable acquisition.

Your most obedient servant and sincere friend,

Dan. Solander.

My dear Sir, London, June 29, 1775.

I am in hopes of being able to wait upon you on Saturday next, and then to give you a full account, both of your paper on the Gorgonias, and also of your Corals in Gray's Inn.

Capt. Cook arrived, the 22d of March last, at the Cape of Good Hope, without having lost a single man by sickness. He has not met with any continent, but found several islands not before seen by Europeans; some above 80 leagues long: most of them near the Tropic. The largest he has called New Caledonia; it is situated a little to the eastward of New Britain, in 18 to 20 degrees of south latitude. He has never been to the northward of the Æquator. He has been in 71 degrees 10 minutes south latitude, in 106° 30′ west longitude; of course much further south than any one before him. He has seen a prodigious quantity of ice, and some ice mountains, whose tops were covered with clouds.

In the South Sea he saw no land in the high latitudes; but in coming from Terra del Fuego to the Cape of Good Hope he fell in with one island in 54°, and another in 59°; both covered with snow and ice.

The Resolution is now expected every day.

Mr. Forster, in a letter to Mr. Barrington, says he has discovered 260 new plants and 200 new animals.

All the Indians they have visited have behaved well; even when he returned to Charlotte Sound, where Capt. Furneaux lost some men, they received Capt. Cook well.

I am ever, &c. my dear Sir, your faithful friend,

DAN. SOLANDER.

MY DEAR SIR, Friday morn, July 21, 1775.

I am afraid it will not be in my power to wait upon you on Sunday next, as I am under an engagement to go down as soon as the express arrives from Capt. Cook, who is now daily expected, and may arrive by Sunday. However, I will soon wait upon you. I have received Mr. Irving's note; and am, for ever, with the greatest regard, my dear Sir, your sincere friend, DAN. SOLANDER.

London, Aug. 28, 1775.

Dr. Solander presents his compliments to Mr. Ellis, and sends him a map with the track of the two last ships; and also a copy of the Jalappa plate. It is exactly copied from the original drawing. If Mr. Ellis wants any more of them, they are at his service.

Dr. Solander was last Saturday at Kew, where he saw Mr. Masson, who is lately come back from the Cape of Good Hope, with a great cargo of new plants, all in perfect health. He has been 800 miles inland to the north of the Cape, and brought from thence a glorious collection.

DEAR SIR, London, May 4, 1776.

The Bread Fruit of the South Sea islands within the tropics, which was by us during several months daily eaten as a substitute for bread, was universally esteemed as palatable and as nourishing as bread itself. No one of the whole ship's company complained when served with Bread Fruit in lieu of biscuit; and from the health and strength of whole nations, whose principal food it is, I do not scruple to call it one of the most useful vegetables in the world. Throughout a great part of the East Indies the same kind is found to grow wild; and I do not doubt that the *Socca*, taken notice of in the eastern

7. Note from Dr. Solander to Mr. Ellis.

Dr. Solander presents his compliments to Mr. Ellis, and will do himself the pleasure of waiting upon him to dinner on Saturday next.

Br. Mus.
Aug. 3. 1775.

8. Revd. Dr. Hales to Mr. Ellis.

Dukes Court March 2. 1752

Dear Sir

I carryed the Collection of red Sea Mosses, which you sent me, to the Excellent Princess, for which she was thankfull & well pleased. And this day showed her the method of spreading them, which she soon practiced her self & liked very much.

9. Earl of Northington to Mr. Ellis.

I thank You for Your Trouble & if You are at Leisure to come down I belive You may be of Use.

I am Yr most Obt
& Hum: Servt
Northington

10. Dillenius to Linnæus, March 30th 1740.

Viro clarissimo & doctissimo,
Carolo Linnæo S. P. D.
Dillenius.

Literæ Tuæ 6. Aug. anno superiori scriptæ, rite ad me perlatæ sunt ex quibus Te & valere & bene agere intellexi & gavisus sum.

R. Martin Lithog. Rochester St. Westr.

The material originally positioned here is too large for reproduction in this reissue. A PDF can be downloaded from the web address given on page iv of this book, by clicking on 'Resources Available'.

part of the East Indies, is in quality equal to that of the South Sea islands. As it undoubtedly must be of the utmost consequence to bring so valuable a fruit to countries where the climate is favourable to a production which cannot bear cold, I think it would be necessary to encourage every body who goes to any part of the world where it is to be met with, to bring it over, either by young plants properly rooted, or by seeds collected in the proper season, and sown during the passage. I am sure no expence ought to be spared, in an undertaking so interesting to the publick.

I am, with great respect, dear Sir, your most obedient servant, DAN. SOLANDER.

THE REV. DR. STEPHEN HALES *, F. R. S. TO MR. ELLIS.

From Mrs. Batcheler's, in Duke's Court, Jan. 27, 1752.

DEAR SIR,

Suspecting that your fine Sea-moss Landscapes would suffer by not being covered with glass in a

* The celebrated vegetable physiologist, one of the eight foreign members of the French *Academie des Sciences*, and clerk of the closet to the late Princess Dowager of Wales, who held Dr. Hales in the highest esteem, and erected a monument to his memory in Westminster Abbey soon after his death, which happened at Teddington, Middlesex, Jan. 4, 1761, in his 84th year. His experiments and enquiries, relative to the theory of vegetation, are the basis of all our subsequent information. The following letters evince his ardour in the prosecution of all kinds of useful knowledge, to the latest period of his life.

frame, I proposed to have them put into a frame with glass, as your's are; which her Royal Highness readily approved of, and which she will pay for. I desire therefore, when you send them here, to let me know the price, that I may acquaint her with it.

As there were not aldermen enough yesterday to do any business, I must be at Guild-hall to-morrow at eleven, and purpose calling on you in my way thither.

I am, Sir, your obliged, humble servant,

STEPHEN HALES.

SIR, Duke's Court, Feb. 3, 1752.

I thank you for sending me the beautiful Sea-moss Landscapes, which I conveyed to the Princess in your name, for which she was very thankful and much pleased, as were the Prince of Wales, Prince Edward, and the Princesses. She desires the favour of you to procure her some varieties of Sea Mosses. I believe Harwich will be one good place, if you know any one there. The Princess will pay any expence it may occasion. I will write to Mr. Pullein at Dublin, to desire him to send me some, and to my nephews in Kent, to desire them to get some on the Dover, Whitstable, and Hearne shore. I conclude, if I direct them to wash the mosses in fresh water, and lay them thin to dry, it will suffice to preserve them well. I shall write also to Dr. Whytt at Edinburgh, to get me some moss. Thus you see a king's or princess's word runneth very swiftly, as Solomon observes *.

The inclosed is for Mr. Pullein, which I have sent open for you to add any thing, or to inclose a letter. Please also to direct it. I am, &c.

DEAR SIR, Duke's Court, March 2, 1752.

I carried the collection of Red Sea-mosses, which you sent me, to the excellent Princess, for which she was thankful and well pleased. And this day I showed her the method of spreading them, which she soon practised herself, and liked very much; my nephew, Sir T. Hales, having got me a few mosses from Dover. But pray let me know how you glue them on with gum arabick. I guess you wet the paper with it, after the mosses are laid in their places dry.

As the Princess designs to put several ladies on this agreeable amusement, it will be well to furnish her with plenty of these mosses. I have desired Dr.

* From these beginnings, recorded by Mr. Ellis in the Introduction to his Natural History of Corallines, and his Dedication of that work to the Princess, the study of these beautiful and curious productions gradually advanced; till this ingenious observer, in the first place, established the animal nature of corals and corallines, and then, in conjunction with Dr. Solander, laid the foundation of that accurate knowledge of marine botany in England, which has finally produced the excellent publications of a Dillwyn and a Turner.

Salter of Yarmouth to direct a parcel of them to you.

I have lately heard from the worthy Mr. Pullein; an answer to which I have here inclosed, that you may inclose a letter to him if you please.

I shall go to Deptford in a few days, to consult about preserving ships by plentiful ventilation, *viz.* at the rate of 28,000 tuns of air an hour into the closed hold of a 20 gun ship; and twice that quantity in larger ships. I could convey 100,000 tuns an hour into these, if needful, but I believe half that quantity will do. I am, &c.

Dear Sir, Teddington, Aug. 15, 1752.

I found the Sea-moss picture, with the *Fucus*, at Kew, which happened very well. I gave it to the Princess, and showed her the manner, which you taught me, to expand the Fucuses, which she was pleased with. This may probably put her upon making some pictures.

On Monday last three Physicians and I, &c. went to the new Small Pox Hospital, where we consulted, and concluded how to draw, with a careful hand, the foul air out of the rooms of the sick. — Since I saw you I have received an account of our success on board the man of war at Deptford. On Tuesday last, when the wind served for 12 hours, to blow at the rate of 18,000 tuns of air in an hour into the closed ship, the force of the air was found so great as to drive the flame of a candle into all the seams of the ship. And four equal-sized linen cloths, which had each four ounces of moisture in them, without dripping, being hung, one at each end of the ship, one in the middle, and the other in the well, they were all very dry, except that in the well, which being hung down near the water, had one fifth of an ounce of moisture in it, *viz.* one tenth part of the whole. So here is full proof that the timber of ships, thus ventilated, will be preserved from decaying much the longer.

As for the parcel from Capt. Ellis to Lord Halifax, as he is at Horton, near Newport Pagnel, Buckinghamshire, where he is like to continue, it will be well for you to write to him, to know where he would have it sent, which I believe will be to Horton.

SIR, Teddington, Sept. 16, 1752.

According to Mr. Pullein's desire, I am getting what mulberry-seeds I can here. He desired me also to apply to you for some. He says that one Perkins in East Smithfield has a large plantation of them, and often squeezes the berries. He says their pulp need not be washed, but only dried in the sun.

SIR, Jan. 7, 1754.

This *Fox-tail,* about four feet long, and in quantity sufficient to fill a three-inch bore, was taken out of the new-laid elm-pipes from Marth-gate to Kew-garden. It is said the like has been found in the old leaden pipe from Richmond park to West Sheen *.

As I am likely to be in London till Monday the 21st, if Capt. Ellis comes, I shall be glad to see him. Saturdays at five, and Mondays at ten, I am for an hour at Leicester-house. You may send a penny post letter to fix a time. I dine at my lodgings at one. I would gladly come to you, but a gravelly disorder prevents me †.

Sir, Teddington, May 29, 1754.

Having just read the Bishop of Clogher's second part of his Vindication of the Histories of the Old and New Testament, which he sent me, with one for the Princess, I find there is something much to the purpose of your coralline affair, page 129. The book is just reprinted here from the Dublin edition, and will be published in a few days, by Baldwin and Cowper in Paternoster-row.

I hope the worthy Captain is well. Pray my respects to him.

Sir, Teddington, June 19, 1754.

Being last week at the Duchess of Somerset's, near Colnbrook, I went into the hot pine-apple

* Perhaps this might be a *Rhizomorpha*, akin to the *R. medullaris*, described in Tr. of the Linn. Soc. v. 12. 272. t. 20.

† The writer was now in his 77th year.

house, where I found the upper air hot, and disagreeable to breathe, as you had observed, but the air near the floor not disagreeable. Upon which I proposed, not only the burning of a candle at the same time both above and below, but also to set, at the same time, pots of divers plants, as mint, &c. both above and below, to see the different effects of those airs on the plants. Here is a fine scene opened to estimate the degrees of unsalutariness to plants, of hot-houses and green-houses, which will lead to use means to refresh them, *viz.* such as I mention in the last leaf of Vegetable Staticks.

Mr. Charles Stanhope, Capt. Ellis's hearty friend, told me lately that the Captain was going another Guinea voyage. If so, and if he is in London, I wish I could see him here before he goes, any day but Sundays, Wednesdays, and Fridays, when I am at Kew; because I want to talk with him about preserving water sweet, both with stone lime and native mineral sulphur, which are very wholesome, and which I am now trying here.

I suppose you know that Mr. Pullein has got a Living, though much disappointed, both in the situation and value of it.

My compliments to the Captain.

DEAR SIR, Kew, Aug. 22, 1755.

I think, with you, that it would be very right in Mr. Pullein to send you a copy of his treatise on Silk-worms, as it would be a probable means to get it published *; and I am persuaded that the book will be very useful, in giving rational hints for the farther and farther improving the method of breeding Silk-worms.

It is a very useful discovery you have made, of the certainty of there being powder blue in refined sugar, which, as you observe, must needs be extremely noxious to us. I had never heard of it before; but as, since the receipt of yours, I have acquainted the Princess and many more with it, so several of them say they have frequently observed the blue sediment at the bottom of tea-cups, but knew not what it was. A gentlewoman of Teddington tells me, that a great sugar-baker said to her, that if some of the ingredients in refined sugar were known, it would not be used; and Dr. Linden, to whom I took the liberty of giving a recommendation to you, says, that the blue smalt powder, used in fine sugar, is highly pernicious, in destroying our smaller vessels, it acting like arsenick; that he has had patients, the mass of whose blood has received a sufficient quantity of it to make them sick; his method of curing whom he told me.

I think therefore that, for the benefit of mankind, it ought to be published in the news-papers; which I conclude you will choose to do without your name, that every one may be satisfied of the truth of it by their own observation; smalt being a composition of

* An "Account of a particular Species of Cocoon, or Silk-pod, from America," apparently by this gentleman, is to be found in the Phil. Trans. vol. LI. p. 54.

cobalt, flint, and potash, melted into a blue glass, and finely powdered, as described in the Philosophical Transactions.

I congratulate you on discovering so fine a blue*. Perhaps it may be more salutary in sugar; or, if no safe blue can be found, the sugar will be good, though not so white as with smalt.

Dr. Linden gave me the following receipt for a blue; *viz.* indigo two drams; lime-water a quart; adding, first, 25 drops of spirit of vitriol.

The patent blue inclines too much to purple; as does also, as you suspected, the *Cyanus* by candle-light.

I have read Mr. Stephens's Account of making Potash. I think, with you, that it might have been easily known, without so great a præmium. I will look into the Transaction which you mention, but have not yet had time to do it.

I have not yet had an opportunity to try to cure the taste of turnip, &c. in milk.

I am sorry to hear of Mrs. Ellis's illness; pray my respects to her. My good parishioners and her acquaintances, Mrs. Dorrells, were very sorry to hear it.

Dear Sir, Teddington, Aug. 29, 1755.

Yesterday Dr. Lewis of Kingston, a skilful physician, called on me; and on my acquainting him

* Probably that made from the *Centaurea Cyanus,* or Corn Blue-bottle, mentioned in the sequel of this letter.

with your important observation, of pounded glass in the finest sugar, he was surprized at it, and said they deserve to be severely punished for putting it in. He says he will get a loaf of the finest sugar, and dissolve it all, in order to discover it. and will let me know the event. Let us defer the publishing it, till we have an account of Dr. Lewis's examination.

Saturday last, in the morning, I took four quarts of milk, just milked from a cow which had fed for 84 hours on cabbage-leaves only, and had drank very little water. It had an ill taste. I put the milk into a leaden vessel, eight inches diameter, and thirty inches deep, which stood in a vessel of very scalding-hot water. I blew through a tin air-box, full of very small holes, showers of air up through the milk, for ten minutes only, which cured the ill taste; and, after standing 24 hours in a broad pan, there was a thick scum, which was half cream and half butter, free from any ill taste. So here is a method to make a greater quantity than usual, of good butter, from ill-tasted milk. But the froth of the milk being so great, by reason of the great heat of the milk, as to make it flow over the vessel at 30 inches height, if it had not been kept down, by constantly lading and breaking the froth; I therefore repeated the like operation with the evening milk of the same cow, but giving it only a heat that I could bear my fingers in for some little time. In this degree of heat the milk did not froth high; but after 45 minutes ventilation, though much mended, yet was not so completely cured as the former milk. Hence we see how necessary heat is to volatilize the rancid oil (which gives the ill taste) to such a degree as to cause it to fly off by ventilation.

My next trials therefore will be to give milk medium degrees of warmth. But this I cannot do till the autumnal leaves fall, and till turnips are more rank; for at present they will not give an ill taste, as I find on trial.

Your fine sky-blue seems to have corroded the paper, the whole length, in the middle.

Please to wipe the blue bottom of a tea-cup with paper, and, when dried, fold it up, and send it to me, to Kew, or by Mr. Goodchild.

Dear Sir, Teddington, Oct. 22, 1755.

I thank you for the copper cuts which you sent me. They finely illustrate the thing, and were well worth giving in the figure. As the book which you gave the Princess is in London, I purposely deferred giving her the cut till she came to reside in town, which will be the 1st of November, when I will fix it in her book.

I think your discovery of powder-blue in the finest sugar is of great importance, and ought to be published in the news-papers, without any one's name, for the benefit of mankind; for finely-powdered glass, which is often given to poison rats, must be hurtful to us.

The coarsest of the powder-blue precipitates to the bottom of the sugar-pot in making the sugar-loaf, and is the reason that the tops of these loaves, being very blue, are broken off. I am told that lemon-juice will make the blue precipitate, but I never tried it. I am obliged to Mrs. Ellis for the specimen of it which she preserved for me on paper. I hope she is well recovered. Mr. Goodchild, jun. tells me that much business has prevented your coming here as you intended.

I believe the method of making potash is sufficiently explained by Dr. Mitchell, &c. There seems to be no great difficulty in preparing it. I have read Mr. Stephens's book, and am of your opinion as to his merit of the præmium.

I am persuaded that much of the expence in evaporating great quantities of lye will be saved, by blowing showers of air up through the lye, from several pipes, full of small holes, laid at the bottom of the evaporating pans. The latter evaporation, to dryness, is to be finished in other pans, without air-pipes.

I am waiting for further trials on ill-tasted milk, from autumnal leaves and turnips, which are not yet rank enough, by reason of rainy seasons. When I can give an account of the event of these trials, I think to lay it before the Royal Society, with the account of distilling sea-water, and then to publish it in a small pamphlet of between 30 and 40 pages, hoping it will, in several respects, be beneficial to the world.

As Mr. Cecil, of Merton Abbey, Surrey, pays me a fee-farm rent, due from the Abbey, I have taken the liberty to desire him to pay it to you, if it suits him better than to send it me here, which I believe he will do. But if he should happen to pay it you, you may, when opportunity offers, pay it to Mr. Goodchild. The sum is £.14. 2*s*. for a year's rent, due at Michaelmas last.

I shall be at my old lodgings in Duke's Court, as usual, from Saturdays to Monday noons, where I shall be glad to see you when you come that way. I long to see or hear from Capt. Ellis. My way of distillation will be of great use in slave-ships.

Teddington — no date — probably written in 1757.

DEAR SIR,

I received the drafts of the Carolina plant *, and am obliged to you for the honour done me therein. I sent one of them immediately to the worthy † Lord Bute; but, as I am named in it, I cannot well convey one to the Princess, but may probably take an opportunity to give one to one of my young pupils.

I wish you could contrive, by means of the Præmium Society, to get the ingenious Mr. Pullein's

* *Halesia tetraptera*; see Ellis's first letter to Linnæus, vol. I. p. 82.

† Such an epithet, from such a man, may defeat a host of slander.

book on Silk-worms printed. Many useful hints will probably thence be taken, from time to time, to improve farther and farther in that affair.

As you are the great promoter of vegetable researches, I must communicate to you the following proposal. Talking the other day with some intelligent persons, on improvements in husbandry, it brought to my remembrance what a Mr. Scott of Montrose, a great improver of lands, told me twelve years since; *viz.* that in wet seasons, when they could not dry their clover hay as it ought to be, they made the ricks of layers of straw and clover; which, he said, not only preserved the clover from spoiling, but also made the straw much better fodder, it being much impregnated with the sweat of the clover. Soon after I had related this, it occurred to my thoughts, that hence a hint might be taken, to try whether, when any hay, of grass, clover, or saintfoin, &c. is about half dried, by evaporation of its more watery part, then to make ricks of it, with layers of straw, which would imbibe the latter richer sweat of the hay, and thereby improve the straw for fodder. By this means it is probable, that a great deal of good nutriment will be saved, which, in the common way of making hay dry, evaporates away in the air. But this should be tried first in little. Pray acquaint Mr. Pullein with this, for him to carry to Ireland. You or he may direct to me at Leicester-house during the summer.

I purpose to try to soften hard water by ventilation. I am, &c.

P.S. Since my sealing of this, I made the following experiments:

1st. My hard pump water.

2d. Some of the same, ventilated 15 minutes.

3d. Some of the same, ventilated 30 minutes.

4th. Long-stagnant rain-water.

Equal pieces of hard soap were put into each, and worked with the hand. That in the rain-water did not curdle; but the other three curdled equally.

As the agitation of ventilation may probably dispose the hardening particles to precipitate, so the experiment ought to be repeated, in order to see whether the ventilated water will grow soft the sooner, by standing some time, compared with the unventilated.

DEAR SIR, Teddington, June 25, 1757.

I thank you for giving me an opportunity to see your Vindication of my friend Monsieur Mazeas, which I have sent to Monsieur Duhamel, with a pressing letter, to take the same care of the English prisoners in France as we do of theirs here, to their great benefit, as the commissioners of our sick and wounded seamen do assure me. The inclosed I had sealed before I received your last letter. I send your Vindication, with my letter, by the help of my neighbour Mr. Hatton, consul at Ostend; so it goes so far post free.

DEAR SIR, Duke's Court, Feb. 25, 1758.

I ordered Mr. Manby, bookseller in the Old Bailey, to send you eight of my second volume on Ventilators, &c. which will probably be published in 14 days; *viz.* for yourself; Mr. Pullein; Capt. Ellis in Georgia; Dr. Garden in Carolina; Dr. Brownrigge at Whitehaven; Rev. Dr. Henry in Kildare-street, Dublin; Dr. Rutty in Dublin; Mr. Lindsay at Lazar's-hill, Dublin.

I hope you will excuse this trouble. As I have seen nor heard nothing from Mr. Pullein, I suppose he is gone to Ireland. Dr. Rutty sent me a present of his large quarto on Mineral Waters. Mr. Lindsay, a new correspondent, wrote to me lately about preserving corn, in several large adjoining buildings, proper for public granaries, which are so situated that they can all be ventilated by one set of ventilators worked by a horse.

The suspending onions in the fumes of brimstone did not spoil their vegetating.

DEAR SIR, Teddington, Nov. 21, 1760.

In answer to yours, which I have just received, I am sorry to hear of the worthy Governor Ellis's ill state of health. Dr. Kirkpatrick, who practised physick many years in South Carolina, told me, that few lived to be more than 50 years of age, and that their children, born there, did not live so long.

As to your ingenious thought, to cool houses by subterraneous pipes, I believe the pipes must be laid in the declivity above the house, because the heavier cool column of air will descend. But let it be tried first with a few yards of pipe. And will it not be cooler, as well as cheaper and more lasting, instead of wooden pipes, to have a channel made, with three or four courses of bricks, without mortar, laid on each other, and covered with bricks? But let things be tried in little first. Ventilators would do well, were it not for the expence of making and working them; for changing the air, though hot, is refreshing.

I have sent 1000 of my book on Ventilators to all our colonies in America, purposely to rouse the nations, not to poison themselves with strong drams, but to make them weak, to the standard of Nature's cordial, wine. And I have sent a parcel of those books to Governor Ellis, to whom pray my best respects. I fear that climate will not agree with him.

I like your different contrivances for bringing seeds from India. As small ventilators are now put into all our transport ships, and into some India ships, and probably will be put into all, so it will be the better for your seeds to have a fresher, cooler air.

The Princess will build a hot green-house, 120 feet long, next spring, at Kew, with a view to have exotics of the hottest climates, in which my pipes, to convey incessantly pure warm air, will probably be very serviceable. And as there will be several partitions in the green-house, I have proposed to have the glass of one of the rooms covered with shutters in winter, to keep the cold out, which will make a perpetual spring and summer, with an incessant succession of pure warm air. What a scene is here opened for improvements in green-house vegetation!

Having been ill lately, though, I thank God, well recovered, I shall not venture to come to London this winter, for fear of exposing myself to the ill consequences of cold to me, who am 81.

Wishing you success in all your laudable pursuits,

I am, &c.

P. S. I this morning received a letter from one Mr. Marsham, who lives at Stratton, near Norwich, a stranger to me, who has been making many observations on the growth of the stems of trees in bulk, and of their lengths, for two years past. He offers to send me his observations, which I shall desire him to do, and I may probably lay them before the Royal Society.

DEAR SIR, Teddington, Jan. 4, 1759.

I have been prevented answering yours of the 23d of December by illness, &c. The progressive motion of the air, through the wooden pipe in the ground, would be descending from the house to the cave, because the long cooler column, of the air in the pipe, would be the heavier. But if the cave was above the house, then it would, for the same reason, descend into the house. — I am sorry to find the heat of the air in the cellar (in Georgia) so great as 81 degrees, and that it disagrees with the health of the worthy Governor, to whom pray my respects. Mr. Martin told me, t'other day, that he has leave to go to another climate, for some time, for the recovery of his health.

The different weight of the air from heat or cold, a merely hydrostatical consideration, whence its specific gravity is varied, I do not remember any author who treats particularly of. There are many instances of the air's being specifically lighter or heavier, according to its different degrees of warmth. Were it not for this property of the air, the lower warmer foul air of the lungs could not be changed, and then we could not live a day.

I thank you for the curious plate of Barnacles. That in the middle looks, *primâ facie*, like a formidable *Hydra*. You have laid open a surprizing scene of Nature.

I thank you also for laying before the Royal Society Mr. Marsham's observations on the growth of trees.

I received very lately a letter from the worthy Dr. Henry, who is now in Kildare-street, Dublin; in which he says he has not received my book on Ventilators; which is, as I guess, that he has been seven months at Arney, near Strabane. I have therefore desired him to enquire of Dr. Rutty, or Mr. Lindsay of Lazar-hill, to whom you sent the book, of whom they received it, which will lead to your agent there.

MR. HOGARTH* TO MR. ELLIS.

DEAR SIR, Chiswick, Nov. 28, 1757.

Being out of town, I did not come by your agreeable present till yesterday, for which I return you my sincere thanks. It must be allowed you print is accurately executed, and very satisfactory too. As for your pretty little seed-cups or vases, they are a sweet confirmation of the pleasure Nature seems to take in superadding an elegance of form to most of her works, wherever you find them. How poor and bungling are all the imitations of art! When I have the pleasure of seeing you next, we will sit down, nay kneel down if you will, and admire these things. I shall be in town in two or three days for good, and will take the first opportunity of waiting on you. In the mean time I am, Sir, your most obliged, humble servant, WM. HOGARTH.

THE DUCHESS OF PORTLAND† TO MR. ELLIS.

SIR, Bath, Oct. 22, 1758.

I received the favour of your letter, with the print of the different kinds of Barnacles, which I think

* The great Moral Painter; whose temper of mind, as displayed in this short letter, may advantageously be contrasted with that of his too celebrated enemies, Wilkes and Churchill. "*Ye shall know them by their fruits.*"

† Margaret Cavendishe Harley, heiress of the Harley and Holles families, married in 1734 to the second Duke of Portland, and long celebrated as the munificent and intelligent patroness of natural history, especially of conchology. Her Grace died in 1785, and her fine collection was afterwards sold by public auction.

very curious, and for which I am very much obliged. Is the fleshy Barnacle with ears discarded out of the Royal Society, and what is become of it?

I am very much obliged to you for the direction to Dr. Pontoppidan. I suppose the letter should be wrote in French, and the direction likewise; and should be glad to know when any ships sail to Norway. I must trouble you with my service and thanks to Mr. Romilly. There are many things in the shell and coral tribe on the coast of Norway, I should imagine, must be very curious.

I am, Sir, your humble servant,

M. Cavendishe Portland.

Sir, Bullstrode, Aug. 21, 1769.

I return you many thanks for the curious shell you are so obliging to lay by for me; and Mr. Lightfoot informs me you have likewise put by some of the blossoms of the Tea Tree, which will likewise be very acceptable. I am very glad the venison was agreeable to you. I hope you enjoy your health.

I am, Sir, your obliged, humble servant,

M. Cavendishe Portland.

I hope you will have made great acquisitions by next winter.

SIR JOHN HILL* TO MR. ELLIS.

SIR, No date — written perhaps in 1761 — see p. 5.

I am very glad to find Mr. Osborn misinformed me, as I never doubted but he did. It gives me particular satisfaction that the plan and design of the undertaking please so good a judge. Linnæus's method has pleased by its novelty; but it is false in the principles, and erroneous in his conduct of it. His discoveries have scarce done more service, than his method hurt, to the science.

There is one expression in your letter so singular that I must desire it may be explained. It is, you are sorry you cannot be acquainted with me. There is something mysterious in such an expression; and you will not wonder it gives me concern. You will easily guess I cannot want acquaintance; but I should not think myself worthy to live, if any thing relating to myself could occasion a person of understanding and integrity to use that form of excuse.

I am, Sir, your very humble servant,

JOHN HILL.

* A man of great pretensions, but of little credit, in botany and physic. He was patronized by the Earl of Bute, and their ideas of botanical arrangement were nearly the same. Sir John Hill published a voluminous Vegetable System, with extremely bad plates, which is probably the "undertaking" here mentioned, and of which we have a character in Mr. Alchorne's letter, p. 5 of this volume. It is not unlikely that Hill might be the anonymous critic, quoted by Collinson, see vol. I. p 36; at least that criticism is worthy of the writer of the above letter. We do not find that Mr. Ellis wrote any answer, or concerned himself with any further disquisition, whether or not the illustrious writer were "worthy to live."

JOHN FORD, ESQ. TO MR. ELLIS.

DEAR UNCLE, Leghorn, July 29, 1763.

Your kind favour dated the 24th of October did not reach my hands till May. I should have answered it then, had I not had hopes of shortly gaining some information in the sea-weed way. To this intent I have been lately near the village of *Santa Margherita,* where the coral fishermen live. But was told the proper season is not till October, when the fishermen retire from the coast of Sardinia for the winter. In the interim I have here met with a branch of a coralline substance, or rather a stony sea-plant, on which the coral grows. I do not pretend to offer this as serviceable in the way of study, being persuaded you have seen many of the kind, but only as an ornament in shell-work, or grotto-work. I beg, for my further information, you will tell me sincerely whether it is worth the money I bestowed on it, the value of a crown English. If you should like such curiosities, or thin branches of Red Coral, I shall do my endeavour to furnish you; but pray limit me in point of price. If I should procure you some sea-weeds or corallines, in which, I must own, I have very little judgment, pray let me know how they should be packed and sent. Without my paying the freight, no captain will take them.

I thank you for your kind offer of the book on Corals, but that is quite out of my way. I assure you I have enough to employ me in studying physick, especially as I have some hopes of practising now and then. I was lately called on to assist a Genoese nobleman, deprived of the use of his limbs by several paralytic fits; and I had the satisfaction not only to see, but also to hear every body say, he was much better in ten days time. I attended a child in a violent fever and swelled belly; and though the father would not believe me at first, he afterwards found my advice to be wholesome. There is a probability of my attending several other obstinate disorders on my return to Genoa, provided the physicians do not oppose me. I endeavour at noble patients, both for their protection and to obtain a certificate of my having performed such cures. The greater the authority, the more credit.

I have often thought that you, who are so great a connoisseur in the productions of the sea, might hit on some secret in the dyeing way. If I be not mistaken, antient books tell us that the *Ostro*, or purple dye, was produced by some shell-fish. This would be worth your inquiry. If that shell-fish were only to be met with in the Mediterranean, I should endeavour to seek it out. An experiment lately made by accident shows me that sea-water has a good effect in dyeing. On washing some red handkerchiefs in hot fresh water, they lost their colour; but by dipping some of the same piece into boiling sea water, and afterwards rincing them in fresh water, their colour was admirably preserved. This may serve for a hint.

Pray give my kind remembrances to all my relations and friends, and believe me, dear uncle, your dutiful nephew, JOHN FORD.

FROM THE SAME TO MRS. ELLIS.

DEAR AUNT, Leghorn, Aug. 12, 1768.

I should readily have complied with your request when at Dieppe, had the town afforded any thing to fill a letter; but I hope you will be satisfied when I tell you it is the chief fishing town in France. Its form is almost square, with handsome streets. The houses are lofty, and make a good appearance, though in fact they are but so many regular hog-sties, bomb proof; and, what is very different from the rest of France, the women there value themselves much for their modest and virtuous education. The harbour is worth seeing, for, besides being naturally a fine bason, well locked, and of a triangular figure, it is vastly improved by art, with strong stone walls, making a canal, which goes to meet the sea.

From hence I took my journey, by the stage, to Paris. We made up seven in number, besides two or three cages of birds; but by good luck we met next day at Rouen with much more agreeable company, two very polite jemmy gentlemen and a trooper, three parrots, two very polite married ladies, a surgeon's daughter, with an old fisherwoman and her daughter. The old woman happened to be my bolster. Indeed, to look at her, she seemed not much less than an Irish heifer; but she assured us she did not weigh above 189 great pounds, though she knew several who weighed 600. In short, from her odd shape, or no shape, her light ness, her silence, her colour, and her long beard, I am apt to conclude her little better than a huge overgrown greasy sponge, moved by polypusses; and if she had but a few cockle-shells stuck up and down to her jagged parts, she might have made an excellent rarity for my uncle's museum. I had almost forgot to tell you, that in going from the inn to the stage, at four in the morning, having near a mile to go, in a most violent rain, and to cross a flooded channel, I and two fat old women got into a cart. There being a hole, my right foot slipped in, up to the hip, and was dragged in the water, while my left remained in the cart. With great difficulty and good luck I recovered myself.

My stay at Rouen was hardly sufficient to see that city, which is very large and irregular, mostly of wood, though the shops are very neat, and richly furnished. It is situated by the side of the Seine, where ships of near 200 tons come up, and has a noble bridge of boats, chained together, paved, and so compact, that it looks like one of stone. The town lies in a hollow, surrounded by hills of easy ascent. The coach for Lyons was so stuffed one could hardly breathe, so I went most of the way upon the top. The prospect was infinitely more agreeable, and the heat not so incommodious. Indeed it is worth while thus to take a view of the finest paved road in Europe, equal to the new-paved streets of London, making almost a straight line for upwards of 60 miles, and continuing, with some little variation, for near 200.

Passing the antient city of Sens, as I entered Burgundy I was delighted with one of the neatest landscapes in France. The town is large, and full of the remains of Roman grandeur. Just out of the east gate is a noble spring, winding about, and supplying many pleasant gardens. To the south extends a beautiful plain, watered by a meandering river, and bordered with smiling hills, of a gentle ascent, admirably disposed. The whole is so industriously cultivated, though without hedge or ditch, or any inclosure, that it looks like one continued farm of wheat, barley, oats, and rye, with shady walnut-trees interspersed here and there, and spots of meadow-ground, where the cows graze with the most peculiar regularity I ever saw. Eight or ten girls hold each a cow by the horns, with a short rope, and lead her round to graze, in a kind of dance, which put me in mind of our Italian balls, with their horned partners; and there is a checquer of untouched meadow left between each company of cows and girls.

After Sens, we meet with Auxerre, a city famous for its antiquity, and still more so for a fountain larger than that at Holywell in Wales. Nothing remarkable occurred till we came to Lyons, where I lodged "*à la rue des enfans qui p——*," and from whence I set out, two days after, for the Alps, in a chaise all alone, though I met with a good deal of company on the road; among the rest, with Mr. Murray, a Scotch gentleman. Here the scene begins to change, from the well-cultivated plains and hillocks of France, to a most romantic country. France and Savoy are divided by a little river, which forms a semicircle, all the inside of which is fine level arable land, belonging to France; the other shore, which is stony and uneven, being in Savoy. On each side of the river is a large village, belonging to the different kings; and on the centre of the bridge stands a pillar, with each prince's arms. After passing this bridge, you wind backwards and forwards, for above an hour, till you come to a most awful sight, a ridge of perpendicular impassable rocks, without the least appearance of entrance or passage. After being driven for some time close under them, you perceive the appearance of a broken cave, and, shortly after, an amazing work of art satisfies your impatience. This is a way cut from top to bottom, through the solid stone, which, on each side, is as high as a steeple. It was begun by the antient Romans, and finished by the Dukes of Savoy. This avenue reaching about a quarter of a mile, you ascend, gradually, to an eminence covered with dwarf hazel-bushes, loaded with nuts. Here a descent begins, with mountains on each side, from the summit of one of which a small river forms a beautiful cascade, wetting us with its mist as we pass by. The country here is rocky, full of walnut-trees, and so singularly romantic, that it seems inhabited by Robinson Crusoe's companions, till we come to Chamberri, the capital of Savoy, where we supped, and had a plate of excellent Savoyard biscuits. We next day dined at the once celebrated fortress of Montmelian, to which our house adjoined, and from whence we had a full view of the village, river, bridge, and a fine straight road beyond it, our prospect terminating in such a beautiful combination of hills, that they appeared all one. At the bottom were hillocks covered with vineyards, backed by high mountains, full of large and shady trees. Behind these were still loftier mountains, clothed with low wood and herbage, and the hoary Alps overtopped all, with their heads half naked, and their summits covered with snow. As I advanced through seven or eight villages, I was lost in a scene of rural delight, amid shady groves, murmuring rills, and verdant meads. I began to be tired of my happiness, for want of an agreeable companion. Yet I met with images enough, as I mused by myself, to call my friends to my remembrance. When I saw great tufts of Sweet-briar growing wild, I recollected how many squabbles I had about stealing it in my sister L's garden. When I plucked the Flowering Willow by the cold streams, I remembered how fond Mrs. M. was of that plant. The majestic Gentian, on the craggy precipices, reminded me of my mother's tea-table cordial; and the sweet Mezereon, sheltered from the sun in the hollows of rocks, brought to my recollection our English or Irish damsels, peeping out of their calashes. But, for fear I should only remember particulars, I was so lucky as to meet with potatoes, scattered up and down in little spots, as we sow onions. You need not doubt that I then thought of you all; and when I met with plenty of Raspberries at the sides of the mountains, I did not quite forget dear self. It is somewhat odd to find the most beautiful plants and flowers sequestered in the most craggy and solitary places, and even the prettiest girls are only to be met with in the middle of the Alps, at a village called Aiguebelle, situated in a deep bottom, and so inclosed with mountains, that it looks as if Nature, jealous of these beauties, had shut them up in a serraglio, for the solace of us travellers, who, Don Quixote like, take pleasure in difficulties for their sakes.

As we jogged on, admiring the great variety of natural beauties, we reached a village called Lanslebourg, at the foot of the great mount Cenis. Here we were accosted by numbers of men, as strong as lions, though as tame as spaniels. Having supped there, with plenty of fresh butter, cheese, and a good roast shoulder of mutton, &c. &c. we set out next morning early, each on a mule, in a grand cavalcade. Winding about for three hours, as the ascent is not rapid, we got almost to the top, having all the way on our right hand a grove of Larch, an ordinary sort of Pine, and a very elegant kind, straight as an arrow, and lofty, with the branches extended in tassels, like the male Cypress; of which I procured some cones, but the seeds were shed *. A noble river, full of trout, washes the

* Probably Spruce Firs, *Pinus Abies*, so picturesque and venerable in these mountain-forests.

skirts of this mountain; but we lost sight of it very soon in the ascent. After having apparently reached the top, we still went on, several miles, over hills and meadows, mounting gradually, but unaccompanied by woods, till at length we got a sight of the great lake, near a mile in circumference, with fishing-boats on it, a convent and two or three small houses near it, as likewise a small lake, or pond. Here we halted, to quit our mules, and take each a rush chair, fixed on long poles, for the descent, which is six miles. We stayed above an hour in the inn, where was plenty of the finest fresh butter, two sorts of excellent cheese, indifferent wine and bread, with fine trouts of the lake. As I sat at table, there stood near me a most curious churn, made of the trunk of a larch tree bored, the top being sunk into it, like a cork, with a hole in the middle, where the staff worked up and down as in ours; but so nicely finished, so neat and clean, as to merit a place in an English lady's cabinet.

From hence one of the most beautiful and uncommon scenes in nature attends us almost to the foot of the mountain. A small river, issuing from the great lake, directs its course into a flowery plain, locked in, almost on all sides, by a chain of very steep and craggy rocks, where, softly gliding over the level surface, its course is scarcely discernible but by the bridge. After traversing this gigantic inclosure, we are conducted by the hoarse noise of waters to a defile, bordered by trees almost regularly ascending on either side, and occupied in the middle by huge stones, through which the river forces itself, with a bellowing noise, in broken cascades. Then hurrying through a village, about half way it meets with more obstructions, spurns against the rocks, and breaks into clamorous waterfalls.

Turning our backs on natural beauties, we now come to the amazing conveniences of art. At a place where nothing but water or goats could pass, we are presented with a surprising zig-zag causeway, like artificial stairs, fixed to the side of an impassable mountain, from whence we have several broken views of the above enchanting river, in various cascades, appearing at a distance like silver embroidery on green velvet. And though it takes a quite different course from the stupendous work of art I have just described, yet we no sooner arrive at the bottom than it meets us again. After rumbling over the stones in a romantic uneven plain, still traversing backwards and forwards, gathering strength from various additional rills, the stream makes a grand effort to render its exit equal to the dignity of its origin, and, collecting all its borrowed strength, throws itself with rapidity into the verdant valleys of Turin, where the thirsty plains divide its refreshing waters between them, and the hay-making nymphs sing its obsequies.

* * * * * * *

FROM THE SAME TO THE SAME.

DEAR AUNT, Dublin, Dec. 25, 1770.

I cannot avoid giving you this trouble, to open my mind to you in my present precarious and uncommon situation. Seldom if ever has there been an instance of a man's being married without his previous consent, either by word or writing, or any intercourse of love, or even a certainty who the object is. Yet circumstances and indications combine to assure me, that all the essential parts of matrimony have been already performed, and that there now remains only an empty ceremony of confirmation. I am even afraid I shall lose the bride cake. I must say my mother has managed wonderfully in this affair; but I cannot say wonderfully well, till I see the lady. Cousin D., the parson, was with her yesterday, and they had a deal of private discourse. In the evening I drank tea with her. She expressed great obligations to him for helping her in a certain affair lately. She afterwards gave me an oblique hint that she was advised by the attorney to prosecute M. for damages; but she looks on the generality of attorney's advice to be very interested, and frequently on a sandy foundation; as a warning to me against any thing I might attempt, if I found myself disposed to resent. So you see I am tied up hand and toe in wedlock, like the parson's coat.

> "Where shall poor needy parson seek for aid,
> When dust and rain at once his coat invade?
> His only coat, &c."

You must know, we soldiers in the wars of

marriage, under our famous Amazonian director and commander, observe the strictest discipline, and most precise regulations, in our progress; having begun with the latter end of the prodigal son's life, who after meeting sharp sufferings, at last tastes of favour, on which account he has the better relish for it. So a veteran soldier, having experienced a famine, thinks himself very happy when a good dish of Irish potatoes falls to his lot. Our next march was into the plain of social duties, where we have performed, and do continue to perform, all sorts of evolutions. We have, at the same time, a constant eye to the great chest, or camp intrenchment, on which are placed an old lease of Mr. Leathley's intermarriage, an old Epithalamium, a love-letter of uncle Roger's at the age of 62, and a copy of verses on the efficacy of being compelled to grow good. All these move forwards and backwards, according as the appearance of our marriage is more or less imminent, or happens to be deferred, by some unforeseen accident. These, with an innumerable company of proverbs, led by corporal trials, all march on to certain victory, like the French to the siege of Dunkirk. But, above all things, I did not expect to be hauled over the coals for a witch; yet such is my luck. She has got some items of it, and I must be one, right or wrong. She frequently points out Manasses to me, and nothing less than his confession, his prayer, and his brass chains, can ever reconcile me to heaven. She has enquired very seriously of me, what region they (witches) chiefly inhabit; if they be regular bred, and learn it by grammar; to all which I must answer in the affirmative, or else I am not believed. You must know we ransack Scripture for incidents of marriage. Nothing so pretty as Jacob's courtship, because he stumbled upon one of his own kidney; except Tobias's taking leave of his father-in-law, who cautioned him to use his daughter well, and of special trust. We are not yet come to the marriage of Cana in Galilee, for we are to have no wine, as that hurts the head.

I must suppose you to be apprized that I am literally between hawk and buzzard; for one of the ladies in question squints, and the other is remarkably quick-sighted. I accidentally met our neighbour the squinter, about duskish, at my mother's, who seemed to stand on tiptoe, and asked me, very cavalierly, how the air of Ireland agreed with me, and seemed to relish my mother's proverb, " when the cart comes to the horse, they must be yoked." I kissed her, and put on her clogs. So you may think that as I took possession of both extremities, the rest must be mine of course. But I rather look upon it in the mercantile style; like a merchant, who having two samples of goods shown him, handles one and takes the other. I should be curious to know the agreeable Miss N's thoughts on this dark affair. They may easily guess mine. They will in all probability laugh and titter, and say, " he has, like Phaeton, driven his chariot too high, and is at last plunged into the bog." Nay, there is an Italian proverb which says, " the Gourd stretches far and wide, but at last comes and ripens at its root."

Mr. B. did not set out for the country on the 16th, being taken very ill. ***** "Butter and Prudence" has a cold, but laughs at such trifles, when business or religion interferes. Mr. M. had a slight fall from his horse, about 12 days ago; but he and mother earth are too good friends to fall out easily.

I am just come upstairs from practising courtship with a lady of 50, who says I am the most polite gentleman in the world. You know I began to learn from 63, and am counting backwards. When I come to 40 I shall stop, as that is the standard of lasting beauty. For all you may think me so hot about matrimony, I assure you I wish I were at Fulham, eating two or three good mince-pies, with cousin M. P., though if my mother were to hear me, she would give me a rap on the knuckles.

I am, in my spiritual wedding garment, dear Aunt, your most humble servant and dutiful nephew, &c.

FROM THE SAME TO MR. ELLIS.

Dear Uncle, Rome, July 8, 1775.

I find myself partly authorised to put you to the expence of this letter, by one lately received from Roger, dated Dublin, May 15th, wherein he mentions having left you in tolerable health at Hampstead, except the complaint in your eyes, for which I fear there is no remedy but patience. It is, however, no small consolation to you, to have your intellectual faculties so improved and acute, that the infirmities of age become less sensible, by your constant occupation.

My brother tells me I should oblige you, by sending seeds or pods of the Locust-tree, and some acorns of the Dwarf Oak. I suppose you mean the Kermes Oak, which is very common in the south of France, but I have not observed it so common here; at least it does not produce the Kermes in Italy. I have no doubt of procuring them, but know not when I can send them to you. I am going from hence to Loretto and Ancona, to spend July, August, and part of September, and from thence to Perugia and Florence, meaning, if it please God, to be in Leghorn by the latter end of October. To save you trouble at the custom-house, I shall direct them to a friend of mine, Mr. James Matthews, medalist at the British Museum, whom I can depend on punctually to send them to you, or to deliver them, for you, to Dr. Solander. This renewal of our correspondence is come very *à-propos,* as I lately travelled over the Bologna mountains, in company with Signor Giovanni Lappi, botanist to the hospital of *Santa Maria nuova* at Florence, who has read your book on Corallines and esteems it much. Nay, I think he told me you corresponded with him. So that, while I stay in Florence, I can make use of his influence, and that of Signor Maffei, Professor of Anatomy, to get plants or seeds from the physic garden. I was once going to send you some fruit of *Tribulus aquaticus**, and some seeds of the *Scio Terebinthus*†. One tree of this species grows in Genoa, but no where else in Italy. Its seeds, or berries, are no bigger than Coriander seeds, but in figure are a kind of wild Pistacia nuts. At a chance time I have met with the Domestic Locust-pod, short, thick, and full of sweet pulp‡; but this year I see only the common sort, with which horses and hogs are fed. I have by me at present a few *Trasi* roots §, originally an Egyptian plant, propagated in the manner of potatoes, but the leaf is grassy. It is cultivated at Verona and Genoa, and tastes like Almonds. The few roots I have, I fear are too much dried to plant. They are called here *Zizole di terra*.

I suppose your friend Roger will have brought you a variety of seeds, shells, and marine productions, as he collected several for you. They doubtless arrived in high perfection, being stuffed into his rattling bomb-cart trunk, which when opened looked like a hard-ware shop. He every where puzzled the custom-house officers to a terrible degree, for they at first mistook him for a contrabandier, but on closer examination, they could make nothing of him but a travelling Robinson Crusoe. It was a regular comedy to me, every time we were stopped

* *Trapa natans.* † *Pistacia Terebinthus.*

‡ *Siliqua edulis, fructu breviori. Tourn. Inst.* 578; a cultivated variety of the Carob, *Ceratonia Siliqua. Linn. Sp. Pl.* 1513.

§ *Cyperus esculentus*; see Sm. Tour on the Continent, ed. 2. v. 3. 103.

to be searched. Roger will, I suppose, have told you of the purchases I have made at Rome, of pictures and antiquities to a large amount. I have been tempted to lay out my money in this manner, from the bad success I have hitherto had in mercantile affairs at Leghorn. I have, besides, the advantage of a friend at Rome, who, though an Italian and a painter, is, to my judgment, as honest a man as most Englishmen, and infinitely more clever in his profession. Perhaps Roger may insinuate the contrary, but I hope the pictures I sent my sister will justify what I advance. It may probably lie in your way to recommend me customers, as you are acquainted with the great.

Pray if you write to me, direct your letter to Messrs. Florence Mac Carty and Son, at Leghorn, who will find me out wherever I may be. My brother says you want to introduce the Dwarf Oak into England. To what purpose? I hope you do not expect it will produce Kermes, for you will be disappointed. I remember you formerly had plants of it, growing in my uncle Neville's garden, which you got from the south of France. If I may be allowed, as one who studies the vegetable productions of the earth, to give my opinion, I think it would be of more consequence to sow seeds of the *Siliquastrum*, or Judas-tree, which bears a beautiful flower, and if in good ground, grows to a very large tree, whose wood is of a fine yellow, and valuable for veneering. The Scio Terebinth-tree is likewise very useful, and would thrive in stony ground, or clay. When full grown, it is the best, and only proper, wood, to make screws for stowing cotton wool; and yields a quantity of valuable turpentine, even in Genoa. The *Ornella*, or Small Ash *, with a stiff leaf, is very beneficial to cattle when sick, and has a more close-grained wood than our English Ash. The wild Mastick † shrub is worth cultivating, on account of its smell, and continual verdure. I know not why we in England do not encourage the sowing of the blue or white Lupines, to fatten the land, and likewise the herb *Eruca* ‡ or *Roquetta;* as also the manner of fattening Oxen with the leaves of the Ypres Elm, as they do at Perugia. In scarcity of fodder they are fed on *Vitalba* §. I believe the Tamarisk-tree would be of infinite use in making hedges by the sea shore, in windy situations, with you, as it proves here. The manner of making a Peach taste of Wormwood, is to set a Wormwood-plant, root by root, near a Peach-tree ||. A large green Plum, grafted on the stem of the long black Fig, has been tried here, and answers extremely well. In the physick garden at Florence, there is a fine collection of forest trees, but not one Velani Oak. I wish you could send me two or three Velani acorns fresh, in

* *Fraxinus Ornus.* † *Pistacia Lentiscus.*

‡ *Brassica Eruca. Linn. Sp. Pl.* 932.

§ Can this mean *Clematis Vitalba?*

|| This notable project is probably founded in vulgar opinion, which no one, but an experimental physiologist, would be anxious to confirm.

wax, for if it once took root here it would thrive prodigiously. A Walnut of my setting bore Walnuts in less than six years, and I have now a fine Pistacia-nut tree, of my own planting and pruning; though the Terebinth seeds which I planted this year have not sprung up. I suppose this is owing to their not being ripe enough *, though I gathered them in December. Yet the tree thrives at Genoa, in a cold, mountainous situation, throwing out many suckers, which I believe would answer better than the seeds. I suppose, after all my trouble in writing, you will tell me, in plain Irish, I have *bothered* you.

The Iris plant † would, I believe, grow to perfection in England, and produce its fine odoriferous root, as it does in Florence, where it grows on the city walls, or chalky ground, with the root almost naked. Caper plants grow on chalky hills, fronting the eastern sun; as likewise Black Mulberry-trees, which then produce much sweeter fruit, and sounder wood, fit for ribbing of boats. Our ponds might be filled with the *Tribulus aquaticus* ‡, as in Holland, where the nuts are eaten instead of Chesnuts. Here they serve to feed hogs. Any of these plants would, I am persuaded, answer better in our climate, than the *Kali,* or Potash plant, encouraged of late in Ireland; or even Flax seed, which only produces a sort of tow, most dextrously converted

* Evidently for want of impregnation, the tree being dioecious.

† *Iris florentina,* now not rare in English gardens.

‡ This has often been tried in vain.

by us into Irish rag. I do not know whether we have the true *Navette* seed, for making oil. It is a species of Rape, but I believe there are different sorts. I should be glad of a few seeds of the Candle-berry Myrtle *, to make a trial of in this warm climate. I suppose you have cultivated it in England, but I never heard with what success. The fruit-trees called *Sorbus, Jujuba,* and *Azarole,* are, I suppose, not esteemed in England, though much cultivated in Italy. But above all things, the Italian poplar is of more general use than any tree we have, as all our cases for merchandise are made of it. This tree might save you a great deal of money, remitted to Norway for deal plank. The tops of turnips, when going to seed, are boiled here as *Brocoli,* and eaten with butter and a shallot.

Wishing you all the happiness you can enjoy, in a double measure, I am, &c. JOHN FORD.

LORD CHANCELLOR NORTHINGTON † TO MR. ELLIS.

DEAR SIR, London, Oct. 30, 1764.

I return you my thanks for your careful remembrance of my Pine Tops, and the ample supply you

* *Myrica cerifera.*

† Robert Henley, Earl of Northington, who succeeded the late Earl of Hardwicke as Lord High Chancellor of Great Britain in 1757, and was advanced to the peerage in 1760. His Lordship took great delight in the cultivation of plants, at his seat at Grainge, in Hampshire, as appears by the following letters.

collected. When I left Grainge all my East wall was planted, and half the South, so that there remain 16 trees to be got for the South wall to finish it. I do not know very well how to get this supply, as business confines me, and I have nobody about me that understands it.

The Pine plants were as useless to Mr. Lee as to me, for his people would not receive them, and they are now in London.

My best compliments, with Lady Northington's, to Mrs. Webb; accept them yourself, and believe me, your most obedient servant, NORTHINGTON.

SIR, Grainge, May 30, 1765.

I have the favour of yours by yesterday's post, relating to the Orange and Lemon trees, among which you say there are some that I want, which must be true, because I have at present in effect none; but you say little in their commendation. However, to be short, my conservatory being begun so far as it must certainly be stocked this Autumn, I think it fit and proper that you should attend this sale, and buy sufficient for one journey, if you can satisfy your own judgment; for I think it will be idle for me to have trees to seek next year.

I have still the gout in my head and eye, but in all conditions, your obedient servant,

NORTHINGTON.

SIR, Grainge, July 25, 1765.

I have yours by yesterday's mail, in which you say, "Mr. Philips, the dealer in glass, desired to know when I would have the glass sent down, and that you told him you expected my orders as soon as I knew what was wanted." I know not the man's name, but the person recommended by Mr. Webb to me, hath an account of the quantity in writing, the size of the different squares, and I suppose according to his agreement they are ready, and you may direct him to send them down forthwith.

The carpenters began putting on the roof yesterday, and I think the trees may be planted in the course of next month. I think we may get water enough for future purposes, for this Summer hath confirmed me in an opinion I always entertained, that if ground be well kept, much water is not necessary for plants.

Your amphibious animal of a new Genus is an Eft, or Οφις, larger than ours in England.

I will give you the trouble of coming down when we are in forwardness enough to go on business; in the mean time, I have only to present Lady Northington, &c's. compliments, and to assure you of my being, your most obedient, most humble servant,

NORTHINGTON.

SIR, Grainge, Oct. 20, 1765.

I received your letter at Bath the day before I returned, and was really concerned that you had been so much out of order, and wish your recovery speedy. I think I am considerably better by my journey, and the waters, and I ought to be so, as the place itself is not suited to my genius. The Governor (Ellis) waited on me, but I did not see him there, or at least did not know him.

I find here every thing to my liking, on my first cursory view, for I came but yesterday. The conservatory in good order, and the terrass and little garden before it, made and planted; the Firs all planted on the Western hill, and the Espaliers in the kitchen garden.

I return you and Mr. Webb thanks for the list of Lord Middleton's plants; but I see very few things there that I have not already to the number I want; for the two *Magnolias* you sent down, take so well to the ground, that they will be soon bigger than his, and are sufficient for me with the two younger plants. And as to all the others, the neighbouring nurseries can supply them better, so I have no wants there.

I purpose being in London about Saturday next, in the mean time remain, your most humble servant,

NORTHINGTON.

DEAR SIR, Grainge, Aug. 20, 1769.

I should have acknowledged the favour of yours before, but I have had my wife and son ill for above this fortnight, and therefore am very indifferent myself, as you will perceive by my pen; I thank God they are now better.

I should have been very glad of your company here, and am sorry that any thing, but particularly illness, prevented it. Howarth confirms your account of Monsieur du Hamel's present; my gardener, like him, found out the mischievous qualities of the *Muscipula*, and therefore killed it. In general my plants are in good order, and my plantations thrive greatly.

I conceive the American animal to be an Otter, and its magnitude owing to its locality; you observe the men in the Southern parts of that continent are giants, and those in the Northern much greater than us Englishmen. When you agents, the Parliament, and the Ministry, have reduced them to a moderate strength and stature, you will have done signal service to your country.

When your business is suspended I shall still hope to see you at Grainge, where you will see one who is very much your friend, and humble servant,

NORTHINGTON.

DEAR SIR, Grainge, Sept. 30, 1770.

I thank you for your several specimens of the Poplars; I shall only take notice of them that are new to me. The specimen from Gray's Inn Garden seems to me the true *Abele*, at a disadvantage, from growing in that poisonous air, and more poisonous earth. It is better in its foliage than the black or the Italian Poplar. The *Populus heterophylla* is quite new. Its beauty must depend on its shape and growth. We have somehow not seen or heard of the Wilmots either through you or otherwise. The Loblolly is much *in statu quo* as to the flower. Both are in health, and the trees in general benefited by change of air. If it blooms I shall not forget your wish.

I am, with our compliments, your most obedient,

NORTHINGTON.

MR. ELLIS TO THE DUCHESS OF NORFOLK*.

London, Oct. 11, 1768.

I am desired by the Right Hon. the Earl of Hillsborough, that if any seeds from America should come to my hands for his Lordship, I should send your Grace a part. These commands I most cheerfully obey, as I have long heard that your Grace is a patron of botany and gardening. What I send your Grace are chiefly Lord Hillsborough's; and though few, I hope they are worth your Grace's acceptance, being rare. As soon as any more arrive, I hope to be so happy as to contribute to so great and elegant a collection.

It is probable that West Florida will soon furnish me with something we have not had before. I daily expect some of the seeds of the *Illicium anisatum*

* Mary Duchess of Norfolk, who died May 27, 1773. The husband of this lady was Edward the ninth Duke of Norfolk, in whom the elder branch of that noble family terminated in 1777.

of Linnæus, of which plant I have already had specimens. John Bartram, in his journey through East Florida, mentions having seen a plant, which I believe to be this. Mr. Clifton, Chief Justice of West Florida, knows the tree, and I am in hopes will procure us the seeds this autumn. It is likewise a Japanese plant, and is called by Kæmpfer, in his *Amœnitates Exoticæ*, *Somo* or *Skimmi*. He has there given a figure of it, with an account of the uses the Japanese put it to. Bartram says it is a hardy evergreen, with bay leaves, and a smell like sassafras. The Loblolly Bay seeds were sent me for this by mistake. The seed-vessel of this tree consists of a circle of pods, with one seed in each, and was used formerly in physick, by the name of *Anisum stellatum* *.

I must further desire your Grace's acceptance of a print of one of the rarest productions of the vegetable kingdom. 'Tis a new Sensitive Plant †, and formed in such an extraordinary manner as if the Great Author of Nature intended it to receive some nourishment from the animals it seizes. For in the internal part of the two lobes of each leaf are three erect little spines, among the little red glands,

* Such were the ideas of English botanists on the first discovery of this Florida plant, which proved, on more accurate examination, a new *Illicium*, with red flowers, now known in our green-houses by the name of *floridanum*. The French have, in Florida, met with a third species, called *parviflorum*, *Ventn. Jard. de Cels t.* 22, whose petals are yellow and orbicular.

† *Dionæa muscipula. Linn. Mant.* 2. 238.

marked with dots in the print. These glands seem to be the irritable part. As soon as a fly or other insect touches these, when the leaves are young and vigorous, the lobes immediately close upon it, and the spines either stick into it, or serve to prevent its escape, where it remains till it dies.

A few of these plants were brought over from Philadelphia this summer, by one Mr. Young, who sold them to Mr. James Gordon, seedsman, Fenchurch-street, and Mr. Brooks in Holborn. I had a drawing taken, and a plate engraved, from a plant that flowered in my chambers last August. This is an entirely new genus. I have sent its characters to Linnæus, our father in botany, which I suppose he will adopt.

THE DUCHESS OF NORFOLK TO MR. ELLIS, GRAY'S INN.

Sunday morning (March 19, 1769).

The Duchess of Norfolk's compliments to Mr. Ellis. She has sent him a branch from Worksop of the Pine, which she takes to be that called (by Mr. du Hamell du Monceaux) the Beaume de Canada *. She has not been able to find it growing in

* *Pinus balsamea, Linn. Sp. Pl.* 1421, now common in our gardens, though its short duration, hardly exceeding 20 years, renders this tree of but little value in ornamental plantations. The balsam, called Balm of Gilead, is very fragrant; and, mixed with some cheap spirit, makes a kind of dram, which is said, as a quack medicine, to have proved very lucrative to its vendor.

any of the gardeners' collections near London. In rubbing the branch with the hand, it has a smell not unlike that which comes from our Broom in England; and she has not observed that any of the other Pines have that smell.

MR. ELLIS TO THE DUCHESS OF NORFOLK.

Aug. 7, 1769.

Mr. Ellis returns her Grace the Duchess of Norfolk ten thousand thanks for her very kind present of venison. It proved, in spite of the hottest day in summer, when the thermometer stood in the shade at 86¾, perfectly sweet and fine, and the fattest Mr. Ellis ever saw.

Lord Hillsborough mentioned to Mr. Ellis, two or three days ago, that he had sent her Grace a present of seeds, which he had received from Michael Collinson. As it may be of use to her Grace's gardener to know from whence they came, Mr. Collinson, before he sent them, told Mr. Ellis they were from the Cape of Good Hope; so that they require a degree of heat.

The specimen of Fir, which her Grace sent to Mr. Ellis on Palm Sunday, is still alive. It was put into a phial of water as soon as received, and about a month ago put forth some young shoots. Mr. Ellis has had various other specimens of Firs and Pines, which he put into water; but all died in a month's time. He acknowledges this differs from the common Norway Spruce, which he thought it at first, and would be extremely glad, if it bears cones, to have a specimen of them.

The sudden death of Mr. Eliot, governor of West Florida, at Pensacola, has disappointed Mr. Ellis in his expectations of receiving many curious seeds from thence for some time; but Lord Hillsborough has been so good as to make Mr. Durnford, the surveyor-general, lieutenant-governor of the province. He is curious and intelligent in the valuable plants of that country, Mr. Ellis being already indebted to him for some of the rarer kinds. He is now here, and will sail this month, so that by next spring Mr. Ellis hopes to be able to present her Grace with some of the productions of West Florida.

Lieutenant-governor Durnford has brought the skin of an animal, of the cat or tiger kind, which exceeds in size any that have been described by the French, and must, when alive, have been a dreadful animal. It was shot within four miles of Pensacola, having just killed a wolf. It is seven feet and a half long, from the nose to the tip of the tail. The tail measures two feet nine inches, and has a remarkable dark-coloured line, ending in a thick tuft of black hair. The back of the animal is two feet six inches high from the ground. The colour is exactly the same as our common water-rat. The fur is fine on the back and sides, and passes insensibly into long whitish hair on the breast and belly. The ears are short, of an oval shape, a little pointed, scarce three inches long. The teeth and claws are formed the same as in all this tribe, strong and terrible to behold. It does not appear to Mr. Ellis to be described in Linnæus *.

JOHN EARL OF MOIRA † TO MR. ELLIS.

SIR, Moira, Feb. 17, 1770.

I received the favour of your's last month, with the seeds of the *Anisum stellatum* ‡ very safe; for which I beg you to accept my most sincere thanks. Any thing of this sort, to one who loves botanical studies as much as I do, is most exceedingly agreeable. I have sown them as you direct, and hope to inform you of their success.

I inclose a sprig of a Dwarf Fir §, which my grandfather imported before the Revolution, among

* *Felis discolor. Schreb. Mammalia t.* 104, B. Black Tiger of Pennant.

† Father of the present Marquis of Hastings, by his third wife, the Lady Elizabeth Hastings, daughter of Theophilus late Earl of Huntingdon.

‡ *Illicium floridanum.*

§ This singular Fir, a dwarf Spruce, of a very compact habit, and extremely slow growth, was first introduced to the notice of the English cultivators by the late Earl of Clanbrassil, who entertained a suspicion of its having been brought from the mountains of Jamaica by Sir Hans Sloane. There is nothing to confirm this opinion in the History of Jamaica; neither does Browne mention any Fir as growing in that island. The tree, moreover, is quite hardy in our gardens, though we never heard of its flowering. Lord Moira's is perhaps the most authentic account of the introduction of this plant into Ireland, though leaving us in the dark as to its native country. Mr. Lambert, other seeds. One tree is alive. It grows in the form of a cone, and is not above five feet high. I do not find it mentioned in any of the books, and should be glad, if you see Mr. Miller, to know his opinion of it. It is perfectly furnished with leaves from the ground, and looks as if clipped into form. I have raised some layers from it.

As to the very important subject you mention, of a continuation of the bounty on American flax, I have mentioned it to the most intelligent members of our Board, as also to the principal drapers, and some manufacturers. I find them *unanimous* in thinking it a measure that may injure us, without a probability of service. It may indeed *serve* the Scotch, and *please* the Americans; but when we are well, why should we risque being worse, in hopes of gaining what, in fact, we do not want? This is the substance of what they all say to me. Lord Chief Baron Forster tells me he will oppose it, as do many others.

I should be exceedingly obliged to you if you could procure me some seeds of the *true* Rhubarb, and the purging Senna from Alexandria; also the Scammony.

I have lately found a curiosity, which I would send to the Royal Society, had I a safe conveyance. 'Tis a honey-comb, perfect and entire, petrified, and

in his Description of the Genus *Pinus*, p. 44, speaks of the Fir in question as probably a variety of his *P. rubra*, *tab.* 28, the Newfoundland Red-Pine. There is a thriving specimen of it at Arley, Staffordshire, the seat of the Earl of Mountnorris.

now an exceeding heavy stone. I found it in a gravel-pit, on the top of a high hill, fourteen feet from the surface. It seems composed of particles of very fine sand, solidly united, and the form of the cells well preserved.

I am, Sir, your most obliged, humble servant,

MOIRA.

MR. THOMAS KNOWLTON * TO MR. ELLIS.

OH MY GOOD SIR! Chatsworth, Oct. 1770.

I had the great pleasure to hear of your health by Mr. Blackburne and Miss, where I have been for a few days. Saw the Tea-tree for the first time, in great good health, with numbers of others, almost as rare; but not your Fly-trap, though he had no less than five of them. I had one, so had Lord Exeter, and Mr. Brownrig of Chelsea, but all suddenly lost them, so I have not seen it to perfection. Your *Gardenia* has, these two years, been nailed up against a south wall, and in such vigour, that I really, at first sight, did not know it, having so shining and florid an appearance. I have it, and the single in your little book, and have another likewise, which is a full flower, and a smaller leaf, different from either,

* Once gardener to Dr. Sherard at Eltham; and, in his advanced age, to the Earl of Burlington, at Londesborough, Yorkshire. He died in 1782, aged above 90. *(Pulteney.)* Several of his letters, full of the gardening history of his time, remain in the MS Correspondence of Dr. Richardson.

and which I never yet saw any where else *. We had it of Clarke, the butcher at Barnet. I have seen the *Uva Ursi*, found on the mountains in Scotland, and the *Betula nana* from that country, which I had not seen before alive, though by the favour of Mr. White †, a surgeon in York, I had a good specimen of it, which he assured me he gathered there. I have the *Chamæpericlymenum* ‡, or Dwarf Honey-suckle, found in the greatest plenty near Salter's Gate and Thornton, three or four miles East of Pickering, Yorkshire, which I esteem as a very great curiosity. It does well with me, as do many of the rare English plants.

I shall, in a few days, go to the Duke of Portland's, Duke of Norfolk's, Lord Scarborough's, and so home. Hope to see the Tea-tree with one of them, as you sent so many seeds to Worksop. I could wish that our young Duke § would, like his father, who every day improved in knowledge, take a turn that way, and that you were acquainted with him who is so good, and I hope amiable, in all things.

It would be a most singular pleasure to have a letter of information now and then from you, as my worthy and good late friend Mr. Collinson is gone, and no more to be seen. I beg it as a favour in you to indulge an old man of 80 years, who is as great

* Perhaps *Gardenia radicans* of Thunberg. *Ait. Hort. Kew. v.* 1. 368, not rare at present in collections.

† Afterwards M. D. a correspondent of Mr. Hudson.

‡ *Cornus Suecica. Linn. Sp. Pl.* 171. *Fl. Brit.* 188.

§ William the fifth Duke of Devonshire, who died in 1811.

a lover of plants perhaps as any that is now living; otherwise would not take such a journey as I have done. If I could be of any use to you by a sketch of any stove I have seen any where, should be glad to oblige or please you in any kind or thing that is in my power, for I do much esteem you for your writings and friendship. I am now told that you are again about to publish, which please to inform me of, and what is the subject, as I could wish to know, and you will much please and oblige your friend and very humble servant, THOMAS KNOWLTON.

At Londesborough, near York.

P. S. I hope your nephew the Captain or Governor is well. I shall be pleased to hear that he is so, as his Hudson's Bay History * is much admired by me, as well as the man.

JOSEPH BANKS ESQ. TO MR. ELLIS.

DEAR SIR, Burlington Street, Nov. 1, 1773.

If you have not at present any particular occasion for that very ingenious young man, whose drawings of Corals you showed me some time ago, I should be much obliged if you would give me leave to employ him a few days to draw for me some subjects which Captain Phipps has brought from Spitzbergen, and Mr. Hunter intends to dissect for that purpose.

* Henry Ellis, F. R. S., published a Voyage to Hudson's Bay, in 1746 and 1747, for discovering a North-west passage; printed at London, 1748, in 8vo. with plates.

I rejoice to hear that your eyes are in a mending state; should be particularly happy if they would give you leave to instruct us a little, as in all matters of that nature even Solander is obliged to allow you the title of master. I am, your affectionate, humble servant,

JOSEPH BANKS.

End of the Correspondence of JOHN ELLIS, *Esq.*

Biographical Memoir

OF

JOHN JAMES DILLENIUS, M. D.

SHERARDIAN PROFESSOR OF BOTANY AT OXFORD;

WITH

His Letters

TO LINNÆUS AND RICHARDSON.

JOHN JAMES DILLENIUS, M. D. whose name is familiar to every student of Cryptogamic Botany, and whose *Historia Muscorum*, published in 1741, still remains unrivalled in that department, with regard to botanical learning and criticism, as well as specific discrimination, was born at Darmstadt in Germany, in 1684 or 1685. He was educated as a physician at Giessen, and while resident there, published several Botanical essays, of considerable acuteness, in the *Ephemerides Acad. Nat. Curiosorum*, as well as a small 8vo. volume, entitled, *Catalogus Plantarum spontè circa Gissam nascentium*, printed in 1719. Being brought to England by the distinguished William Sherard, the greatest botanist of his day, who had been English Consul at Smyrna, Dillenius remained here from August 1721, till his death in 1747. He was closely attached to Consul Sherard, and his brother James, an opulent apothecary, who had a garden at Eltham, of the rare plants of which Dillenius published, in 1732, a splendid history, in two folio volumes; the plates, like those of all his other publications, being drawn and engraved with his own hand. They excel in characteristic fidelity. He also coloured some copies himself. He had previously edited, in 1724, the third edition of Ray's Synopsis, greatly enlarged, if not always improved, by the editor; and he became intimate with many English botanists, especially with Dr. Richardson of North Bierley, Yorkshire, one of the most eminent; as appears from several of the following letters, obligingly communicated to the editor by Miss Currer, the heiress of the Richardson family.

Consul Sherard, in founding his Botanical Professorship at Oxford, appointed Dillenius the first Professor, which place he held, fulfilling its duties, with respect to the garden at least, very assiduously, till he died there of an apoplexy, April 2, 1747, in the 63d year of his age. He was buried in the church of St. Peter in the East, Oxford, where a marble monument, with an ample Latin epitaph, is erected to his memory.

The letters of Dillenius evince great plainness of character, and he lived much esteemed by his contemporaries. If we trace in them nothing of the enthusiasm or elevation of soul displayed by Collinson, any more than the great physiological acuteness, and ardent philanthropy of Ellis or Garden, there is a genuine love of science, and a rectitude of principle, apparent throughout. His temper was not without occasional, though transient, asperity. He disliked forms and compliments, and he expresses his sentiments, without hesitation or reserve, especially in his letters to Linnæus, whom he charges freely with his own fault, an impatience of contradiction, or of criticism. Such is too often the custom of the world. The writer of the last paragraph of Dillenius's *Examen Responsionis Rivini*, could, with a very ill grace, censure any body for irritability, or for the most acrimonious expression of such a feeling. We respect him too much to expose the passage in English, which indeed it would scarcely bear.

The originals of the letters of Dillenius to Linnæus are, of course, in Latin; those to Dr. Richardson are all in English. The editor hoped to have found at Oxford the letters of Linnæus to the Sherardian Professor, which could not but have proved highly interesting; but his present worthy successor, Dr. Williams, has sought in vain for any letters to Dillenius, except a very few from Haller and other botanists. The rest of his correspondence, if not destroyed, may possibly exist among the papers of the Sibthorp family.

Correspondence.

JOHN JAMES DILLENIUS, M. D. TO LINNÆUS, AT MR. CLIFFORT'S, AMSTERDAM. [Latin.]

Oxford, May 16, 1737.

Your letter of December the 18th last, came to my hands in February. What you wrote me on the 1st of March 1736, which we supposed to have been lost, owing to Mr. Dingley's neglect, did not reach me till the 8th of the same month, in the present year. It was accompanied by the *Musa Cliffortiana*, for which I beg of you to return my thanks to Mr. Cliffort, as well as by several plates of your *Flora Lapponica*, and some specimens thereunto belonging.

As we have already conferred together respecting the plants there represented *, I shall say no more about them, except three species of Mosses (Lichens), which you have sent; it being difficult to judge of this tribe without seeing specimens. I would observe, therefore, that *Tab.* 11. *fig.* 1. appears to be a variety of No. 80, Ray's *Synopsis*, ed. 3. 75, for that species varies almost infinitely: *fig.* 2. seems

* The remarks of Dillenius, upon the more difficult Lapland plants, such as *Salices*, are printed in the *Flora Lapponica*, with the sentiments of Linnæus subjoined.

to me No. 85 of the same work, p. 76: and *fig.* 3. No. 88, p. 77 *. This intelligence however comes, probably, too late, the work being already published.

I now advert to your last letter.

The *Arbuscula Ulmi facie* †, &c. *Comm. Hort. Amst. v.* 1. 165. *t.* 85, appears to me to approach the *Azedarach,* though it wants the tube in the middle, and bears more stamens, having above 20. The flower has five petals, each with a nectary at the base, and five permanent calyx-leaves. This tree is different from *Guidonia.* I take Plukenet's *tab.* 221. *f.* 4, and *Hort. Malab. v.* 4. 83. *t.* 40, to be the same as *Muntingia folio Ulmi aspero* ‡, &c., *Plum. Nov. Gen.* 41; though you ask if the latter may be referred to *Ulmus.* I do not like to judge of plants that I have not seen living, but as to the determination of the genera of these, upon your principles, I have ascertained the above plant of Plukenet, as well as *Hort. Malab.* to be no less distinct from this of Commelin, than Plumier's *Muntingia folio Corni,* §, &c., with which it more agrees in leaves and aspect; and I suspect that of *Hort. Malab.* to be a species of *Celtis* ||. As you

* These opinions are certainly all erroneous, the 1st species being *Lichen nivalis,* the 2d *centrifugus,* and the 3d *croceus* of Linnæus.

† *Grewia occidentalis, Linn. Sp. Pl.* 1367, certainly nothing related to *Azedarach* (or *Melia*).

‡ This seems to belong to *Celtis.*

§ *Rhamnus micranthus, Linn. Sp. Pl.* 280, justly referred by Jussieu and Swartz to *Celtis.*

|| *Celtis orientalis, Linn. Sp. Pl.* 1478. *Willden. v.* 4. 995.

were desirous, when here, of seeing the flowers of this last-named genus, I shall now mention what I observed this Spring in *Celtis fructu obscurè purpurascente**.

The flowers appear early in May, as the leaves begin to unfold, disposed in little short clusters, on the lower part of each branch, under the lowest leaf. There are but few situated in the bosoms of the leaves, and those solitary; but they are the only fertile ones, furnished with a perianth, stamens, anthers, and style with its embryo *(germen)*. The rest are merely males, having no style, nor rudiment of fruit; their calyx of one leaf, divided usually into six segments, to the very base, with as many stamens. These flowers soon separate from their stalks, entire, with their stamens; but the upper ones, which alone bear fruit, have the calyx mostly in five divisions, with five stamens. These are what you call hermaphrodite †, which term I am doubtful about, thinking the word androgynous more proper; but this you understand and use in a different sense. The style always terminates in two horns, and is variously incurved, not always in the same manner, downy all over, seated on a roundish-oblong embryo *(germen)*, that springs from the bottom of the calyx, which is greenish, sometimes remaining till the fruit swells, after which they wither, without entirely disappearing. Those of the male blossoms, as I have said, fall with their

* *C. occidentalis. Linn. Sp. Pl.* 1478.

† Now termed *united*, or *perfect*, flowers.

stalks. There are no petals. The stamens are short, their anthers concealed at first; but when the dust is shaken out they become longer. Anthers oblong, thickish, with four angles and four furrows.

Here are a few remarks on the common *Melianthus major*. The calyx is of one leaf, permanent, divided almost to the bottom into five irregular segments; four of which are longer, erect, converging, so as to form a half tube, the two outermost being broadest and longest, the two innermost rather narrower and shorter. The fifth, occupying the under part, is keeled, and much shorter than the rest. The colour of the whole is a rusty dull red. Four narrow tongue-shaped petals lie upon this lower keeled segment, which are reflexed towards the extremity, and connected together by a portion of down. The two uppermost are rather the longest. They are all of a dark red, and deciduous. Under these, within the keel of the perianth, a fifth petal, of a keeled form, and connected with the perianth, is concealed. This is paler and rather more rigid than the other four, divided nearly half way down into three segments, the lowermost cloven. It is not deciduous, but withers as the fruit enlarges. It may be called the nectary indeed, as it abounds with honey. The embryo is quadrangular, placed just above the nectary. At its base, between it and the calyx, stand four, not more, stamens, with oblong, thick, terminal anthers, furrowed along the lower side. From the centre of the embryo arises a simple pointed style, somewhat curved from the first, but more afterwards. As the rudiment of the fruit swells, the calyx becomes inflexed, pendulous, and more spreading, especially its larger segments, which now become inferior, the keeled one being turned uppermost. By the time the fruit becomes the size of a small bean, though the style and stamens, with their dry anthers, remain, the petals fall off, and the nectary withers. The fruit is a four-cornered bladder, to whose furrows the stamens are closely pressed; and at the summit are four projections, each marked with a suture at the inner edge, and bursting with a slight pressure, the style originating from their point of union. When dissected, the fruit exhibits four cells, in a great measure empty, bearing several round seeds at the axis of each cell, though I believe but few come to maturity.

Hermann has described the calyx as the flower (or corolla), taking the tongue-shaped petals for some of the stamens; though he has well observed the keeled petal, or nectary, which he calls a keel-shaped cartilaginous receptacle, &c. The figure of this plant, in the work of Rivinus, is placed among the hexapetalous flowers; the true petals, which stick together towards their extremities, having been, doubtless, taken for one, and the calyx leaves for five more. Compare my account with your No. 324 *, which you describe as if you had seen it, though not indeed fresh; and you will allow there is a great

* *Melianthus. Linn. Gen. Pl. ed.* 1. 114.

difference between the examination of a living and a dead subject.

This makes me wonder the more, at your having admitted, into your *Genera*, so many Indian plants, which you have never seen, and which perhaps are never likely to be seen *.

If the number of stamens be of no importance in *Laurus*, as you say in *Gen. Pl.* p. 382, why do you place this genus, No. 338, in *Enneandria?* Perhaps you will say that *Cinnamomum*, *Camphora*, *Borbonia*, and *Sassafras*, answer to that character. Are genuine species of *Laurus* to be determined by these? And if they be such, why do you deny a calyx to this genus, which these have, though our true *Laurus* has none, the Common Bay being furnished in place of it, with four or five deciduous scales? I have, moreover, noticed monopetalous flowers, in four or five deep segments, in this Common *Laurus*, the stamens being always twice as many, nor have I ever seen nine. I disallow of the next genus, n. 339, *Rheum*, as well as of many others, and I differ from you in many other points. *Trigonella*, No. 880, may be referred to *Fœnum-græcum*, which has, at least the common kind, a similar structure in the flower †, and ought to be

* We may be allowed to wonder, still more, that so few mistakes have been made, with materials so incomplete. This of *Melianthus* is among the most remarkable; yet those who examined it alive, though great botanists, did not avoid errors.

† Linnæus has now removed *Fœnum-græcum* to *Trigonella* accordingly.

distinguished from *Medicago.* The stamens are not worthy of attention in *Sanguisorba*, nor ought *Poterium* to be torn away from that genus. There are more species of *Sanguisorba* with numerous stamens, than with four only. *Cytisus monspessulanus, medicæ foliis*, &c.* as well as some new species which we have from Siberia, have all their stamens proceeding from one common sheath, as is likewise the case with *Coronilla maritima glauco folio* † of C. Bauhin. These plants are therefore not diadelphous, but monadelphous, nor do I doubt that many more will be found so ‡. I could point out many other things, which the limits of a letter will not admit. I must say a word concerning stamens and styles, as being unfit to found a system of arrangement upon; not only because they vary as much as flowers (petals) and seed-vessels, but because they are hardly to be discerned, except by yourself, and such lynx-eyed people §; and in my judgment, every scheme of classification offers violence to nature.

Notwithstanding all this, I applaud and congratulate you, in the highest degree, for having brought your premature birth to such perfection. You have accomplished great things, and that you may go on

* *Genista candicans. Linn. Sp. Pl.* 997.

† *C. glauca. Linn. Sp. Pl.* 1047.

‡ See Linnæus's subsequent distribution of this tribe, in his *Syst. Nat.*

§ A singular objection from the great sharp-sighted cryptogamist!

and prosper still more, let me exhort you to examine more and more species. I do not doubt that you yourself will, one day, overthrow your own system. You see, my dearest Linnæus, how plainly I speak my sentiments, depending on your candour to receive them favourably.

As I am on these subjects, I would observe that *Sherardia* is not distinct from *Spermacoce*. I have said that the *Subylaria* * bears its seeds at the base of the leaves, being clustered in a simple cavity at the inner side of each leaf, which holds good of the first and third species only in Ray's *Synopsis*, p. 306, 307 †. You make *Calamistrum*, No. 800, Gen. Pl. 326, synonymous with Vaillant's *Pilularia*. whereas they are totally distinct in genus ‡.

I received your *Genera Plantarum* through the hands of Gronovius, and thank you.

I have not yet seen the *Flora Lapponica*, though I presume it is finished.

A little parcel of plants, collected in Greenland, which has for many years been thrown aside, among my botanical rubbish, has just now excited my attention. I find in it not a few of your Lapland productions, especially of the Heath tribe, and the *Betula nana*; which last I long ago thought might

* *Isoetes*, Linn.

† Most undoubtedly! for the fourth species is a cruciform plant, the *Subularia* of Linnæus; and the second, if we mistake not, *Littorella lacustris*! Dillenius has here committed greater errors than any that he lays to the charge of his friend.

‡ This is no less indisputable, *Calamistrum*, or *Calamaria*, being the *Isoetes*.

be what Plukenet calls, in his *Mantissa*, 189, "*Vitis idea (forsàn) parvis subrotundis elegantissimè crenatis foliis, Fruticulus ex Groënlandia.*" I conceive it to be also what is meant by *Betula pumila*, in Loesel's *Flora Prussica*, 27.

The sexual differences, in compound flowers, are, in my judgment, altogether useless, superfluous, and mischievous, for botanical characters. What is the object of all this apparatus? These are mere gewgaws. It is quite enough that one botanist, Vaillant, should have had his head turned by them.

To fill up my paper, I will here mention, that I have this spring been looking at the narrow-leaved variety of the Common Rosemary. The lower lip of its calyx has two teeth. The tube of the flower is not longer than the calyx; towards the throat it is shorter; out of which part, towards the under lip, proceed the stamens, which are directed upwards, and applied, with their anthers, to the upper lip, scarcely projecting beyond it; while the style is long, curved, four or five times the length of the stamens, not simple, but most frequently forked, at the extremity*. Such are the characters I have found in a thousand flowers, indeed in all those of a very old bush. But on a younger one, raised a few years since from a cutting, the stamens were longer, falcate, equal to the style.

* These are chiefly intended as corrections of Linnæus, in *Gen. Pl.* and he has partly profited by them: but the relative proportions of the stamens and style, in ringent flowers especially, is now known to be extremely variable.

I am pleased with your remarks on the *Rhodia* *. What we have here is only that kind which does not perfect its seed-vessels. Which is the *major* or *minor* of Tournefort and C. Bauhin, this or the fruit-bearing plant † ? I think, however, the genus ought not to be distinguished from *Anacampseros*, or *Sedum*.

All the species of *Oenanthe* have not the character attributed to the genus; and yet they truly belong to it; for instance, Lobel's *Oenanthe cicutæ facie, succo viroso, crocante* ‡.

Farewell!—do not forget your affectionate friend,

J. J. DILLENIUS.

DILLENIUS TO LINNÆUS, AT HARTECAMP, NEAR LEYDEN. [Latin.]

Oxford, Aug. 18, 1737.

MY DEAR AND MUCH-ESTEEMED FRIEND,

I have seen, I have received, and I have read your *Flora Lapponica*, with great pleasure. Would we had more works of this kind, finished with the same care and attention! You have here well shown your ability. There are few things in which my opinion differs from yours; but you will not, I trust, take it amiss if I point these out. For in-

* *Rhodiola rosea. Linn. Sp. Pl.* 1465.

† C. Bauhin does not make the distinction. The *major* of Tournefort must be presumed to be the fruit-bearing plant, always growing taller and firmer than the barren one; which is also the case with *Valeriana dioica*, and others, for obvious reasons.

‡ *Oenanthe crocata. Linn. Sp. Pl.* 365.

stance: — *Section* 139. Schwenckfeldt is not the first publisher of the *Trientalis*, but Cordus, who wrote long before him, about 1540; and after him Thalius, as well as Francus. — *Sect.* 161. *Vaccinia rubra foliis myrtinis crispis, Raii Syn.* 457, does not belong to this species *, but to the following, *Uva Ursi* of Tournefort, at least if my dried specimens may be relied on, in which the leaves are smooth indeed, though wrinkled, perhaps by drying; scattered without any order, or mostly alternate. I understand your plant to have smooth opposite leaves, which are not veiny. — *Sect.* 176. J. Bauhin's synonym ought, in my opinion, to be expunged †. — *Sect.* 180. The *Lychnis marina repens* of C. Bauhin, which grows in this country, is widely different from the common *Behen* ‡. I keep the former in a pot; the latter is in the open ground. I will take care next year to cultivate both in the same natural or unmanured soil, to compare them better. — *Sect.* 291. *Cirsium alpinum, boni Henrici facie* of Tournefort is abundantly distinct from that with which you have joined it §. Perhaps you have never seen the true plant of Clusius. I can scarcely give credit to what you report at

* *Arbutus alpina. Linn. Sp. Pl.* 566.

† *Saxifraga nivalis. Linn. Sp. Pl.* 573. The same opinion is hazarded by the editor of *Fl. Lapp.* ed. 2.

‡ Undoubtedly. The former is *Silene maritima. Fl. Brit.* 468. Engl. Bot. t. 957.

§ Perhaps Dillenius is here mistaken, not being aware how much *Serratula alpina* varies in its native situations.

Sect. 129 *. I have had the *Lapathum folio acuto crispo* of C. Bauhin, and the *Britannica antiquorum vera* of Muntingius, planted in this garden, and shall observe next year how the matter is. In the copy sent me, the last line of p. 337, sect. 442, is not printed, and I am anxious to know what the words are.

But I feel as much displeased with your *Critica Botanica* as I am pleased with your Lapland *Flora*, especially as you have, without my deserving such a compliment, or knowing of your intention, dedicated the book to me. You must have known my dislike to all ceremonies and compliments. I hope that you have burthened but few copies with this dedication. Perhaps only the copy which you have sent me. If there be more, I beg of you to strip them of this vain parade, or I shall take it much amiss. At least I cannot offer you my thanks for what you have done, though I gratefully acknowledge the favour of the copies you have sent me of the *Critica* as well as *Flora*.

We all know the nomenclature of Botany to be an Augean stable, which C. Hoffmann, and even Gesner, were not able to cleanse. The task requires much reading, and extensive as well as various erudition; nor is it to be given up to hasty or careless hands. You rush upon it, and overturn every thing.

* Linnæus subsequently separated his *Rumex aquaticus*, *Sp. Pl.* 479, from *R. crispus*, 476, with unquestionable propriety, in conformity to the opinion of Dillenius.

I do not object to Greek words, especially in compound names; but I think the names of the antients ought not rashly and promiscuously to be transferred to our new genera, or those of the new world*. The day may possibly come when the plants of Theophrastus and Dioscorides may be ascertained; and, till this happens, we had better leave their names as we find them. That desirable end might even now be attained, if any one would visit the countries of these old botanists, and make a sufficient stay there; for the inhabitants of those regions are very retentive of names and customs, and know plants at this moment by their antient appellations, very little altered, as any person who reads Bellonius may perceive. I remember your being told, by the late Mr. G. Gherard, that the modern Greeks give the name of *Amanita* (ἀμανίτα) to the eatable Field Mushroom; and yet, in *Critica Botanica*, p. 50, you suppose that word to be French. Who will ever believe the *Thya* of Theophrastus to be our *Arbor Vitæ*? Why do you give the name of *Cactus* to the *Tuna*? Do you believe the *Tuna*, or *Melocactus* (pardon the word), and the *Arbor Vitæ*, were known to Theophrastus? An attentive reader of the description Theophrastus gives of his *Sida*, will probably agree with me that it belongs to

* In these sentiments every person at all conversant with the subject must concur; but those who have undertaken to correct Linnæus have often done it without sufficient learning or judgment; so that various names, inaccurately or doubtfully applied by him, are still suffered to remain as he left them.

our *Nymphœa* *, and indeed to the white-flowered kind. You, without any reason, give that name to the *Malvinda;* and so in various other instances, concerning antient names; in which I do not, like Burmann, blame you for introducing new names, but for the bad application of old ones. If there were, in these cases, any resemblance between your plants and those of the antients, you might be excused, but there is not. Why do you, p. 68, derive the word *Medica* from the virtues of the plant, when Pliny, book XVIII. chap. 16, declares it to have been brought from Media? Why do you call the *Molucca, Molucella?* It does not, nor ought it, to owe that name, as is commonly thought, to the Molucca islands; for, as Lobel informs us, the name and the plant are of Asiatic origin. Why then do you adopt a barbarous name, and make it more barbarous? *Biscutella* is not, as you declare, p. 118, a new name, having already been used by Lobel. I am surprised that you do not give the etymology of the new names which you or others have introduced. I wish you would help me to the derivation of some that I cannot trace; as *Ipomœa* for instance. Why are you so much offended with some words, which you denominate barbarous, though many of them are more harmonious than others of Greek or Latin origin? as is the case with several Malabar as well

* This cannot be disputed. Dillenius gave up the very faulty name *Malvinda* for *Abutilon*, which, though Arabic, would surely be preferable to *Sida*, which is false.

as Arabic names. These languages are indeed more antient and estimable than the former.

I could wish you to examine carefully the *Dillenia* of your *Nova Genera* 455, and compare it with the *Clusia* of Plumier, 862. If they prove the same *, you will doubtless leave to this genus the name of an author superior to me in merit, as well as, by a prior right, entitled to the honour; in which measure I shall gladly concur.

You desire me attentively to examine the *Trigonella*. I know the flowers are very like those of *Fœnum græcum*, but the pod is different, being short and compressed, not horn-shaped, though single-valved. But if all the species, as well as the common one, really belong to Tournefort's genus of *Fœnum græcum*, though their pods are not horn-shaped, why do you not refer them to it? Nevertheless, these falcate species of Tournefort's *Fœnum græcum*, which, though twisted like a horn, are not horn-shaped, may be better referred to the *Falcata* of Rivinus. *Lupinaster* of Buxbaum has a pod like that of *Trigonella*, but the flower is differently constructed. I wonder you should join the *Fœnum græcum* with the *Medicæ*. Your characters answer to the *Medicago trifolia frutescens incana* of Tournefort †, but not to the rest.

Calamistrum ‡, at least the first and third species

* They are now known to be very widely different.

† *Medicago arborea. Linn. Sp. Pl.* 1096; the true *Cytisus* of Virgil.

‡ *Isoetes.*

in Ray's *Synopsis*, 306, 307, bears seeds very much resembling those of the White Poppy, in a knob at the root, concealed by the onion-like sheaths of the leaves, and lodged in a simple cavity. Nothing like flowers has been observed. The second species in the *Synopsis* belongs to a widely different genus *. The fourth † bears globular seed-vessels, like pills, upon creeping scions, at some distance from the slender roots. These seed-vessels are hairy, and consist of four cells. The younger foliage is rolled up like that of a fern, and beset with a scaly kind of hairiness. I do not find fault with you for making them one species, but one genus. They differ, however, in species not less than in genus.

I fear I have angered you by saying, as you observe in your last, so much against your system of arrangement. Nevertheless, I could say a great deal more, and should be able to prove to you that you separate and tear asunder several genera nearly related to each other. But this is not my aim, as I value your friendship too much. Your remarks are undoubtedly very excellent, on the alphabetical arrangement of my *Hortus Elthamensis*, and that of my *Catalogus Gissensis*, according to the times of flowering. You desire me to point out any one natural class in Ray's *Synopsis*. What are those of the *Fungi* and *Musci*? What are the plants with Compound Flowers? The *Umbelliferæ*, *Verticil-*

* *Littorella.*

† A strange blunder, this description belonging to *Pilularia*; or *Graminifolia palustris*, &c. *Raii Syn.* 136.

latæ, *Stellatæ*, and *Asperifoliæ*? Surely these are better classes than the loose and sweeping ones of every artificial method, which spring up and pass away like mushrooms! What I am most earnestly desirous of is, that you would re-examine your characters, and note the species they are taken from; without which those characters can never be free from uncertainty. This task should be undertaken in an early stage of the growth of your offspring, which will never be handsome, unless it be moulded into form by constant and repeated polishing. Many of the observations I have made are contrary to yours.

You want to know the flowers of the *Liquidambar*. My specimens are without flowers. You have doubtless seen the seeds, which appear under two different shapes, as they come out of their cases, some being angular and chaffy, others like those of Spruce Fir, with a single wing. These last are truly perfect seeds, as I have raised plants from some of them this spring. You could never have taken the *Palmaria* of Muntingius for the same plant, if you had but read his description. It is a bad way to cast one's eyes upon the figures of a book merely, and, like Vaillant, to form an opinion from them only. We ought indispensably to study the descriptions. I believe the flowers to be apetalous, and perhaps amentaceous.

The *Alaternoides Telephii folio** died with me

* *Cluytia alaternoides*. *Linn. Sp. Pl.* 1475.

last winter, but I have acquired it again by cuttings. I have not made any particular notes upon it.

Acacia africana Abruœ foliis, aculeata, &c. *Pluk. Almag.* 3. *Phyt. t.* 123. *f.* 2 *, is widely different from Catesby's *A. monosperma* †. Catesby is an honest, ingenuous man, who ought not to be suspected of error or fraud, as you seem inclined to do. A figure of Plukenet's species, with the pod, is to be seen in the London Gardener's Catalogue, tab. 21 ‡. Both these, and one or two other distinct species, are growing in the Oxford garden. These plants are not of the same genus with Plumier's *Cœsalpinia.*

I do not wonder at your refusing the appointment of Physician to the settlement at Surinam; and I am very glad of it, the situation being unhealthy. But if I were as young as yourself, I should like to visit the Cape of Good Hope, or North America, instead of returning to your own barren country, which, however, you contrast with the above appointment.

A young man of the name of Snell, who is a distant relation of mine, and lives at Giessen in Hesse, wrote to me a short time since, declaring that he has, from his early youth, had a great taste for Botany, and offering his services to me, or to any one to whom I should think fit to recommend him. I do

* Marked *Mimosa reticulata* by Linnæus. See his *Mant.* 129.

† *Gleditsia monosperma. Willd. Sp. Pl.* v. 4. 1097.

‡ Dillenius is here mistaken; this figure being *Gleditsia triacanthos. Linn. Sp. Pl.* 1509.

not believe that his progress, as yet, has been very great; but he might prove useful to some rich Botanist. I can answer for his being born of honest parents; and, if you think proper, I should be obliged to you to recommend him to Mr. Cliffort; if you have any objection, I would not have the matter mentioned.

I write in haste, but am always most faithfully yours,

J. J. DILLENIUS.

MY VERY DEAR FRIEND, Oxford, Nov. 28, 1737.

Your letter of Sept. 12 is come to hand. From that time I have heard nothing of you, either from yourself or others (indeed it is long since I had any letters from Holland), so that I know not where you may be; and I therefore direct this to Mr. Cliffort's, according to your orders.

I am so far from being angry with you, as you seem, by your last, to apprehend, that I, on the contrary, highly esteem and love you; though, as I freely told you, I did not approve of the dedication of your *Critica*, and it would have been more agreeable to me to have been told beforehand of your intention. I have more reason to apprehend your anger, for not approving entirely of every thing you have done, and for speaking my mind so plainly. I cannot but observe that you are not very patient under the attacks of adversaries. For my part, I am not more pleased with my own opinion, than with that of other people. I am ready to listen to any body's remarks, for the sake of discovering truth; but I have no inclination for the see-saw of altercation.

In your last letter but one, you speak of the *Alaternoides Telephii folio* *, as having precisely the style of a *Ricinoides* †; yet in yours of Sept. 12 you pronounce it a true species of *Cluytia*. But has *Cluytia* such a style? I once saw a plant of this genus in flower, but did not then examine the parts of fructification. I have many specimens of the *Alaternoides* in question, with their seed-vessels, on which the styles, still remaining, are those of a *Ricinoides*. These seed-vessels are three-lobed, or separated half way into three capsules, the seeds being separated from each other by a partially-connected membrane. There is one male specimen, whose flowers are few, and solitary from the bosoms of the leaves, which are not, as in yours, broader than those of the females, but rather narrower. There are so few flowers that I scruple to injure the specimen; but I cannot, with my naked eyes, perceive the stamens. Hermann, who gathered this specimen wild, has written upon it, "*Ricinus floribus herbaceis, foliis oblongis.*"

I by no means take it amiss, that the *Cliffortia* is, as you inform me, engraved for a second time; nor should I complain if any one were to engrave over again all the plants of my *Hortus Elthamensis*,

* See the last letter of Dillenius. † *Croton.*

though I might not think it necessary. That plate of the *Hortus Elthamensis** not only expresses the outline of the plant correctly, but, at the same time, gives its stiff character.

I should like to know whether your female *Rhodiola* be not the *Anacampseros radice rosam spirante minor* of Tournefort.

Your *Sparganium* of *Fl. Lapp. sect.* 345, is *S. non ramosum* of *C. Bauh. Pin.* 15. *S. alterum. J. Bauh. v.* 2. 541. This has often been noticed by me in rather shallow pits, and other stagnant shallow waters; nor does the *S. minimum* of C. Bauhin differ from it, except in size. It is also *S. non ramosum minus* of my *Cat. Giss.* 130. *append.* 58; though there mentioned as different from *S. non ramosum* of Caspar Bauhin †. Whether the difference in the flatness of the leaves be not merely accidental, and whether this plant be specifically distinct from our common *Sparganium*, merits inquiry. Great variations are observable in other aquatic plants; as in *Gramen bulbosum* for example. Pray explain to me what you say respecting the native situation of your floating *Sparganium*, under the paragraph β, p. 272.

Your *Azalea* ‡, sect. 89, appears to differ from

* *Cliffortia ilicifolia. Linn. Sp. Pl.* 1469.

† The learned writer had, most assuredly, no correct idea of these plants. The *stigma* decidedly distinguishes them. See *Engl. Bot.* and *Fl. Brit.* 961—963.

‡ *Azalea lapponica. Linn. Sp. Pl.* 214. This remark accords with the correction of the editor in *Fl. Lapp.* ed. 2, where Bauhin's synonym is likewise objected to.

the plant of Tournefort there cited, in having leaves without footstalks, and fewer flowers. Tournefort's, moreover, has shorter and blunter leaves.

I take the opportunity of this name, *Azalea*, to observe, that many of your newly-constructed names please me much; though I cannot commend your rash mode of applying the names of the antients to our plants, without any discrimination, and even to such as are altogether different from the originals. By this means you daily increase the confusion, which has proved so detrimental to Botany, and which renders a *Pinax* so necessary. Do but read the antient descriptions, and I hope you will in future act with more caution.

When you return home, I shall be obliged to you to send me specimens of all the plants you have described as new; first the more perfect ones; then the rest, such as Mosses. I most wish for your new *Salices*. I have your *S. pentandra*, sect. 370; also your 368 *, and 360 †. I have, besides, your sect. 351 ‡ and 355 §. All that you describe as new besides, I want.

I have no occasion for your *Betula* with orbicular crenate leaves ||, Dr. Ammann having sent me specimens.

If you write to me in future, as I much wish you would, pray leave out all formalities and long-

* *S. lanata. Linn. Sp. Pl.* 1446.

† *S. Arbuscula. ib.* 1445.

‡ *S. phylicifolia. ib.* 1442.

§ *S. herbacea. ib.* 1445.

|| *B. nana.*

winded titles, which are odious to my taste. Write as one friend ought to address another.

If you have no other mode of conveyance, you may transmit any thing for me to Mr. J. Hen. de Sprekelsen.

As to that *Alaternoides Telephii folio,* I once judged it, by the habit, to belong to the genus *Telephioides* *. Is there sufficient reason to separate *Cluytia* from this last?

You quote Ludwig in your *Characteres* †. What is his book, where published, and in what form? We very rarely get botanical books here. I am always obliged to write for them to Holland or Hamburgh.

In your last letter you advert to Catesby's names, as if I were particularly conversant with them, or had perhaps communicated them to that author. I assure you he is indebted to me for but few, and those of the shorter kind. I had no hand in his long descriptive names.

Adieu! J. J. DILLENIUS.

* *Andrachne. Linn.*

† This author is mentioned in the *Corollarium Gen. Pl.;* but no work of his, as far as we can find, is quoted, either there or in *Gen. Pl.* itself.

Oxford, Aug. 28, 1738.

Mr. Seeger, a German, conveyed to me your letter from Rouen, dated June 21st, which informs me of your having spent some time at Paris; though I had previously been told that you had set off for your own country, by way of Germany. My friend abovementioned gives me to understand that you have taken your passage on board a ship bound directly for Sweden. Being ignorant of your present residence, I address this letter to the care of Dr. Olaus Celsius; from whom also you will receive the Catalogue of African Plants, by Dr. Shaw, which you wished for.

At the very time when Mr. Seeger was with me, I received from Mr. Ens, the Prussian, your *Classes* and *Corollarium*, for both which I return you my thanks. You have in these works accomplished a very useful task, though, as I perceive, not without much labour. But how shall any one ascertain the new genera of the *Corollarium*, unless he happens to fall on each identical plant, where no synonyms are indicated*? Nor do I approve of characters taken from dried specimens, as it is easy to fall into errors respecting the stamens and pistils.

* This misfortune attends several new genera, described by the illustrious Schreber in his edition of the Linnæan *Gen. Pl.* he not having lived to publish the *Species*. In this predicament are his *Spartina*, *Schoepfia* (but for Willdenow's discovery), *Ochroxylum*, *Brasenia*, *Meyera*, *Villaria*, *Wheelera*, *Schollera*, *Pappophorum*, and his *Wolfia* alias *Pitumba*.

I understand you have made many observations about flowers at Paris; some of which you have, in your usual manner, communicated to me, and they gave me great pleasure. I wish you had been able, or willing, to make a longer stay in certain countries celebrated for their botanical riches; for in that case I have no doubt you would have furnished us with perfect characters of many genera, now, as you are conscious, but incompletely made out. It would surely have been worth your while to visit Greece, or Asia, that you might become acquainted with, and point out to us, the plants of the antients, whose appellations you have so materially, and worse than any other person, misapplied. You ought to be very cautious in changing names, and appropriating them to particular genera. To mention a few, out of many: why do you retain for the *Jasminoides* the name of *Lycium*, and not that of *Rhamnus?* the *Rhamnus cortice albo Monspeliensis** of John Bauhin being, beyond all doubt, *Rhamnus* of the antients. It should seem better to have called all the species of *Rhamnus*, *Lycium* †. Why do you call the *Euonymoides*, *Celastrus?* as it is any thing rather than *Celastrus* of Theophrastus. Why is *Fraxinella*, *Dictamnus?* as we are in possession of the true Cretan *Dictamnus* ‡. Why is our *Gale*,

* *Lycium europæum*. *Linn. Mant.* 47.

† This is matter of opinion. Linnæus followed C. Bauhin, Tournefort, and all modern Botanists. What would have been said had he gone counter to them?

‡ *Origanum Dictamnus*. *Linn. Sp. Pl.* 823.

Myrica? Why *Alaternoides, Phylica?* the *Phylica* of Theophrastus being an acorn-bearing tree, which, according to Bellonius, is known by that name, at this day, to the inhabitants of Mount Athos *; and so of many others.

I should wish to keep the names which Botanists have sanctioned, unless they are altogether bad; nor would I prefer those formed from the Greek to such as are of barbarous origin, provided the latter be already admitted, and not more harsh in sound than the Greek, as *Ipecacuanha* and *Ouragoga.* But let these matters pass.

I recollect you have made some remarks about *Clusia*, in your letter of December 20th last year, which I have not yet answered. You write thus:

"Are there any species of Plumier's *Clusia* furnished with serrated, and plaited, or undulated, leaves? Certainly there are no serrated leaves in any that I have seen. How then could Plukenet venture to combine my Malabar *Dillenia* with an American plant? Does *Clusia* likewise produce a fruit that is internally fleshy and succulent? Are all the *Clusiæ* of Plumier one and the same species?"

I answer, that I cannot find where Plukenet has combined the *Syalita* of Malabar, with his own American *Cenchramidea*, which is Plumier's *Clusia.* All

* Bellonius merely says the *Alaternus* of Pliny retains its antient name of *Phylica;* and Pliny says his *Alaternus* bears no fruit. More might be advanced against Dillenius here, than in support of him, were it possible, or worth while, to attain any positive certainty on the subject.

Plumier's *Clusiæ* have, doubtless, rigid, thick, smooth leaves, without serratures. I have seen those of but two species, knowing the others from Plumier's drawings only, according to which they are all different, and by no means the same species.

Plumier, as well as you yourself after him, asserts the fruit to be pulpy.

I have thought the *Syalita* and *Clusia* might be reduced to the same genus, so that *Clusia* might absorb *Dillenia;* but it never entered my head that *Syalita* was the same species of plant with any of Plumier's *Clusiæ.* I have merely the leaves of *Syalita,* which differ remarkably from those of *Clusia;* but I do not recollect what differences appear in the *Hortus Malabaricus,* and Plumier's drawings, as to their flowers and fruits.

If you will demonstrate to me any real and essential difference between your characters of *Dillenia,* sect 455, and *Clusia,* sect 862, I will allow the genera to be distinct. I wonder, however, that you ascribe deciduous petals to the *Dillenia,* when the contrary is evident from the *Hortus Malabaricus.*

Whether Plumier's *Clusia* be monopetalous or polypetalous, and whether the flowers be deciduous or permanent, does not appear. In his figures the calyx remains with the fruit, but the flowers (petals) do not appear with it, nor do they, as in *Syalita,* invest the fruit.

Clusia flore albo, fructu coccineo, of Plumier *, has middling sized flowers, of four petals, or at least

* *Clusia alba. Linn. Sp. Pl.* 1495.

divided to the base into four roundish segments; and seems the same as *Terebinthus folio singulari*, &c.*, of Sloane's Catalogue, and Hist. v. 2. 91. t. 200. f. 1. *Cenchramidea, Pluk. Phyt. t.* 157. *f.* 2.†. *C. flore roseo, major, fructu subviridi, Plum.* bears large flowers, in seven segments, of the shape of *Nymphæa lutea,* divided to the very base, if indeed they have not separate petals. *C. flore roseo, minor*‡, of the same author, has smaller flowers, of six blunt petals, or segments. *C. alia minor, flore albo, fructu virescente, Plum.* has also moderate-sized flowers, but divided only half way down into five blunt segments. All these, nevertheless, appear, by his drawings, to be genuine species of *Clusia.* To the last of them I refer *Coapoiba,* species 1. of Margrave, book 3, chap. 17. The *C. flore roseo major fructu sub-viridi,* I consider as the second *Coapoiba* of the same writer.

You enquire the number of stamens, in the male flowers of *Cynocrambe* §. What I have observed is as follows. The flower, supported by a slender, perhaps tubular, stalk, is of one petal, naked, tender, greenish, divided almost to the base into two, occasionally three, flat, equal, revolute segments. There are two flowers at each joint, with or without leaves. The stamens are 12, or more, springing from the bottom of the flower, very short, with oblong, slen-

* *C. flava. ibid.* certainly different from *alba.*

† *C. rosea. ibid.* different from both.

‡ *C. venosa. ibid.*

§ *Theligonum Cynocrambe. Linn. Sp. Pl.* 1411.

der, pendulous, tufted anthers. The female flower, or rather the embryo *(germen)*, grows on the same branches, opposite to the foregoing, in the bosoms of the leaves, in pairs likewise, round, marked with a scarcely visible point, which is the stigma, without any style, and destitute of calyx as well as petals, for the little scales, delineated by Tournefort, belong rather to the footstalks of the young leaves; as appears by the larger scales, attached to those of the large leaves.

Tournefort represents five stamens, with short round anthers, which are really as above described, if his species be the same as mine. Indeed one only has hitherto been discovered, in which Columna found numerous stamens. My specimens were raised from seeds sent from Petersburgh, and exactly resemble Columna's plant, except their leaves being larger, owing perhaps to culture *.

Allow me to make a few remarks on your *Lycopodium, Fl. Lapp.* 324. *sect.* 417 †. Boccone's synonym ought to be expunged, his plant being a *Sphagnum*. Ray's also, taken from C. Bauhin, should be struck out, if your species be truly distinct, and the synonyms under your *sect.* 416 ‡ be rightly applied. Ray meant the same plant as the two Bauhins, Tournefort, and Ruppius. It is to be observed that he describes it with " long flower-

* Columna represents the whole plant, leaves and all, much smaller than it is found wild.

† *L. alpinum. Linn. Sp. Pl.* 1567.

‡ *L. complanatum. ibid.*

stalks, clothed with small scattered leaves, and bearing spikes, &c." I have gathered this moss in England, without any stalks, the branches being beset with numerous little leaves, close to the spikes. But I find in my notes, that such is the case at an early period only, the stalks growing out afterwards to the length of half an inch, or more, bearing sharp scattered scales; and I have noticed spikes with and without stalks, on the same specimens. Perhaps therefore your two species, 416 and 417, differ in age only. The distinction you derive from the leaves is nothing; for the plant you delineate, t. 11. f. 6, which is certainly the English one, has both kinds of leaves; those shoots which have no spikes being flat at one side, leafy at the other; while the shoots that bear spikes in the present, or perhaps ensuing, season, are encompassed with leaves on every side, and seem square, though not really so. Whatever the truth of the matter may be, I wish you would send me some good specimens of your No. 416. I want also No. 426, 427, 442, 443, 450, 454 (this *Lichen* * is very desirable), 456, 468, 527, and 528.

Have you ever seen the *Muscus Norwegicus, umbraculo ruberrimo insignitus* of Petiver? *(Cent.* 1. 11. *t.* 1. *f.* 70.) Some have suspected it to be a kind of *Fungus*, but Buxbaum, in the Petersburg Transactions, asserts it to be a real Moss. Pray enquire †.

* *Lichen velleus. Linn. Sp. Pl.* 1617. *Gyrophora vellea. Achar. Syn.* 68.

† It is now well known, though not the less admirable, as *Splachnum rubrum. Linn. Sp. Pl.* 1572.

I have not yet seen the *Hortus Cliffortianus* and *Viridarium Cliffortianum.* Perhaps they are not published.

I enclose flowers of the *Sinapistrum ægyptium heptaphyllum, flore carneo, majus, spinosum. Herm. Hort, L. Bat.* 564 *. The flower is white with me. Also, I send those of *S. triphyllum fœtidum, siliquâ crassâ,* of Boerhaave. You will thus be enabled to see how they differ from the rest.

I am sorry Mr. Von Spreckelsen has not yet forwarded the *Hortus Elthamensis* to you, as I trusted he certainly would. I will mention it when I write. I have had no letters from him these eight months, though I have twice written to him. Farewell.

Linnæus has left the following minutes of his answer; which appears, by the ensuing letter of Dillenius, to have been written Aug. 6, 1739.

With regard to unoccupied names in antient writers, which I have adopted for other well-defined genera, I learned this of you. You, moreover, long ago, pointed out to me, that your own *Draba, Nova Pl. Genera* 122, is different from the plant so called by Dioscorides.

What are the stamens of the *Siliqua,* or *Ceratonia?*

What is *Bonduc duplicato-pinnatis foliis, inermis?*

* *Cleome heptaphylla. Linn. Sp. Pl.* 937.

How many are the stamens in the unknown genera *?

DILLENIUS TO LINNÆUS.

Oxford, March 30, 1740.

Your letter of Aug. 6, 1739, reached me in due course, and I am rejoiced to perceive by it that you are in good health, and usefully employed.

You wish for a description of the flowers of *Ceratonia;* instead of which I send you a flower or two from a dried specimen, having never seen the live plant in blossom.

You ask also for the synonyms of the *Bonduc* which is at Paris. I know none. There is a specimen of this plant, sent by Vaillant to William Sherard, under the following appellation. *Arbor canadensis, foliis Bonduc, vel Androsœmi majoris.* Guarded enough!

You desire, moreover, a flower of Tournefort's *Thalictrum montanum præcox* †. I have none. There is a flower or two, resembling Clusius's figure, on Sherard's specimen, firmly glued to the paper, which I cannot conscientiously take off, having promised to preserve all these things entire. If there were plenty, and loose, I might have been able to spare one.

* This question seems as difficult to understand as to answer, and it is probably a concise allusion to what was more fully expressed in the letter. Dillenius passes it over.

† *Isopyrum thalictroides. Linn. Sp. Pl.* 783.

I am neither able nor willing to say any thing about the *Pinax*. As to Mosses you see the state of the case. If you would, within six or seven months, send me those I have asked for, and you have promised, I shall be able to introduce the new ones into their proper places, and more certainly to fix the synonyms of the old ones. I hope you will not forget that curious *Muscus licheniformis* found in Smoland, about whose fructification you relate wonders, such as have never occurred to me.

I have this winter seen the *Blasia* of Micheli (not very accurately described and delineated by him), which agrees in many points with your description, though it differs in several others. The whole upper side of the prostrate leaves is thickly covered with undivided tubercles, or warts, and from the middle of each leaf arises a short stalk, bearing an ovate powdery head, like that of a *Mnium;* which stalk, when the powder is dislodged, appears hollow or tubular, as it seems to me. At the time of my present writing, this plant is almost withered and dried up, as if it were annual, the tubercles being already opened and bearing seed. The hairs have disappeared. While the plant is growing, the tubercles are inseparable from the leaf, being solid, and not hollow *.

The Mosses, or whatever you may have for me,

* The accurate observations of Dr. Hooker, now Professor of Botany at Glasgow, have proved this plant a real *Jungermannia*, whose calyx and veil are imbedded in the leaf. See his excellent Monograph on this genus.

may be sent to London; either to the care of your Swedish clergyman, Mr. Biorke, or to Mr. Peter Collinson, Mercer, at the Red Lion in Gracechurch Street, from whom I should be sure to receive them safe, even if they were a large sum of guineas, for the printing of my *Historia Muscorum.*

Farewell, my excellent and highly-valued friend, and when you write to Upsal, present my compliments to Dr. Olaus Celsius, with the enclosed sheet *.

My much valued friend, Oxford, April 30, 1741.

Your letter of October 11th last year, full of much botanical news, was highly acceptable, and I return you my thanks for your communications. The maritime plant found by you in a certain island, is new and puzzling to me.

Our friend Peter Collinson sent, for me to peruse, the Transactions of the Upsal Society for 1740, which book was, I think, given him by Mr. Biorke, and therefore I returned it to him. Another volume has, doubtless, come out by this time, which I should be much obliged to you to send me, with the abovementioned, and I will pay Mr. Biorke or Mr. Collinson for them on your account.

My first volume of the Swedish Transactions for 1720, wants, at page 97, the figure of *Dortmanna*

* Probably the proposals for the *Historia Muscorum.*

laçustris: and the fourth volume, for 1730, wants a sheet, trom p. 65 to 72, which defects I should be particularly obliged to you to make good.

You some time ago wished for the Catalogue of the plants about Nottingham, and I sent it; but I think I have been told the ship by which it went was lost, so I now send you another copy.

I am happy that Celsius is so well. I also grow an old man, but have not enjoyed good health for some years past. Pray remember me to him. Believe me, at the same time, your most devoted,

J. J. DILLENIUS.

Oxford, Oct. 15, 1741.

MY EMINENT AND LEARNED FRIEND,

Having at length got rid of the burthen of my *Historia Muscorum*, I proceed to reply to your letters of September 11, and October 13, 1740. When I name that work, I hope it will obtain your pardon for my silence. All the plates, amounting to 85, are printed, as well as 61 sheets of letter-press: 12 or 14 more remain to be printed, which will be finished before I can receive an answer to this letter. I am sorry that it must be the middle of winter before the book can come out, nor can it therefore reach you so speedily as I wish. You say you could get it better from Amsterdam than from London, but I beg to be informed to whom the copies should be directed. I am to send eight copies to Wetstein and Schmidt, to which I propose to add yours, which you shall have so far free of expense, only you must let me know who will take charge of the parcel for you. Mr. Collinson has received, in my name, the first subscription from Mr. Waesberg, of two guineas, one for the copy on fine paper, and one for the two copies ono rdinary paper.

Micheli's ***Blasia*** is surely a species of ***Mnium.*** You seem to me to have mistaken for it some species of a true *Lichen* *. The figure in Micheli, which you reckon excellent, is not without fault. The Moss you observed in Smoland, and have sent me, is *Lichenastrum*, No. 3, in Ray's *Synopsis*, p. 111, the 41st species in my *Hist. Musc.* † The leaves differ widely from the *Marchantia* of authors.

The other doubtful Mosses of the *Flora Lapponica*, which I asked for, and you mention having sent (but when, by what ship, and to whose care, I know not), have neither come to my hands, nor to those of Biorke or Collinson. I have therefore been obliged to leave them doubtful in my *Historia*. It is proper that you should explain them, either referring them to some of my species, or pointing them out as new, which you will easily do when you see that work.

Enclosed are the female flowers of *Ceratonia*, which grow on a separate tree from the male, from the stem itself, in numerous crowded spikes, each

* The writer means what is now called *Marchantia*.

† *Jungermannia epiphylla. Linn. Sp. Pl.* 1602.

an inch or inch and half long. Every flower consists of a simple-leaved calyx, in five small fleshy divisions, like so many warts, with a simple pistil. The latter is at first short, and purple; then longer, and of a hoary green, springing from a green fleshy receptacle, which is small at first, but becomes more conspicuous afterwards. These flowers are well represented in Camerarius, except that the segments of the calyx in mine are smaller, and the spikes shorter. I have not found any petals.

I hear Rosen is dead, and I hope you are by this time his successor. Adieu. J. J. Dillenius.

Oxford, Sept. 23, 1742. O. S.

I replied on the 15th of October 1741, to your letters of September 15, and October 3, 1740, informing you that the *Historia Muscorum* was likely to be completed in a few months, and requesting you would say to whom you would have your copies entrusted. If you had given me this information within the time limited, they would have been sent with others, to Wetstein and Schmidt at Amsterdam, early in February last. But as your letter of the 19th of February did not reach me till the 13th of March, the books have remained with P. Collinson, at London; more especially as Mr. Waesberg, who is in possession of the receipts for the first subscription, has returned no answer to Mr. Collinson's repeated letters on the subject.

As soon as I received yours of the 19th of February, I charged Mr. Collinson to send, as soon as possible, your copies above mentioned, to the Widow Schmidt, at Amsterdam, who had mentioned that the concluding subscription money would be paid, by some merchant or other, to Mr. Gronovius; and when I was in London last June, I found the books had, accordingly, been sent about the end of May. Whether the Widow Schmidt has received and forwarded them, or whether Gronovius has received the money for me, I know not; for since this friend of ours has devoted himself to public business, he writes but seldom, and very concisely. I am sorry I can give you no better account, but you see the delay is not owing to any fault of mine, and I hope to hear, by your next letter, that you have got your three copies of the *Historia Muscorum*.

There were in your letter of February 19th four specimens of plants, but so broken and bruised, that what they were I could not make out. You desire me to enclose my letter in a cover to the Royal Society of Upsal; but I cannot comply with this, as I must pay for every such cover, double postage here.

In your last letter of all, I find a plant gathered in Charles island (Stora Carlsöen) on the coast of Gothland *, which you judge to be *Polygonum majus erectum angustifolium, floribus candidis*, of Mentzelius, *Pugill. t.* 2; and *Caryophyllus saxa-*

* *Gypsophila fastigiata.* See *Fl. Suec. ed.* 2. 145.

tilis, foliis gramineis, umbellatis corymbis, C. Bauh. Pin. 211; nor do I object. But it is by no means Tournefort's *Lychnis alpina linifolia multiflora, perampld radice* *, whose flowers are more scattered, and the leaves broader in the middle, though narrower at the end.

I have no seed of the *Thalictrum montanum præcox* of Tournefort †, nor have I ever seen that plant growing. Your conjecture appears to me just, that *Helleborus fumariæ foliis, Amman Stirp.* 74. *t.* 12 ‡, belongs to the same genus. But can you tell me whether this *Helleborus* of Amman, which by the fragment enclosed, I perceive you have seen, be a true species of *Helleborus*, or of *Thalictrum*, or perhaps of *Aquilegia?* You may ascertain this by its nectaries.

I shall be happy to send you any seeds that I can, but you had best let me know what you want. I am accustomed to send a printed list of my wants to the lovers of Botany. You are however so rich already, having two thousand plants in your garden, that I doubt whether I shall be able to contribute much; especially as you are about sending collectors to America. The unfavourable seasons, for three years past, have robbed me of many plants; and the few seeds sent hither from America, are generally either decayed, or not well ripened. You may

* *Gypsopila repens. Linn. Sp. Pl.* 581.

† *Isophrum thalictroides. Linn. Sp. Pl.* 783. *Jacq. Austr.* 105.

‡ *J. fumairoides. Linn. Sp. Pl.* 783.

expect better success, your people being more diligent, as well as more learned and curious, than ours. Moreover, you must recollect that the situation of Oxford is low and watery, the neighbouring meadows being overflowed, close to the garden, after the autumnal rains; so that the dews, and heavy morning fogs, will scarcely ever allow the plants of hot climates to bring their seeds to perfection.

Present my best compliments to Dr. Olaus Celsius, and let him know that I have duly received his letter of the 25th of September, 1741, by Mr. Bæck, who was lately here, and is now at Chelsea. I agree with him as to the enclosed plants, one being *Clathroidastrum obscurum majus et minus, Mich. Nov. Gen. p.* 215. *n.* 1, 2. *tab.* 94. *f.* 1, 2 *; the other *Muscus capillaceus aphyllos, capitulo crasso bivalvi, Buxb. Cent.* 2. *p.* 8. *t.* 4. *f.* 2 †; which last I have described in my *Hist. Musc. p.* 477. *t.* 68. *f.* 5; and which, whenever you meet with it growing, I wish you would examine with your usual care and accuracy. Adieu.

Oxford, Feb. 4, 1743. O. S.

My very dear Friend,

I received your letter written September 18th last (though without a date, as is sometimes your prac-

* *Clathrus nudus. Linn. Sp. Pl.* 1649. *Stemonitis fasciculata. Persoon Syn.* 187.

† *Buxbaumia aphylla. Linn. Sp. Pl.* 1570. *Engl. Bot. t.* 1596.

tice), by which I am happy to learn the safe arrival of your copies of the *Historia Muscorum*. Be so good as to refer to their proper species, those doubtful names of yours, which I have here and there mentioned at the conclusion of each genus; as well as some other uncertain matters which you will meet with; and if you have specimens to spare, I shall thank you for them. Send me likewise that Moss which you take for my first species of *Lichenastrum*, and which you mention as bearing the heads of a *Mnium*. Your specific character applies to some others of this genus likewise.

The *Absinthium** enclosed in your letter is not *A. ponticum repens vel supinum* of C. Bauhin, according to the specimen in Sherard's herbarium, but a new species. I have Siberian specimens also from Amman, but without a name. Is it not his No. 197 of *Stirp. Ruth.?* But I beseech you, in future, not to stuff your letters with plants or seeds, as they cost me too much for postage. Without such, your letters will always be acceptable to me, especially if they are not enclosed in a cover.

On the 24th of December 1742, I sent to Mr. Biorke, the minister of your Swedish church at London, through the hands of P. Collinson, the Catalogue of Nottingham Plants, with a parcel of seeds. On the 31st of January 1743, I sent several things, through the same channel, for you, among which you will find 36 different seeds, lately brought

* *Artemisia palustris. Linn. Sp. Pl.* 1185.

from Pennsylvania. I wish they may reach you speedily, and not too late for sowing, as the ships for Sweden are tardy in sailing.

I want the Swedish Transactions for 1733, and all the following that have come out. Pray buy them for me, and I will either return you some other book, or pay the money in London to your order. Direct the parcel to Mr. Collinson, as before mentioned, but inform me when, and by what ship, it is dispatched; or you may send it to Leyden, to the care of Dr. J. Fr. Gronovius; or to Hamburgh, to Mr. J. H. Van Sprekelsen. I applied to the latter, several years ago, for these books, but he informed me they were not to be had there.

Compliments to Olaus Celsius. Believe me your devoted, J. J. DILLENIUS.

London, Nov. 29, 1743.

MY MUCH-ESTEEMED FRIEND,

I write in haste from Mr. Collinson's, where I am staying for a few days, before I return to Oxford. I have written to Mr. Biorke, to call for this letter and a parcel of seeds, at our friend's house. They consist of African Geraniums, which I have collected for you this summer, according to promise; about 13 species. I wish they may speedily come to hand, and find you in good health.

Do not forget your faithful friend,

J. J. DILLENIUS.

Oxford, Feb. 10, 1746.

Having heard that the ship, by which I sent Deering's catalogue, was lost, I send you another copy, as you have expressed so earnest a wish for it. I might perhaps have rendered you an acceptable service in transmitting a few other works, of a similar kind; for books of a small size are seldom exported or imported, as I find by experience, being very rarely able to procure such from abroad.

A great part of the plants you ask for are natives of hot countries, of which I have but few. Miller could supply you better, though I do not believe that more of these plants ripen their seeds with him than with me. As to those cultivated in the open ground, one of these is the *Carum*, Caroway, which I send you. The *Gratiola*, like other creeping plants, seldom bears seeds. The *Chionanthus* is a small tree, which flowers with us every year, but never brings its seeds to perfection. I have several times had them from America, but none ever came up, as they were generally pierced through and through by maggots. I have a notion this plant might be referred to some common genus. I gathered four or five seeds of *Morina* * last Summer, but committed them all immediately to the ground. The *Viburnum* is so common here that nobody thinks of collecting its seeds. I obtained a root or

* Hence it appears that this interesting plant survived at Oxford the hard winter of 1739-40, which destroyed it in most gardens; and that it perished there from subsequent neglect, after the time of Dillenius.

two of *Collinsonia* last year, but seeds are to be had from very old plants only. The *Euonymoides* *, *Sassafras, Sanguinaria, Tetragonotheca, Amorpha*, and American *Gale*, are to be enquired for from your countrymen in America. *Patagonula* died with J. Sherard. Who could ever suppose that the Tamarind tree, the *Rauwolfia*, or the *Plumieria*, &c. bore seeds with us? and yet you ask for these! I have the *Turnera foliis oblongis* †, which bears plenty of seeds, but as they scatter themselves about the pots, and thus produce plenty of young plants, they are never gathered. I will take care of some for you next season.

I now however send you seeds of 140 annuals, which grow here freely, without attention, and these are more in number than I ever received from any body at once. Many of them indeed are common things, and you may perhaps say, as others have said to me, either that you do not care for them, or that you have them already. I shall henceforth deal with you as I do with others, not sending a single seed that you do not expressly ask for. Nor will I accept in return either books, empty titles, gold, or silver, and least of all children ‡. You must send me other seeds, according to this rule; that if

* *Celastrus.* † *T. ulmifolia β. Linn. Sp. Pl.* 387.

‡ The original is "*minimè verò natorum.*" This strange passage may perhaps be explained by Dillenius's letter of August 18th 1737, where he recommends to the care of Linnæus a young relation of his own, with whom he evidently did not wish to be encumbered himself.

you obtain any from abroad, you shall divide them with me; or if yon have none of these, I wish for any of Swedish growth that are not in Ray's *Synopsis.* What have I received in return for all the seeds sent you, but those of one plant, the *Helleborus Fumariæ foliis* * ? I have never seen one of the Lapland seeds you promised me. But this I can excuse, knowing very well the difficulty of procuring them; nor am I ignorant that the seeds of alpine plants do not prosper in low and warm situations, any more than the plants themselves, which are accustomed to a very cold climate; as I have learned from some of the productions of Greenland.

I cannot however forgive you for not sending me, after so long a period, those doubtful Mosses, of which I think I sent you a list; or if not, you may see what they are by my *Historia Muscorum.* I formerly asked you to get me the figure of *Dortmanna, Act. Suec. v.* 1. *p.*'97, and sheet I of *v.* 3, both wanting in my copy. I now repeat this request, in complying with which you may, with little trouble or cost, render me an acceptable service. Farewell!

* *Isopyrum fumarioides.* See, hereafter, Professor Amman's letter of Nov. 18, 1740.

DR. DILLENIUS TO RICHARD RICHARDSON, M. D. NORTH BIERLY, NEAR BRADFORD, YORKSHIRE.

[English. Slightly corrected.]

London, Aug. 1, 1724, from Mrs. Allandson's in Barking Alley.

HONOURED SIR,

The *Synopsis Stirpium Britanniæ* being lately finished, I intended to send you a couple of books; but Consul Sherard having some other things to send you, I defer, and leave the sending of them to him. I had resolved to do myself the honour of dedicating it to you and Mr. Sherard, two persons who have contributed the most to its perfection. But having some apprehension, being a foreigner, of making natives uneasy if I should publish it in my name; and considering that a dedication without a name could not be very acceptable, I must deprive myself of that honour. However, under the name of Editor, I have dedicated it, in the form of an inscription, to all those Lovers of Botany who have contributed the most to this edition and augmentation of the work; in hope that the two principal persons concerned will take it no less kindly than if it had been directly addressed to them.

Perhaps I may be so happy as to find another and a better opportunity for such a compliment; I mean the History of Mosses, if I could find time to finish it, and if a simpling voyage to America, which, having lately quitted home, I am resolved on, when a good opportunity offers, does not prevent the publication. I wish with all my heart I could finish this work before I leave England, as nobody will ever again take so much pains, nor will have so much encouragement and assistance as I have had from you and Dr. Sherard. But as long as I live with him, and am employed in a more necessary work, I doubt very much whether I can find any leisure time, unless some good friend of his would please to persuade him to let me have one day in a week for this purpose.

I am, with all respect, honoured Sir, your most obliged and obedient servant, Jo. Jac. Dillenius.

P. S. Dr. Sherard hath not been well these five or six weeks, and hath now almost lost his stomach. The worst is, that he refuses all physick, and neither Dr. Mead, Dr. Grew, nor his brother, can yet prevail with him to take any thing. But I beg the favour of you not to take any notice of this particular especially. I hope it will go off again, since the only fault seems to lie in the stomach. Please to excuse my bad writing.

Honoured Sir, London, Oct. 8, 1726.

I spent two months in my botanical journey, and intended giving you a short account of what I observed, as soon as I returned home, which was on the 7th of last month; but my plants being dispersed, and left partly in Wiltshire, partly in Shropshire, I was obliged to defer it till I got them together.

From Trowbridge Mr. Brewer and I went to Mendip hills, where we could not find the *Muscus denticulatus* of Clusius (1), mentioned by Lobel as growing there; but instead of it we saw the *Muscus lanuginosus alpinus* (2), and a new Mushroom, of the *Fungoides* kind, very tender, of a straw-colour, and ending in sharp points, not branched (3). These hills are of great extent, and at one end of them, near Chedder, is a remarkable place, as well for its singularity as for the plants there growing. We saw there several Welsh plants, not known to grow in England, as *Papaver luteum perenne* (4); *Sedum alpinum trifido folio* (5); and several Welsh Ferns; also a new *Lichen* (6), with very delicate bright-green leaves.

From hence we walked to Brent Down, a peninsula not noticed by geographers, though as remarkable as any of the Holms islands, over against which it lies. Here we found in plenty, on the top and about the middle of the hill, *Chamæcistus montanus polii folio* of Plukenet (7); and an unknown Grass, *Spicâ Sparti, foliis reflexis angustis glaucis striatis, radice crassâ et fungosâ.* A little lower, *Lychnis*

(1) Or rather of Gerarde. *Lycopodium denticulatum. Linn. Sp. Pl.* 1569.

(2) *Trichostomum lanuginosum. Fl. Brit.* 1240; see *Dill. Musc.* 372.

(3) Apparently *Clavaria fastigiata. Linn. Sp. Pl.* 1652; figured by Dill. in *Raii Syn. t.* 24. *f.* 5.

(4) *P. cambricum.*

(5) *Saxifraga hypnoides.*

(6) This should be some *Marchantia, Jungermannia,* or *Riccia.*

(7) *Cistus polifolius.*

maritima, Behen dicta, flore et folio majore (1), first observed, after the *Synopsis* was printed, by Mr. Brewer, and sent to Mr. Sherard's garden, where I believe you have seen it. The place mentioned for the *Polygonum maritimum, longiùs radicatum,* &c. of Dr. Plukenet, is but two or three miles from hence, and we could not miss it, being of no great extent; but we searched in vain. Over against Brent Down, on a rocky hill, where Uphill church stands, I gathered seeds of *Peucedanum minus* (2), and sent a few by post to Mr. Sherard, who raised them all, and you may have plants or seeds from him next year; which I mention, having lost the rest that I gathered. I have seeds of the *Cistus* for you, and a few others, which I will send the first opportunity.

From these parts we set out for Bristol, and from thence travelled through Gloucestershire, Worcestershire, and Shropshire, to meet Mr. Brown at Bishop's Castle; he being desirous of going with us to Snowdon, but he went only as far as Cader Idris. We observed little remarkable by the way. *Alcea tenuifolia crispa* (3) of John Bauhin is pretty common that way, and no other. In a hilly wood near Worcester we observed a species of *Campanula*, with scattered flowers, on long slender spreading stalks, a square upright hairy stem, upper leaves very nar-

(1) *Silene maritima, With.* 414. *Fl. Brit.* 468.

(2) *Pimpinella dioica. Linn. Syst. Veg.* and *Fl. Brit.* 332.

(3) *Malva moschata.*

row, lower broader, almost of the shape of *Veronica officinalis,* slightly hairy, minutely and elegantly crenate; the root short, annual, with few fibres (1). I take it to be new. In boggy meadows here, as well as in other counties, I have observed, this year and the last, a *Gramen junceum* with jointed leaves and black shining heads, a root more fibrous and creeping than the common kind, the whole plant of more humble stature, and earlier (2). This is as common, if not more so, than the other. It is one of Micheli's, in the *Hortus Pisanus.* Along the Severn, to a great extent, grows wild the *Brassica sylvestris, rapum radice oblongâ* (3), and *Sinapi siliquâ latiusculâ glabrâ,* &c. (4) of J. Bauhin. We saw here and there, in Shropshire, *Sphondylium foliis angustioribus* (5), which I believe to be a different species.

Near Norbury, four or five miles before we came to Bishop's Castle, grows *Pimpinella tenuifolia* of Rivinus (6), *Pentap. Irr. t.* 83.

Travelling from Bishop's Castle into Wales, in boggy ground upon the downs of Montgomeryshire, we observed *Gramen miliaceum exiguum palustre, paniculâ e locustis globularibus minimis constructa,*

(1) *Campanula patula.* See *Hort. Elth.* 68. *t.* 58.
(2) *Juncus lampocarpus. Davies, Tr. of L. Soc. v.* 10. 13.
(3) *Brassica Rapa β. Fl. Brit.* 720.
(4) *Sinapis nigra.*
(5) *Heracleum Sphondylium β. Fl. Brit.* 307.
(6) *P. saxifraga β. Fl. Brit.* 331.

new, as I think. Towards Llanydlos, in the hedges, *Oxyacantha folio et fructu minore*, noticed, if I mistake not, by Pontedera.

Betwixt Llanydlos and Dolgelle, and between the latter and Carnarvon, we observed several new Mosses of the *Pulmonaria* kind; viz. *Pulmonaria arborea minor*. Micheli *Nov. Gen. t.* 45 (1). *Lichenoides arboreum, foliis lætè virentibus latis, scutellis fuscis*, nondescript (2). *Lichenoides arboreum foliosum, ex cinereo glaucum, infernè scabrum* (3). The best country for Mosses, that I ever was in, is between Dolgelle and Carnarvon. We might have found a good many new ones there, had not the violence of the rain and wind prevented us.

We had only one fair day at Dolgelly, on which we ascended the hill of Cader Idris, and found there many of the Welsh plants; but Snowdon has still the preference, above this or any other mountain I have visited. *Campanula alpina, foliis imis rotundioribus* (4), grows there, as well as on Snowdon; but I think it only a variety of the common one. About the cascades, in ascending the highest part of the hill, I found a *Lichenastrum*, with round, silvery, densely-fibrous shoots, not described (5), which I saw afterwards upon the Glyder; and a

(1) A narrow variety of *Lichen pulmonarius, Linn.*

(2) *Lichenoides n.* 98. *Dill. Musc.* 195. *t.* 25. *Lichen lætèvirens. Lightfoot Scot.* 852.

(3) Perhaps *Lichen caperatus Linn.*

(4) *C. rotundifolia β. Fl. Brit.* 235.

(5) *Jungermannia julacea.*

very elegant *Muscus coralloides, facie Corallinæ marinæ* (1), growing out of the slate rocks. This I did not observe on Snowdon. Between Carnarvon and Dolgelle, amongst ferns, in heathy ground, I found a very elegant upright Vetch, with pointed glaucous leaves, pods like those of the Lentil, growing many together on a long stalk, no tendrils. I had no time, nor would the rain permit me, to look after the root, whether it were that of an *Orobus*, but the leaves do not agree with the *O. sylvaticus nostras* (2). Here, as well as in other parts of Wales, along the banks of rivers, grow two *Salices*, one with a sage-like rugged leaf (3), the other with an obtuse, somewhat glaucous, leaf, *neque compacto, neque laxiore, sed medio* (4), which I take to be different from all the rest of the English Willows.

The weather being so bad, we resolved to go to Carnarvon, and to spend some time there and in the island of Anglesea, till it should settle fair, before we visited Snowdon. In the Carnarvon river, which runs down from Llanberis, I met with the seeds of *Subularia repens, folio minùs rigido* (5). It has a

(1) *Lichen fragilis. Linn. Sp. Pl.* 1621.

(2) This could scarcely be any thing else than *O. sylvaticus*.

(3) Perhaps *Salix cinerea*.

(4) Possibly *S. Lambertiana. Fl. Brit.* 1041.

(5) *Dill. in Raii Syn.* 306. Nothing is more certain than that this plant is *Littorella lacustris*, mentioned as a *Plantago* in the same work, 316, n. 11. Whether insects caused the appearances described by Dillenius, and exhibited in his *Hist. Musc. t.* 81, we can but conjecture. They seem to have been found only once.

naked seed, contained in a calyx cut into four segments. There is never more than one seed upon each little stalk or pedicle. Along the leaves come out, here and there, small horns beset with four or five marginal teeth, which may probably contain a dust, like the *apices* (or anthers) of perfect flowers. I was too late to ascertain this with certainty. The *Subulariæ rigidæ* (1) are of a quite different character, for they bear at the bottom of their leaves, within, numerous seeds, like those of a Poppy in shape and size, which you may find, I believe, in your dried specimens, if you cut them, in a sloping direction, just above the tuberous root. Some leaves contain nothing but dust, like what is in the head of a Moss. I know not whether this be unripe seeds, or a fecundating powder. It appears at the same time with the seeds formed.

Anglesea is in its soil very like England, and, except some marine plants, has no great variety or diversity. *Echium marinum* (2) does not grow near Trefarthen; but we found it afterwards plentifully by Llyffny river, where we went in search of it, three or four miles before we got to the place mentioned (Clynog). *Pneumonanthe* of Cordus (3) grows plentifully on some boggy commons in Anglesea. In a wood I found *Fungi digitelli* of Par-

(1) These are the *Isoetes*.

(2) *Pulmonaria maritima*.

(3) *Gentiana Pneumonanthe*.

kinson, never seen by me before; and a new *Agaricus globosus anthracinus,* destitute of either pores or gills (1). Two new Sea-mosses over against Prestholm island, where we found also, in plenty, *Chamæfilix marina anglica* (2). In a small river that runs out of a pond near Esquire Baly's, I observed a *Spongia fluviatilis,* a soft, unbranched, very elegant species, of a bright green; and a *Potamogeton,* with oblong flat leaves, the lower ones alternate, the upper opposite. *Plantago marina,* the same with that found in Durham, having thinner and more carinated leaves (3), a variety of the maritime one, grows all over the inland part of the island (of Anglesea). *Odontites* (4) with a white flower, in some pastures. At Llandwyn near Newborough, besides other marine plants, grows the *Chamæfilix marina* (abovementioned); *Thalictrum minus; Anonis maritima procumbens* (5), &c. of Plukenet; *Vulneraria flore coccineo* (6); Mr. Stonestreet's *Tithymalus* (7), but rarely, on a small neck of land running into the sea; *Viola alpina lutea, cum flore minore* (8), a variety of the larger Welsh; at Abermeny ferry, *Cakile marina* (9), which I believe has

(1) *Sphæria maxima?* (2) *Asplenium marinum.*

(3) *P. maritima β. Fl. Brit.* 184.

(4) *Bartsia Odontites.*

(5) *Ononis arvensis γ. Fl. Brit.* 758.

(6) *Anthyllis vulneraria β. Fl. Brit.* 760.

(7) *Euphorbia Portlandica.*

(8) *Viola lutea.* (9) *Bunias Cakile.*

been mistaken for *Leucojum marinum* (1); and *Eruca monensis* (2); a plant different from Boccone's, but the same with Plukenet's, though very ill figured by him. The flower is pretty large, like *Rapistrum* (3). I could find but very few specimens, and no seeds; but brought some young plants with me, which grow well at Mr. Sherard's.

After a week's stay in this island, we got a fair call for Snowdon, for the wind turning north-east, cleared all the Welsh hills, so that we left Holyhead, and the northern part of Anglesea, unsearched. We had pretty fair weather most of the time we were at Llanberis. There grows here and there, in wet places amongst the rocks, a *Bryum* or *Hypnum*, of a deep shining purple colour (4); and a green one, pointed and pungent at the extremities (5); which I remember in the Consul's collection, probably sent by you, but not taken notice of in the *Synopsis*. I could not find any heads on either of them. We found most of the Welsh plants then in season; but missed some upon Clogwyn y Garnedh, viz. *Filix pedicularis rubræ foliis* (6); *Salix pumila, folio rotundo* (7); *Cirsium*

(1) *Cheiranthus sinuatus*, certainly found in Wales.

(2) *Sisymbrium monense.*

(3) *Raphanus Raphanistrum.*

(4) *Bryum alpinum.*

(5) *Sphagnum alpinum;* see *Dill. Musc.* 245.

(6) *Woodsia hyperborea, Brown, Tr. L. Soc. v.* 11. 170. t. 11. *Compend. Fl. Brit.* 158.

(7) *Salix reticulata.*

humile montanum, cynoglossi folio, polyanthemum (1). At the very top of Snowdon I met with *Muscus islandicus purgans* of Bartholin (2); and at the bottom of it, on the east side, in a meadow, *Campanula foliis cymbalariæ* (3), in plenty. In a lake at the foot of Gribgoch I found *Potamogeton lapathi minoris foliis pellucidis*, D. Llwyd (4). On the green pastures near the top of Gribgoch I could find nothing like a *Bistorta folio vario* (5), but an *Acetosa lanceolato folio glabro spisso, obtuso, et vix auriculato* (6), in great plenty, which I have seen on other hills in Wales, and found only one specimen in flower. The lower leaves are very small, and roundish; that on the stalk broad at the base, long and tapering to a sharp point (deltoid and pointed, by the sketch in the letter). Whether Parkinson mistook this for a *Bistorta*, I cannot assert. His figure does not agree with my specimen. I brought plants with me, and shall see next year what they come to. The *Hieracium latifolium, uno vel altero flore* (7), is only a variety of the common *Pulmo-*

(1) *Serratula alpina.* (2) *Lichen islandicus. Linn.*

(3) *Campanula hederacea.*

(4) *Raii Syn.* 150, *n.* 10, possibly *P. heterophyllum. Engl. Bot. t.* 1285.

(5) A variety of *Polygonum viviparum*, found here by Parkinson.

(6) May be a variety of *Rumex Acetosa*, a very variable plant, if there be not more than one species confounded under it. (7) *Raii Syn.* 170, *n.* 13.

naria gallica (1). Not far from Llanberis church, along the road, grows a *Gentianella pilosa, flore semper quadripartito* (2), very different from *pratensis flore languinoso* of C. Bauhin. I find specimens of one amongst Consul Sherard's, gathered near Malham, which agree with this, except that the Malham ones seem to have the flowers divided into five segments. If I had a loose specimen or two I could better determine the difference. Pray, Sir, when does that at Malham blossom, in spring or autumn?

Our guide not being so well acquainted with the Glyder as with the hills on the other side, we could not get to the place where the *Bulbosa alpina juncifolia* (3) grows. Nor could we find, on the south side of Llyn y Cown, the *Hieracium* mentioned to grow there (4); nor the *Virga aurea montana, flore conglobato* (5). There grows one on all the hills about Llanberis, and on other hills in Wales, which is indeed nothing but the common one. I am sure we were at the right place, for we found there *Lycopodium foliis juniperi* (6). In the lake at Cown

(1) *Hieracium sylvaticum.*

(2) This may have been a four-cleft variety of *G. Amarella.*

(3) *Anthericum serotinum.*

(4) See *Rau Syn.* 168, *n.* 7. Gerarde's plant is *Cineraria integrifolia.* Dr. Richardson's, found at Llyn y Cwin, appears by his own specimen, shown to the writer of this by Mr. Hailstone, to be *H. sylvaticum β.* Tr. L. Soc. v. 9. 240.

(5) *Raii Syn.* 177, *n.* 4. Dillenius was surely more correct here than afterwards, when he published the Welsh *Solidago* as distinct.

(6) *L. annotinum.*

I found the common *Subularia folio rigido* (1), mentioned to grow only in Phynon Vreech, and the *Juncifolia cochleariæ capsulis* (2) pretty plentifully, which relieved me very much of our disappointment of not seeing more Glyder plants. In the lake near Llanberis, a little further on, where you found the *Subularia fragilis, folio longiore et tenuiore* (3), cast out of the lake, I pulled off my shoes and stockings, and found it growing there in great plenty. If any body had the means of fishing out plants from the depths of these lakes, I am inclined to think he might find strange things. Near this place, about three years ago, Mr. Evans, coming home late from a christening, in stormy and rainy weather, was drowned. His corpse could not be found by any means used for fishing. There being no parson living at the place at present, it is almost impossible for any body to go herborizing thither. We had very hard and uncomfortable lodging at the alehouse, and with difficulty got a young man to be our interpreter and guide. At last young Mr. Evans, of Bangor, gave us leave to lie at his house, and sent us provisions from Bangor.

If some rich botanist, that has no relations or children, would build a house there, and buy some land to it, which might be done with little money, it would be a very kind invitation for botanists to

(1) *Isoetes lacustris.* (2) *Subularia aquatica.*

(3) *Isoetes lacustris,* the long-leaved variety, described in *Dill. Musc.* 541. t. 80. *f.* 2.

visit these strange places, and be an inducement for making a collection of Welsh plants, as you proposed. Without such a fixed place of abode, it seems to me impracticable.

I am afraid, Sir, I have tired your patience. I give you many thanks for your directions, which were very useful to us. I do not know whether you make any collection of (dried) plants. If so, any of my duplicates that you may want are at your service.

Dr. Sherard arrived safe from Holland last Wednesday, and is at present at Eltham. Mrs. Wansell, who gives her humble service to you, told me something about Oxford, which I am glad you are sensible of; especially if you should think it proper to advise the Consul, and to tell him your opinion. I beg and hope you will excuse the trouble of this, from your most obliged and most humble servant,

J. J. Dillenius.

Dear Sir, London, Jan. 30, 1727.

I received both your kind letters in due time, and delivered the money inclosed, to Consul Sherard immediately. I am extremely obliged by the communication of your observations and additions to the *Synopsis*, which I shall take care to enter, and to preserve. Since you promise me an account of the times of flowering of some other plants, I have drawn up a list of most of the northern plants.

leaving room to put the month down, to save you the trouble of writing it over. I should not have made it so large, were not Mr. Ray deficient in this respect, even in his History. I might have left out several of them, which I have seen, either abroad or in gardens here; but thinking the time of flowering in their native places in the north, might differ somewhat from what they observe abroad, or about London, I indulged myself in making it perhaps too large. I do not, however, design to give you the trouble of putting down the time of all of them, but only of such as easily occur to your remembrance.

I have had frequent letters from Mr. Brewer, and his last is so melancholy I am afraid the poor man will die. He tells me he will send my plants next week (his letter is dated Jan. 21); but if he should drop before he had finished and sent them, Dr. Richardson would take care of them. The return of his son from the Indies, who was dismissed the service, brought great trouble upon him, and he has been ill ever since. I wish he may not be troublesome to his landlord. I believe he has given directions that nobody should know where he is, or what may become of him. I wish with all my heart he might live, and get over this trouble, for he has done me a great deal of service, and I am sure I shall never meet with a better searcher, especially for Mosses. When we travelled together in Wales, he would stop in the worst weather and most violent rain to pick up Mosses.

Mr. Brown, an ingenious young clergyman, last year found the *Muscus absinthii folio* (1) in flower, about Bishop's Castle. It flowers in April; has a pretty thick and long stalk, almost like *Lichenastrum, n. 3. Raii Syn.* 110 (2); divides into four at the top, and is a true *Lichenastrum.* The stalk comes out of a hairy tube or *vagina,* somewhat longer than in other species. He also found in head the *Bryum roseum* (3), which ought to be removed to the tribe *capitulis pendulis;* another new *Bryum, foliis juniperi* (4); some crustaceous Mosses (Lichens); and a new Willow; with peach-like leaves, hoary beneath (5). The mischief is, he gathers so very few specimens, and has displeased Mr. Brewer by not sending him some things of his finding, though the latter was so kind as to leave plants for him at Bangor. Dr. Sherard desires me to give his service to you, and thanks for the pot of fowls. He will write himself as soon as he has any thing to communicate.

I am, &c.

DEAR SIR, London, April 29, 1727.

I had the favour of your last letter, and am very glad to hear of your good health. As soon as you gave me an account of the abundance of Mosses that grow about Settle I determined with myself to go thither, and am still in the same mind, in order to

(1) *Jungermannia tomentella.* (2) *J. epiphylla.*

(3) *B. roseum. Fl. Brit.* 1370.

(4) Probably *Tortula rigida;* see *Dill. Musc.* 388.

(5) *Salix mollissima, Fl.Brit.* 1070, answers to this description.

finish my History of Mosses there. I intend to stay there a whole season, but find it impracticable this year. When I last heard from Mr. Brewer, almost two months ago, he told me he designed to take a ramble to Holyhead. Whether he still continues in Anglesea or in Wales, I do not know. He then had thoughts of going over to Ireland. He has a particular genius for plants, and had he been bred a scholar, he would have done great things.

I am sorry Mr. Wood disappoints you as well as Consul Sherard. I heard somebody last winter complaining that they had sent him books to the value of £.8, but had never heard of him since.

Your remarks for the *Synopsis*, though an Appendix is not yet in any forwardness, will always be very acceptable, if, at your leisure, you will be so kind as to communicate them. There is indeed a good number of the imperfect tribe, but in the perfect ones we are deficient.

Dr. Jussieu the younger, second Professor of Botany at Paris, has been here for eight or ten weeks, seeing the gardens, and looking over Sir Hans Sloane's and Dr. Sherard's collections. He intends to return home the latter end of next week. He would assure me that the *Muscus folio absinthii* * is a *Hypnum*.

Dr. Sherard designs going to Oxford next month, in order to settle his collection. He joins with me. &c. &c.

* See the preceding letter.

Dear Sir, London, Aug. 13, 1728, in haste.

I had the favour of yours by the hands of Mr. Foxley, to whom I will deliver whatever comes to me for the use of Mr. Brewer, as soon as I receive it, according to your desire and his orders; but have heard nothing yet of any money stirring. I wish it may come soon, as I shall be in the country most of the remainder of the summer, which will hinder my taking care of Mr. Brewer's letters and affairs. When the Consul stayed at Eltham, I was obliged to be often there, as I told Mr. Brewer; but since he came to town, I remained with him, attending him continually till the moment he died, which happened last Saturday, between one and two in the morning, of a *marasmus*. He is to be buried next Monday at Eltham, from his house here. He has settled all his affairs, and has left his collection to the University of Oxford, if they please to find a fund for the garden in six months. If not, the executors, Mr. James Sherard and Sir Richard Hopkins, are to take care to find a place for it. He has been so kind as to nominate me his first Professor, for life, enjoying the yearly revenue from the present time, in order to take care of the collection, and to carry on and finish his *Pinax*.

I have laid by a matter of 100 English plants for Mr. Brewer and others, and can furnish him with some more, if of any service. I am, &c.

P. S. by the Rev. Mr. Brown. — Honoured Sir, I am very confident that by this time the Doctor has the money or a good bill in his hands; and I hope I shall have an account of it by this post or the next. Your obedient, LIT. BROWN. Service to Mr. Brewer.

DEAR SIR, London, May 26, 1730.

When I gave your son Buxbaum's *Centuriœ*, I could not tell the price; but my friend abroad has since sent me word that it costs nine guilders. I presume they have no paper-mills yet about Petersburg, and must pay dear for foreign printers, which is the reason of this exorbitant price. However, it is better to pay a pistole for 200 plants, than a guinea or two for a dozen or score of plants, which the new writers oblige you to do in London. I hear two *Centuriœ* more are coming out, if not out already, and have sent for them.

I hear also that Micheli's book is come out at last, but cannot tell when and where the subscribers will have it. Mr. Sherard designs to write about this.

At your convenience I hope you will return those plates of the said Micheli, belonging to the Consul's library; and at your leisure I hope you will favour me with your observations and emendations of the English plants, as you gave me hopes formerly.

I remain, &c.

DEAR SIR, London, July 21, 1730.

I have been much in the country, else I should have returned a more speedy answer. The parcel and money came safe; and I have directed that Kolben's Present State of the Cape of Good Hope should be sent for you to Mr. Bartlett, of Bradford, as soon as it comes out. Twelve sheets are already printed.

When in Wales, I made particular observations on the *Subulariæ*, and am pretty well assured the *repens, folio minùs rigido,* is of a quite different genus from a Plantain. What I observed was a single naked seed, though enclosed in a calyx, on a single footstalk, of which I have still specimens by me *.

The Vetch you raised proved the same at Eltham, two years ago. When I saw it, it grew upright amongst *Filix fœmina;* but now it lies all flat on the ground, and is the *Orobus* you mention †. At your leisure, I hope you will collect your observations, which will further oblige, dear Sir, yours, &c.

London, July 11, 1734.

Some weeks ago, when I had the pleasure of seeing your sons in Oxford, I promised the elder, to whom please to give my humble service, to enquire whether the *Commercium Literarium,* an account of books and other news relating to physic and natural history, printed since 1731, at Nuremberg, in 4to, was to be had in London. After my return I found the same at Mr. Vaillant's, over against Southampton-street, Strand. He has the

* This is correct. See letter of Oct. 8, 1726, note, p. 136.

† *Orobus sylvaticus.* See p. 136.

years 1731, 1732, and the first seven months of 1733. The price is 18 shillings in sheets; the two first volumes bound 16 shillings. If you have no correspondence with this bookseller, and like the book, I will take care that it be sent; &c.

Dr. Sherard * designs to go to Oxford next week, to give directions about the green-houses now building. I believe I shall bear him company thither, but hope to be in London again at the end of next week. Mean while, I remain, &c.

Sir, London, Sept. 28, 1734.

The gentleman you ordered to call for the books not coming, I have sent them to Mr. Bartlett, of Bradford. You will find with them *Columna's Ecphrasis,* which is a favour not from me, but from Dr. Sherard.

Dr. Sherard has lost several of the Welch and Northern plants, as for example, *Saxifraga ericoides* †, *Lunaria contorta major* and *minor* ‡. The *Alsinanthemos,* or *Pyrola alsines flore* §, he never had. If it be not too much trouble, I accept your kind offer of such of these as are

* James Sherard, younger brother of the late Consul, and proprietor of the garden at Eltham; who received the degree of M. D. at Oxford, in consequence of his brother's noble benefaction, and in the hope that he might follow it up with further contributions.

† *S. oppositifolia.*

‡ *Draba incana,* α and β. *Fl. Brit.* 678.

§ *Trientalis europæa.*

growing with you, for the Oxford garden; there being a gardener for the present, and besides him, Mr. Bobart, who will take care of them. I hope to remove thither in February, or the beginning of March next.

If your son be with you, please to give my service to him, and the same please to accept, Sir, &c.

DEAR SIR, Oxford, Aug. 25, 1736.

I received your favour of the 21st, and accordingly shall send to-morrow, by our Northampton carrier, the *Ficoides* * you named, and we have to spare, viz. *Boerhaave Lugd. Bat. v.* 1. 289. *n.* 3, which is *Hort. Elth. fig.* 272; *N.* 18, is *H. E. fig.* 265; *n.* 19 *is fig.* 275, 276; *n.* 21. is *fig.* 245—247; *p.* 291. *n.* 2, and 290; *n.* 1. are *fig.* 263 and 262; *p.* 291. *n.* 8, is *fig.* 277 and 280. To these I have added *H. E. fig.* 270, 239, 282, 233, 256, 259, 241, 242, 267, and 268. Of Boerhaave's *p.* 290. *n.* 11, 12, and 14, we have but single plants, which will not bear cutting at present, but hope to supply you next year. His *p.* 290. *n.* 13. and 291. *n.* 4, are dubious to me. I do not know what he means by them, nor perhaps does any one know but himself. Before I printed the *Hortus Elthamensis*, I sent a coloured set of the figures of these things to him,

* *Mesembryanthema.* It appears by this letter that Boerhaave's *Hort. L. Bat.* served the collectors at this period as a catalogue for correspondence.

desiring him to fix his synonyms to them, but could not obtain that favour.

If I remember right, your son had, amongst the rest, *Geranium. fig.* 152 *H. E.*; and another which is very like *G. africanum frutescens malvæ folio laciniato odorato, Hort. L. Bat.* but differs in having larger flowers, the leaves more fragrant, like roses, and the whole plant taller.

You say you have all Boerhaave's *Gerania,* except *n.* 8, 9, 17, and 21. We have lost *n.* 17 last winter; but of 8 and 9, which are but one, and of 21, I can gather you seeds. Now if you have *n.* 14, 23, 26, and 62, I should be obliged to you for them, also for the following North-country plants.

(Here follows a long list, of no importance at present.)

P. S. A new Botanist is arisen in the North, founder of a new method, on the stamens and pistils, whose name is Linnæus. He has printed *Fundamenta botanica, Bibliotheca botanica, Systema Naturæ;* and is now printing in Holland his *Characteres,* and his *Flora Lapponica.* He is a Swede, and has travelled over Lapland. He has a thorough insight and knowledge of Botany, but I am afraid his method will not hold. He came hither, and stayed about eight days, but is now returned to Sweden.

Several letters, relating merely to the interchange of plants, are here omitted. They evince the ardour with which the new Professor laboured to enrich the Oxford garden, and to make up for some losses which had happened to the Eltham collection, removed thither; but they are chiefly lists of names, and contain nothing relative to the introduction of any species from abroad.

Sir, Oxford, Sept. 8, 1737.

Your letter, and the box of Northern plants, came safe to hand. Several of these have been lost here, and some we never had before. I find these plants do not thrive with us, as they do in their native places, though planted under a cold Northern wall. Our gardener has, no more than yours, been able to make the cuttings of the double striped Oleander take.

Whether I myself shall live to see W. Sherard's *Pinax* finished, though I am in good health, I do not know. I have but three months in a year to myself, to do any thing with it, and have within these two years finished near half the *Dendrologia*. The garden and other avocations fill up all my time in Spring, Summer, and Autumn. The provision for a Professor is scanty, and that little is not settled. £.3000, left by Consul Sherard, has been placed in South-sea annuities, and being bought at 104, may be sold cheaper, and not fetch £.3000 again. Besides, there is always a loss whenever 6 *per cent.* comes to be paid off and bought in again. Nay, at present, as James Sherard's executors are going to law, there seems to be a stop put to the Professor's salary, and I have had no money these 11 months. It is to be laid out in land, and when that is done, perhaps it will not be above 80 or £.90 *per annum*. I ought in so long, tedious, and difficult a work, to have had an assistant or two. But what can a man do with so small an allowance, in a place not cheap to live in?

James Sherard has spoiled it (the *Pinax*). After William Sherard's death, he took me off, and set me to work in his gardens, to make himself known, promising to do great things, that is, to pay for the plates, paper, &c. But when it came to the performance, he did nothing, and not to lose so many years' labour, I undertook it (the *Hortus Elthamensis)* at my own expence. All his kindness ended in an offer to lend me money, and to take 30 books (copies), which he did to prevent my taking subscriptions. I finished the work without his money, and instead of 30, he took only 10 copies; for which I had 30 guineas; and besides gave him one gratis, and some time after, another, which he in a manner begged of me. This is all I ever had from him. Besides time and labour, I lose by him at least £.200; for it is a book of but few people's buying, and therefore I do not think it safe to go through the whole impression. Five hundred copies were printed, at his desire, I would have printed but half as many, to which I got 145 copies of the plates printed off. The rest I do not design to complete. If he had let me go on my own way, I should have made a book in 4to or small folio; but he did not like this, and made me draw over again 50 or more of the plates, to make it look bigger and more pompous, and persuaded me to use royal paper, but gave me not a penny towards buying it.

If the time spent in composing this work, had been employed on the *Pinax*, I dare say the latter would have been finished, for I had nothing else to do, and was free from all avocations for six or seven years.

This is the character I have to give of our old friend. I might and should have quarrelled with him three or four years ago, and should have brought him in a bill. But I did not, because it should not be construed as if I had been the occasion of what has happened now. For he had given himself airs of doing great things, and raised great expectations in every body, especially the university; but I knew him too well, &c.

You will excuse my freedom, and excuse the trouble of reading my scribble. I am, &c.

SIR, Oxford, March 31, 1738.

I sent yesterday, by the Northampton carrier, Palmer, a box of seeds directed for you, in which you will find some perennial umbels. Our collection of *Umbelliferæ* is but small, and we have not been at all successful in raising this family. The difficulty is, I could never get the seeds early enough to sow in Autumn, and have found, by my small experience in gardening, that it is throwing seeds away to sow them in Spring. I therefore design never to sow any more but in August or September, though the seeds should grow older by doing so.

But the major part of the seeds in the box came from Eltham, and were sent me after Dr. Sherard's decease. How old or fresh they may be I know not, being without date.

Mrs. Sherard has offered to the university such plants as we may want, and we have received a good many stove plants, some hardy ones, and a few for the green-house. The hardier and more ornamental green-house plants she designs to keep, as our gardener tells me; and therefore he was reserved in asking for any of those.

We have but one small plant of the Hedge-hog Aloe, and an old one of the larger Pearl Aloe, without any offset. The Dog's-chap *Ficoides* * we lost this Winter, but recovered it from Eltham. I hope to supply you with this and the Caltrop *Ficoides* † this Summer. Pray put me in mind of it.

As you encourage me, I will mention some of the Northern plants we have lost. (Here follows a long list.)

If *Linum n.* 4, *Syn.* 362, be the same, as Mr. Ray suspects, with *n.* 3, it is wrongly called *perenne.* At least what we have for it is but biennial.

* *Mesembryanthemum caninum, Ait. Hort. Kew. v.* 3. 218.

† Perhaps *M. deltoides.*

I am likewise in doubt about the 1st and 2d *Lunaria, Syn.* 291, 292. We have had both from you. The 1st has disappeared; the 2d we have. It is not perennial, but biennial. I save the seeds, and it also usually sows itself. If the 1st be really different, please to send it once more; or both sorts together, for I may mistake the 2d, if there be really a perennial sort *.

My service attends all friends, to whom I have the honour to be known.

The *Specimen Phytographiæ Africanæ* is a part of Dr. Shaw's Travels, which work is now finished, and will be published, as he says, in three weeks time. I am, &c.

SIR, Oxford, Dec. 7, 1738.

Your son Henry told me this Summer of Furze being made use of instead of tanner's bark, but could not inform me how long it would keep the heat. I should be glad to know whether it be preferable to bark, how it must be used, whether fresh, half or quite dry. Bark being dear with us, this might be an improvement for our stove.

I have taken the Mosses in hand this winter, to settle their *Synomyms* for the *Pinax;* and examining Vaillant's and Micheli's works, I find there remains a good deal which they have not touched. I observe Micheli has no descriptions at all, bad names, and

* *Draba incana,* certainly biennial. Varying much in size, it was mistaken for two different species.

his figures, for the major part, are more ornamental than true *. Vaillant has no great number, has not described the whole of them, and most but imperfectly. His figures are more artificial than natural; several of both these authors not being so good as those in the *Historia Oxoniensis.* This almost induces me to finish what I had formerly begun; I mean a History of Mosses, in which I have already made some progress, and done about 150 species. In the whole tribe of *Lichenoides,* the most singular always appeared to me *Syn. ed.* 3. 74. *n.* 71 †, found first by yourself. and by nobody, that I know of, afterwards. If you have any samples of this by you, I beg the favour of you to send me one or two in a letter, and to give me what information you can about it. The specimens in Consul Sherard's collection are pasted down, and I therefore can neither see them on both sides, nor soak them for full examination. If you have any Mosses not in the last edition of the *Synopsis,* by communicating them you will much oblige, Sir, yours &c.

* Poor Micheli falls under this heavy censure from botanizing in Italy, where the plants in question are finer and more luxuriant than Dillenius ever saw!

† *Lichen ampullaceus, Linn. Sp. Pl.* 1613, found since by Mr. Menzies, who ascertained it to be only an accidental state of *L. Glaucus.*

SIR, Oxford, Aug. 22, 1740.

To-morrow will be a week since I had the pleasure of seeing your son. He paid me the subscription money, as you mention, two guineas and a half, for which I have given him a receipt.

I proposed to have sent some plants by him, but on Wednesday evening, when they were to have been gathered, we had a great rain, and the next day being too late for the London carrier, I have sent them by our Northampton one, to the care of Mr. Bartlett. At the top of the box you will find three proposals, which I beg of you to send to any noted bookseller at York, when opportunity offers. On the other side you have a list of plants. We escaped very well last winter as to green-house plants. Our method was to put half a dozen large garden pots, with lighted charcoal, into each green-house. But in the open ground we lost abundance of plants, particularly all from the South of France, Italy, Spain, and Africa; and yet those of Virginia, Maryland, and Pennsylvania stood the frost. The *Jacea lusitanica sempervirens* * we had great plenty of in the ground, as well as in pots, but all are dead. *Jacea argentea ragusina* † we lost two years ago. It grew well in the open ground, but never increased, and our wet Autumns always rotted the heads. We never had the *Jacea sphærocephala tingitana* ‡.

If you have *Amaranthoides perenne, floribus*

* *Centaurea sempervirens.* † *C. ragusina.*

‡ *C. sphærocephala.*

stramineis radiatis, Hort. Elth. t. 20 *, I should be glad of some cuttings.

(Here follow various notes, relating to the interchange of plants, with lists, of no moment at present.)

DILLENIUS TO DR. RICHARDSON, JUN.

SIR, Oxford, Oct. 14, 1741.

I received yours of July 1st, and give you many thanks for the enclosed ring. I could have wished your father had lived a few years longer. I flatter myself I should have had his approbation of the work in hand, my History of Mosses, in which he was so eminently skilled. The manuscript being much interlined, and having in some places many additions, I find myself mistaken in the number of sheets, of which 61 are already printed, and I compute that there will be at least a dozen more. The plates are all printed off. Towards Christmas, at furthest, it will be finished, and delivered here in London by Mr. Manby, on Ludgate-hill, which you will please to notify to Sir John Kaye and Dr. Stanhope.

I enclose a list of plants which we have lost. If any of them be in your collection, you will oblige me by sending cuttings next season, and you may be sure of having any we have, that you may want. I am, &c.

* *Gomphrena perennis.*

MR. JAMES PETIVER, F. R. S.* TO RICHARD RICHARDSON, M. D. NORTH BIERLEY, YORKSHIRE.

[No date. Written in 1702.]

HONOURED AND WORTHY SIR,

I received yours about the middle of the month, and your enclosed Catalogue of the Leyden plants was highly acceptable to me, because I find in it a great number that I was dubious of, many of which the sight of will presently set me to rights. I question not but I have already several of them by me, which I have yearly gathered at Hampton Court, Fulham, &c. Nevertheless, not having seen them from Holland, under their undoubted names by Dr. Hermann's authority, we dare not be positive in several of them, especially since his death has deprived us of expecting them from him.

I have annexed a catalogue of such as I am positive of, which indeed are but a few, though there may be many more that I need not scruple, yet, since the sight of them from his authority will so easily remove these doubts, I hope you will give me that opportunity, and pardon my scrupulosity. It is only my aim to do each man justice, in preserving his title to his own discoveries, and by doing that I prevent multiplying species, which has always been my greatest care.

* An apothecary and celebrated naturalist at London, who died April 20, 1718.

I have this morning, by John Hall, the Yorkshire carrier, sent you my three first Centuries, which I remember, in some of your former letters, you hinted to me you wanted. To them I have added the first Decade of my *Gazophylacium Naturæ et Artis*, which I finished but yesterday, so that you have the first I have yet parted with. I desire your free thoughts of it; and let me know my errors, that I may mend or avoid them in my next, for I intend to proceed as I shall meet with encouragement; since I have many things by me, and daily receive more, which have never yet been figured in any author. I have engraved four or five more plates, in which are many new and strange things, and hope to finish my second Decade by Christmas. I should be glad to sprinkle some formed stones, and other fossils, in my future tables; although I should incur the good-natured Dr. Woodward's displeasure, who, I hear, suddenly designs to proclaim war, and damn all such as have meddled with his province. However, though as yet a novice in the knowledge of them, yet I will venture to figure some, if you and my other kind friends will assist, and help me to them.

I thank you for your insects; and though they were but five or six, yet two or three of them were very rare. I have sent you four tables of foreign Butterflies, several of which, for largeness and beauty, are of the first rank, and therefore, I hope, will be acceptable to you. I dare not say so of the two tables of English ones, though some of them are scarce about London, viz. *Mus. nostr.* 301 *, 303 †, 326 ‡. My chief design in sending them was to show you what we have about us, with references to my *Museum*. I expect an addition from you next summer, and hope you will begin your collections and observations of them early in the spring. I doubt not but you have some with you that we have not yet seen. My next Decade will have the figures of at least half a dozen English Butterflies, not yet mentioned in my Centuries. The eleventh in my second table is a small brown one, wholly green underneath §, very rare about us.

I know not whether you will like my method, of sending the Butterflies in quarto tables, but I chose that size, that if you were minded, or thought any of them worth putting into frames with glasses over them, which you may cheap and easily procure in the country, they will keep many years. If at any time you find lice or worms in them, you may safely take out the glass and clean them.

My next collection for you shall be shells and plants, of either of which if there be any you particularly desire which are in my Centuries, *Gazophylacium*, of Philosophical Transactions, I will endeavour to send them, if you will please to mention them.

* *Papilio Cratægi.*

† *P. Sinapis;* figured in Petiver's *Papil. Brit. t.* 1. *f.* 23, 24.

‡ *P. Cardui.*

§ *P. Rubi.*

I have a great itching after the knowledge of Fossils, and must therefore beg of you to send me some, though never so few, as soon as you can, with your remarks on them, where they are found, &c. Dr. Sloane shewed me the designs you gave him. What you can spare of them will be very acceptable, there being several among them very curious. I find Dr. Sloane has a mind they should be figured, and I doubt not but he would gladly let me sprinkle them in my tables, if you would be pleased to give him leave. I should esteem them no small ornament to my second Decade.

I shall conclude, fearing I have tired you with my long scribble. At your leisure I should be glad to hear that you have received what I sent you safe, which is hoped by, worthy Sir, your much obliged and very affectionate friend, JAMES PETIVER.

WORTHY SIR, [No Date, Written in 1703.]

Your very acceptable letter, and collection of Leyden (garden) plants, came safe to hand, for which I had returned you sooner thanks, but was willing they should be attended with somewhat more than bare words. Your kind encouragement, and expectations from other friends, heartens me to go on with my Tables, having finished six more, and herewith send you them. I intend some of your designs in the remainder, having a promise of them from Dr. Sloane, since the receipt of your letter. I think every day more than two, until I receive some fossils as you promised; and your account of them will be very instructive to me, being as yet a novice in the knowledge of them. Your notions of spars, marcasites, pyrites, &c. with samples of them though never so common, will be inexpressibly welcome to me, and give me great opportunity to procure the same from abroad.

I this week finished the seventh and last book of Mr. Samuel Brown's collections of Indian plants, which will be published in the next month's Philosophical Transactions. They chiefly consist of Corns, Cyperuses, Rushes, and Grasses of each kind to the number of threescore.

Before Christmas I hope to have printed the 9th and 10th Centuries of my *Museum*, which I will send you as a new year's gift, with a collection of shells, which I have hitherto deferred, because of some I have therein mentioned; as also other Indian ones, which will be in this account of Mr. Brown's seventh volume.

Before I had herewith sent this small footboy, and made you acquainted with what is in embryo, and what I designed to send you, I durst not presume to tell you that your two remaining bundles, as you please to call them, of the Leyden plants, will be very welcome to me. However, expose them not to a rainy season, without better covering. A settled frost will be less prejudicial.

By Christmas next, I hope to have finished my second Decade, and I will then send you a large collection of those East India plants I have given an account of in the late Philosophical Transactions, and whatever in the interim will be acceptable to you, besides what I have promised, pray let me know, and I will endeavour to obey your commands, who am, honoured and worthy Sir, your very affectionate, and most highly obliged, friend to serve you, whilst JAMES PETIVER.

P. S. The cold and winter season may give some friends of yours, that are good marksmen, frequent opportunities of shooting several sorts of birds, which do not appear in our parts. These may be preserved, the great ones, by taking out all their entrails, cutting them under the wing, and then stuffing them with tow, mixed with tar and a little unslaked lime. The little ones scarce need any other care than drying gently in an oven, after the bread is taken out, and so sent us.

I observe several small birds in Mr. Ray's Ornithology, communicated to him from your parts. I should be glad to figure, or give an account of, whatever uncommon birds you can procure me. But I fear I have already too much troubled you with my desires, and will therefore humbly beg your pardon, and conclude.

WORTHY SIR, London, Sept. 19, 1704.

It is now some months since by Signor Vigani * I sent you my second and third Decade; the printed part of which last I have not till now finished, for want of subscribers. My fourth is also done, and stays but for three or four promoters; however, in the interim, if you desire it, I will send it as it is. I hope, Sir, this summer has given you an opportunity of collecting some of your more rare northern plants; and what specimens of them you can spare will be very acceptable to me, particularly your *Salix rotundifolia* † with its *juli* or palm. I would give a figure of it, if I could get good samples, with my conjectures of it, concerning which I gave you some hints formerly. You will find in my third and fourth Decades several English Insects, which I design a continuation of, and I doubt not but your northern parts may afford us several strangers, which I should be glad to see. As for minerals, ores, and other fossils, *viz.* formed Stones, &c. I am but a novice in them, and therefore whatever you send me of these will be all discovery. I could procure several from abroad, if I had but samples of them to send as patterns to procure others by. I remember in your letter to Mr. Llwyd you have vast quantities and varieties of the *Entrochi* particularly, and other formed stones, in your parts. I should

* This gentleman, though a foreigner, was elected Professor of Chemistry at Cambridge, on account of his superior knowledge.

† *Salix reticulata.*

think it no difficult matter to find some poor fellows, belonging to your coal-pits, quarries, lead-mines, &c. to lay by whatever of these things may come in their way, which for a shilling or two I believe they would do. I would now and then bestow half a crown on any poor man who would himself lay by, or procure from his fellow-workmen, whatever were not cumbersome, or overloaded with worse matter, and pay for the carriage besides. A plenty of these will also enable me to make some returns to several German correspondents, who are very curious this way, and mightily desire to see what our island in this kind produces. Amongst these, I have lately received a very curious collection of plants, many of them such as Caspar Bauhin in his *Prodromus* has described from those parts, which were not only highly acceptable to me, but will give me an opportunity to clear the obscure Botany of that time. These were sent me from Dr. Scheuchzer, a physician at Zurich, who has printed a treatise on the formed stones of that country, with their figures. If you have not seen the book, I will send it for your perusal, having it by me, as also whatever you will desire out of my ten Centuries, and what I have figured in my Decades, being very desirous of meriting a commutation with you, who am, Sir, most respectfully, &c.

JAMES PETIVER.

WORTHY SIR, London, Sept. 11, 1712.

I am just returned from making a trip to the Bath and Bristow, which was the reason I answered not your very kind letter sooner. I am very much obliged to you for your very generous proffer, of assisting me with dry specimens towards the completing my *Icones Plantarum Britanniæ*, which will be highly acceptable to me. I have therefore annexed a catalogue of such as are most peculiar to your parts, and Wales, and duplicates of such of them as you can spare will be very welcome.

I should be very glad to take a herbarizing journey with you into Wales next summer, if I do not make a tour into France, which I have some thoughts of doing, if health, peace, and plenty, crown my wishes. I was this day at Chelsea garden, where we dined at the Swan, it being our last herbarizing till next spring. I am now finishing my cognizance of the most curious plants we have raised there this summer, which I design for the ensuing Transactions, and think to give the figures of a dozen of the most rare, in a table, after the manner of my Herbal. And if there be any amongst them, or those I have already published, that you desire specimens of, if you please to let me know, I will gladly send them, or whatever else you shall desire, being very ambitious to approve myself, kind and worthy Sir, your much obliged, &c.

JAMES PETIVER.

KIND AND WORTHY SIR, London, Dec. 20, 1712.

I am very sensible you may justly blame me for being so rude as not sooner to acknowledge the very acceptable present of dry specimens which I have so long received. I was in hopes Mr. Buddle had in part done it for me, as I desired him, for which reason I did design to defer it till after Christmas, this being a very busy season with us, and then to have sent you what you want of my Gazophylacick Tables, with twenty others of East India Shells, which I shall have ready for you as soon as the holidays are over; and then I design to write more largely to you concerning what you have sent, and other things. In the interim I heartily wish you a merry Christmas and a happy new year, who am, worthy Sir, your very affectionate friend and humble servant, JAMES PETIVER.

End of PETIVER'S *Letters to* RICHARDSON.

11. Petiver to Dr. Richardson.

Kind & Worthy Sr.

I am very sensible you may justly blame me for being so rude, as not sooner to acknowledge yr very acceptible present of dry Specimens, which I have so long receiv'd: I was in hopes Mr Buddle had in part done it for me, as I desired him, for which reason I did design to deferr it till after Christmass.

James Petiver.

Dear Sir

12. Gronovius to Dr. Richardson.

My last letter was dated the 7th of Decemb. By these I have the honour to present to you an Index of all the Stones, I have collected since the Systema Naturæ of Linnæus was printed.
Lately is at the Hague printed Seguieri Bibliotheca Botanica in Quarto, being the price 4 gilders.
Wishof is printing Halleri observationes in plantas Helveticas, with several good plates. Rumphy his work go's well on at Amsterdam.

13. Amman to Linnæus, Novr. 15th 1737.

Viro doctissimo atque clarissimo
CAROLO LINNAEO
Med. Doct.
S. P.
Ioannes Amman.

Gratissimæ mihi fuerunt literæ Tuæ ante aliquot menses e Musæo Cliffortiano ad me datæ. Pro transmissis autem hactenus a Te editis Scriptis Botanicis maximas ago gratias.

Martin, Lithogr. Rochester St. Westmr.

The material originally positioned here is too large for reproduction in this reissue. A PDF can be downloaded from the web address given on page iv of this book, by clicking on 'Resources Available'.

MR. ROBERT FOULKES TO DR. RICHARDSON.

SIR, Llanbeder, near Ruthin, Nov. 7, 1727.

I am to beg your pardon, which I do heartily, for informing you that *Alsine baccifera** grew in Anglesea, which I did from the account of it from one who pretended to know plants very well, and had often attended Mr. Edward Lloyd. But I could find no such plant. * * * * * * &c.

JOHN FREDERICK GRONOVIUS, M. D. † TO R. RICHARDSON, ESQ. [English, corrected.]

DEAR SIR, Leyden, July 22, 1738.

I yesterday received, with great pleasure, your letter of the 20th of June. On receipt of it I went to the gardener, Jacob Haze, who told me it would

* *Cucubalus baccifer.*

The only authority for this plant being reckoned a native of Britain is the above Mr. Foulkes. Dillenius has first introduced it, under the name of *Cucubalus Plinii,* into the third edition of Ray's *Synopsis*, 267, as having been gathered "in hedges in the isle of *Mona,* Anglesea, by Mr. Foulkes, of Llanbeder, near Ruthin, and communicated to Dr. Richardson." Nobody, as far as I could learn, has ever met with the plant since, except in curious botanic gardens, in any part of the British isles; and accordingly I was obliged to be content with a garden specimen, for the figure in *English Botany*, *t.* 1577. I am therefore under the necessity, however unwillingly, of excluding the *Cucubalus baccifer* from our British Flora. The Rev. Hugh Davies, who is so intimately acquainted with the botany of Anglesea, could never meet with this plant.

J. E. SMITH.

† He died in 1762, aged 72.

be almost three weeks before the bulbs could be sent, several of them being still in flower. I shall let them remain till I hear of a ship going to Hull. The price of a Camphor-tree is 25 guilders, which is not much more than the two guineas you sent. Haze is not sure whether he has a good one or not; but he means to-morrow to cut off all the layers from the old tree; so we shall see in a week whether they have made good roots. If so, you shall have one safely packed. I sent one, two years ago, to Lord Petre, which arrived in good condition.

It is very strange that Dr. Boerhaave's death has been reported in England, Switzerland, and France, which is a great mistake. A month ago he was so much recovered as to walk from one room to another, and finding his strength improving, he resolved to go to his country-house, where he rides out every day in a *chaise romaine*, having his wife with him, and being his own coachman. My brother-in-law coming last Thursday by the boat from Haarlem, met him riding in his chaise, about half way, where the boats commonly stop for a few minutes. Since which, my own brother met him at Warmond, and spoke to him. He looks so well, that a stranger would not believe he had been so ill; and I cannot but hope he may resume his labours after the vacation.

The public garden is now enlarged, as far as the ramparts of the town, so that its size is more than doubled; for what purpose, except the pleasure of the Professor, I cannot say. Nobody is permitted to go into the new part; and this is all I can tell you on the subject. Dr. Van Royen has no extensive correspondence, to procure new plants.

You have, no doubt, heard of the great advances made in the science of Botany, within these three years, by Dr. Linnæus, whose method I like very much. That man has worked prodigiously. Dr. Lawson and I were the first who supported him here; and on seeing his Tables, we resolved to print them at our own expense *. Afterwards we resigned him to Mr. Cliffort, to make a Catalogue of his garden, according to the sexual method, which is now printed, though not yet distributed, in 500 pages folio, with about 40 exceeding curious plates. I assure you it is one of the most curious books ever printed. But it is strange that Mr. Cliffort, who is at the whole expense, will not sell a single copy, being resolved only to make presents of the book, or to exchange it for other curiosities. He is a great lover of petrifactions; therefore if you could procure a collection of these things, named, I think I could obtain for you a copy of the above work and of his *Viridarium*. Almost all my new plants from Virginia are curiously described there, which were formerly under very bad names in Ray, Plukenet, &c. You will also see there a genus called *Richardia*, after your father. Linnæus has printed, besides, *Characteres Generum*, with a *Corollarium; Critica Botanica; Methodus Sexualis; Flora Lapponica; Classes Plantarum;* and *Artedii Ichthyologia;* all

* The first edition of the celebrated *Systema Naturæ*, in eight folio sheets, now very rare.

very learned books. Last winter we had a very excellent club, or society, which met every Saturday, composed of Lawson, Van Swieten, and some other gentlemen, Linnæus being our president. Sometimes we examined minerals; on other days flowers of plants, as well as insects or fishes. We have made so much progress, that by his Tables we can refer any fish, plant, or mineral, to its genus, and, subsequently, to its species, though none of us had seen it before. I think these Tables so eminently useful, that every body ought to have them hanging in his study, like maps. Boerhaave highly esteems all these performances, and they are his daily recreation. If you want any of them, please to let me know by the first post. Albinus *de musculis*, &c. shall be sent you with the roots, as well as a small work of his printed in colours. His Eustachius is in the same condition it was when your brother left Leyden. The second volume of Swammerdam is not yet printed. It is necessary your brother should send the paper of subscriptions. When the work is finished, I will do my utmost to get it, but I wish to know whether you have the first volume. You shall also have the copper plate of Boerhaave with the roots, and a further answer to your letter; when also I shall write at large about Lawson, who is now in Saxony.

Burmann at Amsterdam has printed a *Thesaurus Plantarum Zeylanicarum*, with very good plates, and four *Decades Plantarum Africanarum*, in 4to; but the price of the *Thesaurus* is extravagant, sixteen guilders (about 32 shillings). If you have any more orders, you may send me a letter by post before I send the above articles.

I am, dear Sir, your most obedient servant,

J. F. GRONOVIUS.

DEAR SIR, Leyden, July 29, 1738.

In my last letter I promised to write to you about Lawson, and to answer the rest of your wishes.

When the second table of the *Systema* was printed, containing plants, we undertook to examine all the flowers we could get, finding it very easy, by the number of stamens, to make out the classes. This charmed us so much, we were desirous, by a like method, to determine the characters of stones; after which we came to animals. It was not enough to know the names; we endeavoured to get the specimens themselves, and to examine them by fire. Thus our club became a forge of Vulcan, where we went through all the assaying processes, in the course of a whole winter. At the conclusion our assay-master was so kind as to cheat us, giving my own manuscripts to Professor Gaubius, who was proceeding, without my knowledge, to print them. This I put a stop to. On which our assay-master himself went to work, and still continues printing. I assure you it will be one of the best books of its kind.

Lawson was so taken up with these experiments, that he resolved to give up attending lectures. He took his degree, and engaged a master to teach him high Dutch, in which he became a great proficient.

speaking it like a native German. He went last March from Norden to Osnabrugge, having only one companion in the coach, whom he discovered to be a servant of the hangman at Haarlem. He next visited Clausthal, the *Sylva Hercynia*, and Berlin. His last letter, received a fortnight since, was dated Fribourg in Saxony, when he was not sure whether or not he should go to England.

It is now fifteen years since I gave a commission to a merchant at Paris to send me the portrait of Monsieur Tournefort in his græcian habit and order of St. Louis. He, like many other friends to whom I have given the same commission, could never meet with this print; but he discovered that the copper plate was in the hands of Tournefort's relations, to whom I wrote for half a dozen impressions, which at last I got, but the price is extravagant, twenty-two stuyvers each.

I have again enquired about the Camphor-tree. There are five young ones, to be taken off in September, when it will be seen whether the roots are good. The 28th of August I must go to Middleburgh in Zealand. On my return, about the 20th of September, I will send you a good plant, except you order the contrary.

Respecting the print of Boerhaave, you must know that Wandelaar desired the Doctor to find leisure to spend half a day with him, that he might correct the old plate. This Wandelaar has done so well as to gain great credit by it, and the Doctor took the plate into his own hands. Every person, therefore, except two or three booksellers, is obliged to send to his house for the print. Besides this, there is another portrait of the Doctor, by a painter named De Groot, which is not at all a good likeness. I therefore send but one of them.

A few weeks ago, Dr. Linnæus printed his *Classes Plantarum*, which I think you cannot yet have got, being so recently published.

I add to this parcel three copies of the *Systema Naturæ*, besides four plates of the Classes of Linnæus, which is of great use, as giving an idea of all his Classes. I do not doubt you will find friends enough to dispose of it *. I forgot to tell you, in my last, that besides the books there mentioned, the Doctor has printed a *Bibliotheca*, and *Fundamenta, Botanica*.

Among the works of Albinus you will see a new one, *de Colore Æthiopum*. If this or any other book that I have sent does not please you, send it freely back.

In the new garden are two new green-houses, and foundations are laid for a third. You may be sure of having, in due time, any catalogues of good books that may appear. Professor Fabricius died yesterday. He has, no doubt, left a fine library. You may depend on having the catalogue when printed.

Dr. Van Swieten gives his humble service to you. He is obliged to you for Miller's Appendix. He has

* The writer means, perhaps, four copies of the book thus named.

still a great deal of practice. My brother and Dr. Van Royen send also their humble service.

A year ago I received a letter from Dr. Johnston, of New York, who seems curious about plants. He desired me to send him Plukenet's works.

I have this moment a letter from Dr. Lawson, who is inclined to go to Schneeberg; from thence to the baths of Carlsbad, and so on to Prague, Vienna, and England.

Boerhaave is not so well as when I wrote to you last; but he takes an airing every day.

I am obliged to you for your kindness about my *Hortus siccus*. Some of the *Musci* I want. I shall take the liberty of writing about them another time.

Van Hazen has collected all the roots you want. At the end of his list you will see a note, which I interpret as follows:

"Every thing packed in sand or moss must be planted immediately. The tulips and hyacinths in September. *Narcissi* in November. *Anemones* and *Ranunculi* in the latter end of January."

I hope all these things will reach you in good order. Mr. Hudig wrote me word that a ship would be ready to sail in eight days, and I have therefore lost no time. I am, &c.

J. F. GRONOVIUS.

Dear Sir, Leyden, Sept. 2, 1738.

Mr. Hudig having informed me of a ship going to Hull, I have endeavoured to execute your orders in your letter of the 22d of July.

Dr. Boerhaave is so much recovered since I wrote last, that he walks in his garden every day, and is glad to see a friend; the dropsy and all other inconveniences being quite gone. There is even a report that he drinks wine, and intends to give lectures in person.

You will receive, by this occasion, the best Camphor-tree I could procure, and a dozen roots of the double Tuberose.

The *Laurus Aldini* * is not to be had; but Haze has put in some layers, which, if they take root, will next year be good trees.

I now send you all the rest of Linnæus's works, in which you will find great learning, and many curious remarks. Most of your triflers in Botany dislike his method, and particularly his *Critica*, because they do not understand him; especially some of his characters, of which a great number are illustrated by examples to be found in his *Hortus Cliffortianus*. I assure you, Sir, it was by his principles alone that I could reduce several of your Virginian plants to their proper genus, which till now could not be referred to any class, as you will see in the *Flora Virginica*. This work will go to

* *Laurus indica, Linn. Sp. Pl.* 529, figured in *Aldini Hort. Farnes.* p. 60.

press as soon as the *Hortus Cliffortianus* is published, which I expect every day. I shall do my utmost to persuade Mr. Cliffort to sell some copies of his *Hortus;* otherwise I know not how to get one, except in the way before mentioned.

I send you also two copies of Albinus *de vasis intestinorum,* and *de cute.* But I find in the booksellers' shops three more similar pieces, of each of which I now send you a copy. If you want more, they may go with the second volume of Swammerdam, which I hear will be printed in a month hence. Copies with coloured plates grow dearer every day, I know not for what reason.

Whilst writing this, I have sent to van der Aa, for the first volume of Swammerdam. He has kindly sent the second also, with a note, in which he says that though they do not yet distribute this volume, he has furnished me with a copy. He makes an apology for the second volume being much thicker than was expected, its price being 11 guilders 5 stuyvers. So you see how we are imposed on by subscriptions! I inclose his own paper, with his receipt for the whole work of van Swammerdam, 16 guilders 5 stuyvers.

You have also the *Thesaurus Zeylanicus; Artedii Hist. Piscium,* by Linnæus; *Linn. Fl. Lapponica, Critica, Characteres, Corollarium, Methodus sexualis, Bibliotheca Botanica,* and *Fundamenta Botanica;* for all which I paid 24 guilders. These books of Linnæus go off very well, and will be scarce in a short time. I beg pardon for not being able now to send Burmann's *Decades*, which I confess is my own fault, having quite forgot them, as well as Commelin's *Plantæ rariores & exoticæ*. I shall remember them another time.

Verbeck does not recollect the book of buildings and ruins, printed by Mortier. I intend, before winter, to go to Amsterdam, when I shall enquire for it myself.

You see here Lawson's dissertation. He arrived yesterday, and intends going to-morrow by the sloop for England.

The children and executors of Seba are still greatly at variance, and have involved each other in so many law-suits, that I believe his book will never be finished. Two volumes are published, which come to about 100 guilders, or rather more; I forget the exact price. The first volume is very well done; but when I saw the second, I immediately sold my subscription paper and the first volume for a ducat loss, being very glad to save 90 guilders, or thereabouts. If I remember right, a coloured copy would come to 500 guilders. There are, doubtless, several fine things in the work, but bad are every where mixed with good. There are several common things, not worthy to be named, but less to be exhibited in figures. The two volumes may perhaps be procured by chance, at some auction, for 50 guilders.

I have not been to Zealand, my son having taken the measles a few days before the time fixed for my departure; and afterwards my wife, my daughter, and maid-servants. Myself and the footman only escaped. I thank God this storm is over.

I have no catalogue of any consequence now; but there will be one of van der Aa's, as well as that of Fabricius's library, which I will send when printed.

I hope, Sir, I have executed your orders to your satisfaction, except the two books I forgot, for which I beg pardon. I am, &c.

J. F. GRONOVIUS.

P. S. Having yesterday morning written the above letter, I by chance, in a bookseller's shop, met with a catalogue of the library of a friend at Amsterdam, of whose death I had not heard. If you wish for any thing in it, you can write to me. The possessor of these books was Mr. van Meel, author of the *Epistolæ obscurorum virorum.* There is one remarkable book, which I have seen only twice before, de Bry, being what Caspar Bauhin in his *Pinax* always calls *Historia orientalis et occidentalis.* There are two editions, one in Latin, which I never saw but at the public library; the other in German, which I never knew to be sold under 50 guilders. When I hear of any considerable auction, I will send you a catalogue, that you may order any book you like; but you must always put the price you will go to.

DEAR SIR, Leyden, March 11, 1739.

Last September I sent you a Camphor-tree, which I hope came safe. I now send Boerhaave's catalogue, and shall be ready to do you any service respecting any articles in it. The *Hortus Cliffortianus* is not yet to be bought. I send you some printed sheets of the *Flora Virginica*, which I hope may be acceptable. Mr. Seguier, a learned botanist, who resides with Scipio Maffei at Verona, is printing at the Hague a *Bibliotheca Botanica*. Targioni of Florence is preparing for the press a second volume of Micheli *. Burmann is printing *Decades Plantarum Africanarum*, which is really a fine book, with very good plates. Ten of them being completed, he is going on with the *Herbarium Amboinense* of Rumphius. Logan's experiments on the Generation of Plants are in the press. Van Swieten means to publish Boerhaave's lectures on the practice of physic, and Haller at Gottingen those on the theory. Gaubius has printed his book *de formulis Medicamentorum*. I am, &c.

DEAR SIR, Leyden, June 2, 1739

Last Friday I received your letter of the 5th of May, which gave me great pleasure, especially when I learned that the Camphor-tree and other things arrived safe. I have often been anxious to hear

* This never appeared, but the plates are in the Linnæan and Banksian libraries.

from you, and, as you desire, I take the first opportunity of acknowledging the receipt of your letter. I shall execute your commission at Boerhaave's auction as if it were my own.

I have sent this morning to van Haze, to enquire for a Camphor-tree. He reports that he has three exceeding good ones. I shall send the best of them to you next week, by Mr. Hudig.

As you are so kind as to offer me some English beer, it will be very acceptable. The surest way is to give the captain a letter to Dr. Abraham Gronovius, at Leyden, my brother having the freedom of the custom-house there. As soon as the captain arrives at Rotterdam, he must send that letter, either by the post or market-boat, to Leyden. Be so good as to mention therein how many gallons the barrel contains, with the names of the captain and of his vessel. I shall then send the "bill of freedom" to the proper officer at Rotterdam, and get your present without further difficulty.

Gaubius's work shall not be forgotten. I know not whether I ever informed you of the association I had, two years ago, with van Swieten and Lawson, to make experiments upon minerals. Our master was one Cramer, a man very skilful in that science. But after we had finished, he robbed us of our observations, which he sent to the press. The work is printed in duodecimo, and in all respects well done; the plates very exact. I think its price is 2 guilders 10 stuyvers.

I think your intention of sending Mr. Cliffort some fossils, will be the best and surest way of acquiring a copy of his *Hortus*. I am, &c.

R. RICHARDSON, ESQ. TO DR. J. F. GRONOVIUS.

DEAR SIR, No date.

I sent you a parcel of strong beer, packed up in a cask, which went from Hull, on board Mr. Parver's ship, the beginning of August, with a letter to your brother, according to your directions. I should be glad to know that it came safe to your hands, and proved good. I received the books you bought for me at Boerhaave's auction, along with the Camphor-tree, about eight weeks ago. I am very much obliged to you for your care about them. The tree lost its leaves in the voyage, but is now, in the stove, putting out new ones.

You mention having some cuts of Tournefort, in his Turkish dress, at 22 stuyvers each. I must desire you would send me one. I should be glad to see Mr. Seguier's *Bibliotheca Botanica*. As Targioni is about publishing Micheli's second volume, I conclude the latter must be dead, which I never heard of before. Please to send me Burmann's *Decades*, which were forgot. I have proposed, for some time, to send Mr. Cliffort some fossils, but my father has been either indisposed or engaged, so that we could not look them out. I hope we shall shortly, when I shall take the liberty to direct them to you. When you have an opportunity pray enquire of Mortier, at Amsterdam, about the book of ruins and buildings in Italy. I do not know whether it may not be called *Theatrum Italiæ.* It was printed I suppose 30 or 40 years ago.

DR. J. F. GRONOVIUS TO R. RICHARDSON, ESQ.

DEAR SIR, Leyden, June 23, 1739.

This is the list of the books I bought for you at Boerhaave's auction. The rest went extravagantly dear. The king of Portugal has given several very great commissions. You ought to have a catalogue marked with all the prices. The whole amounted to 13,000 (guilders?). I have no time to write to you fully, because I must pack up, in order to go to Middleburg to-morrow. When I return, you shall see the terms of subscription for *Rumphii Plantæ Amboinenses;* also the rest of the first volume of *Flora Virginica.* At my return I shall go on with the second. After the vacation there will be an auction of Boerhaave's Corals and Minerals, also of his mathematical instruments. I will take care you shall be duly informed of it. The *Hortus Cliffortianus* went for 31 guilders; really too dear. I think some fossils will do very well. Dear Sir, I must finish this, and remain, &c.

DEAR SIR, Leyden, Aug. 11, 1739.

This is to acquaint you that I am returned from my voyage to Zeeland. I hope you received the box of books, and the Camphor-tree, in good order, which I shall be glad to hear.

I had supposed the catalogue of Boerhaave's Corals and Insects would have been printed, but Gaubius, who has undertaken this, is gone to Germany, and the catalogue he has made cannot be printed before he returns. Besides this, there will be a catalogue of several chemical preparations, made by Boerhaave himself.

Mr. Cliffort is not disposed to sell his *Hortus*, but he has been so far persuaded, that, of his few remaining copies, he has given some to a fellow who sells them at 28 guilders 10 stuyvers.

Breynius of Dantzic has reprinted some botanical books of his father's, with very good notes, and some works of his own, viz.

Prodromus primus et secundus, of his father's.

Fasciculus rariorum plantarum. This was never printed before.

Vita et Effigies Jacobi Breynii.

Dissertatio de radice Ginsem et Herbâ Acmellâ, cum additamentis, in quarto, 1739. This is indeed a very good book.

And now, dear Sir, I hope to hear from you, being &c.

DEAR SIR, Leyden, Dec. 7, 1739.

I was mightily pleased with your letter of the 1st of October, and I do not doubt your having, in the mean while, received the catalogue of Boerhaave's preparations, Corals, &c. which I sent for you to Mr. Hudig, the day it came from the press. The chemical preparations did not sell dear, the Insects sold very dear, and the Corals as they were in your time, with the chests, went for 330 guilders. So there is an end of Boerhaave's collection! But I must tell you one thing more. I no doubt informed you that Boerhaave left no legacy to the University, but bequeathed all his manuscripts to his sister's sons, both of them physicians at the Hague. The eldest married afterwards a lady of no good character, and was soon obliged to seek an asylum at Cuylenburg, a congregation of thieves, murderers, &c. who never dare return to Holland. All his goods are now confiscated by his creditors, and if the legacy of Boerhaave's manuscripts were in his house before he went off, they must all be sold, which you shall be informed of by letter when I know it.

I hereby acquaint you that I have sent the *Decades* of Burmann to Mr. Hudig's care. Van Royen has printed his *Floræ Leydensis Prodromus*, really a very good book, which I also send. If you do not like it, you may send it back. I am sorry you have not the *Hortus Cliffortianus*, because you will see, by this *Prodromus*, that the said *Hortus* is a standard for specific characters. But I do not doubt Mr. Cliffort's giving it to you in return for fossils. You may depend on seeing Seguier's *Bibliotheca Botanica* when printed. It is about three years since Micheli died.

I have now finished my catalogue of all I possess belonging to the *Regnum Lapideum* of the *Systema Naturæ* of Linnæus, reducing them to Classes, Orders, and Species, and observing the same laws in the specific definitions as Linnæus has prescribed in his *Critica* for plants.

I believe you were told, when in Holland, that the heirs of Mr. De Wild, of Amsterdam, had a mind to sell that gentleman's collection, consisting of an unrivalled series of Greek and Roman coins, with antique gems, of which he printed, in his lifetime, an account in two or three volumes quarto. These things will all be sold next year.

I have enquired after the Italian book you wanted. There is a work printed 40 years since, by a bookseller named Overbeek, entitled *Reliquiæ Romæ Antiquæ*, in large folio. The price at that time was 70 guilders, but it is not now to be had, except at auctions, and that very seldom.

Dr. van Swieten is printing Boerhaave's lectures on the practice of physic.

Mr. La Court, who had that curious garden near this town, died a few months since. Two years ago he printed his observations on gardens, trees, and flowers, with explanatory descriptions of his stoves, &c. a large book in quarto, but in Dutch only, and without his name. I remain, &c.

DEAR SIR, Leyden, Sept. 26, 1740.

My last letter was dated the 7th of December. With this I have the honour to present you with an *Index* of all the stones I have collected since the *Systema Naturæ* of Linnæus was printed *.

Seguier's *Bibliotheca Botanica* is lately printed at the Hague, in quarto, price four guilders.

Wishof is printing Haller's observations on Swiss Plants, with several good plates. The work of Rumphius goes on well at Amsterdam.

Dr. van Swieten gives his humble service to you. The first volume of his Commentaries on Boerhaave's aphorisms will be published the beginning of next year.

A new edition is now printing of the *Fundamenta Botanica* of Linnæus, in large octavo, with several augmentations and corrections.

His *Characteres Generum* are also to be reprinted, with several new *genera* of plants, which I have lately obtained from Virginia. These will form the second part of my *Flora Virginica*.

I remain, dear Sir, your most obedient servant,

JOHN FREDERICK GRONOVIUS.

* *Index Suppellectilis Lapideæ, quam collegit, in classes et ordines digessit, specificis nominibus ac synonymis illustravit, J. F. Gronovius. Lugd. Bat.* ed. 1. 1740. ed. 2. 1750. The dedication of the latter to Linnæus commemorates those mineralogical studies, in company with him and Dr. Lawson, which are mentioned in the above letters. Lawson had now been dead two years.

DR. JOHN AMMAN*, PROFESSOR OF BOTANY AT PETERSBURGH, TO LINNÆUS. [Latin.]

SIR, Petersburgh, Sept. 1736.

I have received your letter of the 26th of June, addressed to me at Amsterdam, as well as one from Dr. Gronovius, accompanied by your systematic tables †. I have long wished to see these tables, of which I find such frequent mention in the literary news of the day, and am much obliged to Dr. Gronovius for giving me an opportunity of perusing them. They evince great genius and knowledge in the study of nature. Your new method of arranging plants, by the number and situation of the stamens and anthers, appears to me very serviceable in defining the inferior genera; but less so, with regard to the superior ones ‡, than others already contrived.

The character of *Ammania*, as to its stamens, pistil and calyx, is unknown to me, as I have never seen the plant in flower. I have indeed a most elegant dried specimen from America, but it shows the fruit only. The petals and stamens are said by Houstoun to be so fugacious as scarcely to be detected, even in the living plant. I am possessed of many new genera, from India and America, as well

* He died in 1741, at the early age of 34.

† The first edition of Linnæus's *Systema Naturæ*. See the letter of Gronovius to Richardson, p. 173, of this volume.

‡ This distinction seems not very intelligible, but the next letter explains it. By "superior genera" are meant classes or orders, especially natural ones.

as from Tartary, Siberia, and Davuria, but I have not at present sufficient leisure to give you their characters. I long very much to see your *Flora Lapponica*, as I am certain it must contain many of our native plants. It is some time since I saw Mr. Siegesbeck, on account of the distance at which we live from each other; nor have I, as yet, been able to send him your publications. The garden of which he has the care, has nothing to do with the Academy, but belongs to the Apothecaries' department. In this country there is no communication between the Academy and the Medical College, to which last the subject of the *Materia medica* is attached. Whether Siegesbeck has written a catalogue of the Apothecaries' garden I am entirely ignorant. Farewell.

Petersburgh, Nov. 15, 1737.

Your letter dated from Mr. Cliffort's Museum, which I received a few months since, afforded me great pleasure. I owe you many thanks for your kind present of all your Botanical publications; and I beg of you also to thank that great cultivator and patron of Botany, Mr. Cliffort, for his present of the description of the *Musa* which flowered in his garden. I remember to have heard from Dr. Gronovius, whilst I was studying medicine at Leyden, how much expense and attention he bestowed on the cultivation of all the rarest plants that could be procured, in his delightful garden near Harlem. We have therefore much information to expect from your intended catalogue of this collection.

I perceive you take somewhat amiss what I wrote to Gronovius and Dillenius about your new method founded on the stamens and pistils. I was merely joking, in my letter to Gronovius, if I mistake not; and I could not suppose you would seriously be displeased, at my remarking the great concourse of husbands to one wife, which often happens, and which is so unsuitable to the laws and manners of our people here. I was not speaking of those natural laws of the vegetable kingdom, instituted by the Creator of all things. I observed to Dillenius that your system was excellent for establishing and defining the genera of plants, though scarcely of any use as to classes. I continue of the same opinion; for according to your method, plants which agree in the number of their stamens and pistils, though totally different in every other particular, are placed in the same class. What affinity is there, except in the number of their stamens, between *Valeriana* and *Cyperus*, *Persicaria* and *Campanula*, *Gentiana*, *Ribes*, and *Angelica*, &c.? I have read your *Flora Lapponica* quite through, and have found many plants mentioned and described therein, which grow wild about Petersburgh. For example, your *Betula* with round crenate leaves *, a figure and description of which I communicated to our Academy two years since. You omit among your

* *B. nana.*

synonyms *Betula pumila*, of the *Flora Prussica*. Your *Rubus* with a solitary flower and ternate leaves *, is the same as Buxbaum's *R. humilis flore purpureo*, *Cent.* 5. *t.* 26, elegantly and accurately described 15 years ago, in Siberia, by Messersmidt. The *Struthiopteris* † of Cordus, and *Thalii Sylv. Hercyn.* which grows abundantly in boggy places about this town, is the same plant as the *Filix palustris maxima* of Caspar Bauhin's *Prodromus*, and the *Struthiofera* of Muntingius; being likewise *Lonchitis norwegica major* of Petiver. Muntingius and Petiver saw the seed-bearing leaves only, which are totally unlike the barren ones. This plant has led me to write a dissertation, to be printed in the *Commentarii* of our Academy, upon all the species of Ferns, taking that term in an extensive sense, which bear two kinds of leaves, one barren, another fertile, in which I shall describe and figure some new kinds from America.

I have no doubt of your genera of plants having been accurately determined from living specimens; and yet the characters of some of them appear to have been taken from dried ones only, or from figures or descriptions not altogether exact, such as those of the *Hortus Malabaricus*. Hence it is possible that no small number of errors may have crept in.

You promise to account, in your *Critica Botanica*, for your numerous alterations of names. I presume you have followed the rules laid down in your *Fun-*

* *R. arcticus.* † *Osmunda Struthiopteris* of Linnæus.

damenta Botanica. Still many of those rules may, perhaps, not be universally approved, any more than your changes of names. I beseech you to consider what would be the consequence, if every body were to lay down such laws and regulations, at his pleasure, overturning names, already known and approved by the best authors, for the sake of making new ones. Would it not lead to worse than the confusion of Babel? I write with candour and sincerity, not from a spirit of wrangling or contradiction.

Siegesbeck, a man of a singular character, is printing a critical dissertation, in which your writings are, harshly enough, found fault with. I have had an opportunity of perusing it, as Dr. Fischer, chief physician to the Empress, has obtained leave to have it printed at the Academical press. The work is very short, but its brevity is, in my opinion, counterbalanced by spite and arrogance. The same may be said of the fundamental principles of Botany, prefixed to this treatise.

I have never seen *Ceratocarpum* in flower; but I here enclose its capsules, taken from a specimen gathered in the furthest part of Siberia, where this plant serves as food for a kind of jumping Rabbits, with a very long tail. If you will let me know what you want of the plants described in Buxbaum's *Centuriæ*, or in the Transactions of our Academy, I hope to be able, by next spring, to send you some of them, along with my treatise on the rarer plants of Tartary and Siberia, illustrated with descriptions and figures. Farewell, and continue your friendship for me.

Petersburgh, Jan. 23, 1738.

I send, according to your desire, the critical dissertation of Siegesbeck on the works you have hitherto published, as well as the *Botanosophiæ verioris Sciagraphia* of the same author. You will judge for yourself what is to be thought of them. As to the price of postage for any thing you may send me, I beg you will never think about it. The packet which you told me I should receive from Gronovius, is not yet come to hand. When you return to Stockholm I shall write more at length, directing my letter to the care of my countryman Mr. Hedlinger, to whom I beg of you to present my best compliments. When you see the great Boerhaave, the learned Van Royen, the distinguished Gronovius and his son, or any others of my Leyden friends, be so kind as to wish them, on my part, health and prosperity. Adieu — believe me ever yours.

Petersburgh, Dec. 2, 1739.

Your very kind letter, dated Stockholm, the 12th of October, came safe to hand. I learn with joy that you have reached your own country in health and safety, and that the highly honourable appointments, conferred upon you, have determined you to settle there, on which I congratulate you most sincerely. My little work, upon the rare native plants of the Russian empire, as well as the performances of Siegesbeck, his first attempts, I should have sent you by a ship going to Stockholm; but it happened that the lakes and rivers, and even the sea itself, became frozen almost entirely over, soon after I received your letter, so that I was unable, as I still am, to find any mode of conveyance to you during winter, unless you can suggest any.

I have written no treatise on the Russian Ferns, unless you are pleased to give that name to an Academical dissertation, upon all the Ferns hitherto known, that bear two sorts of leaves, one producing flowers and fruit, the other barren. I hope this will appear, next year, in our Commentaries, illustrated by a few figures *. Gronovius has, this autumn, sent me the *Flora Virginica*. He promised me the *Hortus Cliffortianus*, but it is not yet come. I have not yet seen Van Royen's *Prodromus* of the Leyden Flora. Pray send me whatever is published by the newly-instituted Society of Sciences at Stockholm, as it cannot but be highly interesting. Farewell, and grant me a continuance of your regard.

* See *Commentarii Acad. Scient. Imp. Petrop. v.* 10. *ad annum* 1738, *J. Amman de Filicastro, p.* 278—302. *tab.* 18—23.

Petersburgh, Sept. 7, 1740.

I wrote to you, my dear Sir, several months since, at Stockholm; but your silence leads me to doubt whether you ever received my letter. I therefore beg to know whether or no it ever reached you. I would gladly have sent my work on Russian plants, had I last year met with a good opportunity. The same was the case in the ensuing spring, and, now almost departed, summer. This must be known to you as well as to myself. I will, however, this autumn, or next winter, send you that work, with the only copy I have of the beginning of the Petropolitan Flora *. I hope you received the *Botanosophiæ verioris Sciagraphia*, which I sent you by post, during your residence in Holland. We this year had in flower, in the garden of the Academy, the *Lapathum orientale frutescens*, &c. † This elegant plant does not properly belong to *Atraphaxis;* having eight stamens, and three stigmas. I have given a figure and description of it to the Academy, to be published in its Commentaries. Two species of *Persicaria*, from Siberian seeds, have also flowered. Each of them had eight stamens, with a pistil terminating in three stigmas. The flowers of both were fragrant. In one they formed dense clusters; in the other they were mostly axillary. We also raised a very curious whorled plant, with labiated flowers ‡; the upper

* We know not what work is here alluded to.

† *Polygonum frutescens, Linn. Sp. Pl.* 516.

‡ *Amethystea cærulea, Linn. Sp. Pl.* 30.

lip cloven, the lower in three nearly equal segments; the stamens but two; seeds four, naked; calyx deeply divided into five equal, very sharp, spreading segments. The flowers are by no means spiked, but grow on little branched stalks, from the bosoms of the leaves. The leaves, moreover, are three-cleft, much in the manner of *Moldavica americana trifolia, odore gravi, Tourn. Inst.* 184 *. Does not this plant form a new genus? At least I am not able to refer it to any known one.

I this summer examined three different species of Jussieu's *Corispermum*, but could not discern petals in any of them, even with magnifying glasses. I have also a very extraordinary plant, which seems hitherto unknown †. It is furnished with two kinds of flowers; the male ones growing at the summits of the branches, and consisting merely of barren stamens, forming very dense heads, disposed in long reflexed spikes. The female flowers are concealed in the bosoms of the leaves, and so minute, that I have not been able to investigate them fully. The seeds are contained in little scaly and leafy heads, solitary, compressed, oblong; but have not yet arrived at maturity. Pray direct your answer to the Academy, with my name at the corner. Farewell!

* *Dracocephalum canariense.*

† The description answers in some points to *Liquidambar styraciflua*, or to *Comptonia asplenifolia, Ait. Hort. Kew.* v. 5. 254.

Petersburgh, Nov. 18, 1740.

Your kind letter from Stockholm, dated Oct. 16, came safe to hand. I am surprised to learn by it that you had not then received one which I wrote to you in the course of this year. I entrusted my book, on the rare plants of the Russian empire, to an eminent merchant of this place, who undertook to convey it to you at Stockholm. But if this also is lost, pray let me speedily be informed. I wrote to you, a few weeks ago, by the post, with observations upon several uncommon plants. I trust you have received that letter.

I now proceed to notice what you say in your last, respecting some doubts and *desiderata* among the plants mentioned or described in my work. I begin with the *Argusia* * of Messersmidt. This pretty and singular plant flowered last summer, in the Academic garden. You enquire concerning its stamens, style, and stigmas, which are as follow. The style, or, to use Tournefort's language, the pistil, is conical, short, and thick, concealed within the flower, and inserted into the calyx, having a very thick, almost hemispherical stigma. The stamens are usually five, sometimes six, very short, closely attached to the bottom of the flower, and bearing white *apices* (anthers) about a line long, cloven, discharging a very white powder, which nearly fills the whole bottom of the flower. The corolla is white, monopetalous, tubular in the lower part, divided upwards into five spreading, acute,

* *Messersmidia Argusia, Linn. Mant.* 42.

equal segments, undulated at the margin. The calyx is small, of one leaf, likewise divided into five very narrow segments. You will find this plant among the dried specimens which accompany the present letter.

The flower of the *Helleborus Fumariæ foliis* * has five nectaries, of a yellow colour, as well as five petals. The stamens are as many as the little seed-cases (germens), generally twelve, having closely attached to their backs very small and slender yellow anthers. I send you likewise a specimen of this, with its seeds gathered last summer in the garden of the Academy. It is certainly a very singular, as well as elegant, plant. The figure in my book is not very accurate.

The *Valerianella nemorosa repens*, &c. of Messersmidt, I confess, altogether agrees in its creeping root, procumbent stems, leaves, flowers, number of stamens, and mode of growth, with the *Serpyllifolia* of Buxbaum. But if we consider the shape and structure of the calyx and fruit, as described by Messersmidt and yourself, there appears some difference. You in your *Genera Plantarum* describe a double calyx, one of which is seated on the germen, the other embraces the fruit, which is an ovate smooth capsule, of three cells, covered with this calyx of the fruit. Messersmidt, however, describes the calyx of the flower as being deeply five-cleft, seated on the top of the roundish-oblong hairy seed, and remaining permanent after the

* *Isopyrum fumarioides, Linn. Sp. Pl.* 783.

flower is fallen. You see, therefore, that, according to your description, the plant of Messersmidt differs from yours in its simple calyx, and naked hairy seed, not contained in a capsule. He brought neither seeds nor dried specimens with him from Siberia, the examination of which might remove our doubts; but I am almost convinced that this *Valerianella* of his is Buxbaum's *Serpyllifolia*. I have given his account of it in my work on Russian Plants, for the use of those who may visit the place where Messersmidt found the plant *.

The dried specimen of my *Lathyroides* †, now sent, will abundantly show how different that plant is from your fourth *Orobus* ‡ in the *Hortus Cliffortianus*. Several roots of this species are growing in our garden, from Irkutensian seeds, but have not yet flowered.

Besides the leguminous plant, whose seeds I first sent to Dillenius at Oxford, and from which you constituted your new genus *Trigonella*, by the form of the corolla, I have in the garden of the Academy another very rare species, described in my work on Russian Plants, under the name of *Melilotus*, No. 159 §.

* The plant in question is *Linnæa borealis*. The description of Linnæus is correct, and that of Messersmidt, as well as his generic name, most unfortunate.

† *Orobus Lathyroides, Linn. Sp. Pl.* 1027.

‡ *O. hirsutus*, ibid.

§ This appears to be no other than *Trigonella ruthenica. Linn. Sp. Pl.* 1093. *Herb. Linn.*

*Chamæjasme** has a dry turbinate fruit, almost like the common *Valerianella*, so that it cannot belong to the *Thymelææ*†. I have never seen any thing of *Menispermum* but the seeds. I have sown the seeds of *Cymbaria* four times, in wet and dry, hot and cold situations, rich and poor soil, but in vain. They have never vegetated. I never saw the flowers of *Coris juncea*‡. You will find flowers of *(Isopyrum) Thalictroides*, and seeds of *Hypecoum* among my dried specimens.

You mention having seen something cultivated by Jussieu, at Paris, under the name of *Thalictroides*, which you think scarcely different from mine. I believe them to be very distinct. The plant of Jussieu bears berries; mine has a dry, horn-shaped fruit, full of hairy seeds. Although the flowers agree, the different fruits must make them distinct genera.

There is found in Siberia a red-berried kind of *Christophoriana (Actæa)*, which may probably be the same with the American one in Gronovius and Morison.

I think I have now answered all your enquiries. Adieu therefore, my distinguished friend; continue your kindness to me as before.

[This appears to have been Professor Amman's last letter, as he died in the following year.]

* *Stellera Chamæjasme, Linn. Sp. Pl.* 513.

† It certainly does, having no affinity nor resemblance to the plant mentioned, *Valeriana Locusta* of Linnæus.

‡ *Hypericum Coris. Linn. Sp. Pl.* 1107.

DR. HERMAN BOERHAAVE TO LINNÆUS. [Latin.]

Leyden, Jan. 13, 1737.

I return you my thanks for your letter, as well as for your book *. The former evinces your strong, though unmerited, attachment, your kind and liberal friendship. I wish I had any thing to deserve your good will, or the means of making any adequate return!

Whoever reads your book must be struck with it, as a work of infinite attention, singular perseverance, and unrivalled science. Nor can I fully calculate the result of so valuable an undertaking. Ages to come will applaud, good men will imitate, and all will be improved.

While in the most classical and correct manner you trace out your botanical characters, with so much felicity, in every instance, I know not how it has happened that you are incorrect on the subject of my character alone. I never have deserved, nor can I hope to deserve, the high botanical encomiums which you are so kind as to bestow upon me. I am well aware that my work † abounds with errors. I hastened its publication for the sake of the academical students. This haste, and the very numerous occupations which distracted my attention, have caused but too many oversights in my performance.

* *Genera Plantarum*, 1st *ed.* dedicated to Boerhaave.

† *Index alter plantarum, quæ in Hort. Acad. Lugd. Bat. aluntur*, 4to, 1720.

You, in every instance, write nothing but what announces a man of experience, and a profound critic.

Nevertheless, as you have been pleased to inscribe to me this immortal work, I shall be anxious to seize every opportunity of showing how highly I esteem such an honour.

May God grant you health of body and mind, to be Nature's historian, for many years to come!

DR. BOERHAAVE TO BARON J. B. BASSANDO, PHYSICIAN TO THE EMPEROR, AND THE GRAND DUKE OF TUSCANY. [Latin.]

Leyden, March 16, 1738.

A British gentleman and very learned physician, Dr. Lawson, having finished his scientific education, and travelled over the most cultivated parts of the world, is desirous of studying the subject of mining and mineralogy.

I beg the favour of you to give him all the assistance you can in the prosecution of this object, particularly as to the Imperial mines. He is certainly entitled to such a favour, on account of his knowledge and respectable character; and I know how desirous you always are of rendering any service to those who are worthy of your notice.

I labour under a vomica in the lungs, which for the last three months has greatly oppressed my breathing, on the slightest exertion, and has been hitherto increasing daily. If it goes on enlarging, without bursting, it must suffocate me; if it bursts, the consequences are uncertain *. Whatever may be the event, it is at the disposal of the Supreme Governor of all things. What therefore can I have to fear or to desire? We must submit to the divine will. Yet I am not the less desirous of neglecting no probable remedy that may alleviate my pain or bring the abscess to maturity, though I am perfectly easy as to the final result. I have lived more than 68 years, with unabated cheerfulness. May you, my worthy friend, enjoy a long and happy life!

ANTOINE DE JUSSIEU, M. D. REGIUS PROFESSOR OF BOTANY AT PARIS †, TO LINNÆUS AT HARTECAMP NEAR LEYDEN. [Latin.]

SIR, Paris, July 1, 1736.

I received, with much pleasure, your work on the *Musa*, which I immediately read through with avidity, and no less satisfaction; not only because of

* The tumour, whatever it was, continued to enlarge, and terminated the valuable life of this great and good man about six months after the above date. Dr. Lawson never had an opportunity of delivering the letter; but he allowed Dr. Richardson to take a copy, of which we here give a translation.

† He died in 1758, aged 72. This gentleman was elder brother of the celebrated Bernard de Jussieu, and likewise uncle of the present Professor Antoine de Jussieu, so well known by his *Genera Plantarum*, published in 1789.

the singularity of the plant itself, but for the sake of your remarks. I never suspected that this plant, which I had seen bearing flowers and fruit in Spain, could produce any in Holland, as we have never had an instance of the kind in the royal garden at Paris, where it has not even flowered. None of the other works, mentioned as having been published by you, have ever reached me, and I shall be greatly obliged by your ordering them to be sent hither, at my expense. I long very much to see your *Hortus Cliffortianus* and *Flora Lapponica;* especially the latter, as the king has recently sent some of our academicians towards the most northern parts of Europe; to whom, in their search after plants in those countries, your book would be a guide, instructing them what seeds, or dried specimens, to send us. If therefore you are likely soon to complete this work, I request the favour of two copies, which shall be paid for with the above-mentioned publications. If you know of any thing, issuing from our Parisian press, likely to be worthy of your notice, nothing will give me more pleasure than to procure it for you.

Be pleased, Sir, to accept the respects of my brother and myself.

I remain, &c.

ANTOINE DE JUSSIEU.

BERNARD DE JUSSIEU * TO LINNÆUS AT HARTECAMP. [Latin.]

SIR, Paris, Oct. 9, 1737.

I am favoured with your two letters, as well as with your *Critica Botanica*, or fourth part of the *Fundamenta*. The other things which you mention having sent, the *Genera* you have established, and the *Flora Lapponica*, I have not yet seen, at least they have not yet come to my hands. Nevertheless, I return you my best thanks for these marks of your favour, and beg of you to accept a few botanical publications, though of small moment, which have appeared at Paris. I should not mind the expense, if you could purchase for me any works on the subject of botany that I have not yet met with, such as Siegesbeck's *Hortus Petropolitanus*, Ludwig's *Characteres Generici*, *Plantarum Indices circa Aboam*, Tilland's *Flora Uplandica*, or *Bromelii Chloris Gothica*. Your friend Lithenius promised me this last, when he was at Paris. We have, as yet, no tidings of my younger brother, who sailed for Peru last year in July, on a botanical expedition. If, as I hope, he should return in safety, laden with plants, I shall be happy to supply you with some of his dried specimens. I rely on your

* Regius Professor of Botany at Paris, the original author of the celebrated System of Natural Orders, since improved and published by his distinguished nephew, Antoine de Jussieu, the present professor, who is also nephew of the writer of the preceding letter.

14. Bernard de Jussieu to Linnæus, Oct 9th 1737.

Clarissimo Viro

D.D. Carolo Linnæo doctori medico
et Botanico illustrissimo, doctissimoque

S. P. D.

Bernardus de Jussieu. Prof Bot. Reg.

Tuas quas dedisti binas epistolas accepi, simulque criticam Botanicam seu fundamentorum botanicorum partem quartam caetera quae scribis misisse, genera atq constituta et floram Lapponicam non vidi huc usque.

15. Haller to Linnæus, Septr 12th 1737.

Experientissimo Viro
Botanico Summo DD
Carolo Linnaeo
Spd
AlbHaller.

Epistolam Tuam cum Critica & Musa recte accepi, mittam vicissim specimina Botanica de Pediculari, et Veronica. Fuerunt autem Tuae literae similimae dissertationis, mihi oppositae, sed id omne ab animo amicissimo profectum esse certus confido, neque renuo doceri meliora.

16, Haller to Linnæus. Octr 12th 1748.

Quid me moves ad pacem Linnaee celeberrime, quo nemo avidius persequitur! quem nisi Tuum [illegible] animum semper demonstrasse, et quem [illegible] hominum injuria ita in ordinem coegerit, ut non potuerim contentionem vitare, qui multa tibi [illegible] [illegible] teneri vel tribuli a multis [illegible] possem [illegible]

Martin, Lithogr. Rochester St. Westmr.

The material originally positioned here is too large for reproduction in this reissue. A PDF can be downloaded from the web address given on page iv of this book, by clicking on 'Resources Available'.

kindness, in return, to furnish me with the plants of Lapland, Sweden, and Norway, if you will so far favour me. Farewell!

My brother desires his compliments.

TO THE SAME, AT STOCKHOLM.

Paris, July 20, 1740.

MY DEAR AND MUCH-ESTEEMED FRIEND,

I have received two letters from you, since your prosperous return to your native country, by which I am informed of your happy marriage. Nothing gives me more pleasure than to congratulate you on your fortunate lot, after so many labours and exertions, in being quietly settled under your own roof, enjoying well-earned fame in medicine, and especially that eminent botanical appointment to which you are so fully entitled. At all these events I most sincerely rejoice; not that I ever doubted of your success, but the news of it is most welcome to me. You promised me all your various publications, but they, by some accident, have never reached me. I have long looked for them; but the great length of the voyage may, probably, be the cause of their delay. Whenever you send me any of your works, pray think, at the same time, of the books you spoke of, for our friend Danti (d'Isnard). They may be entrusted to our embassador at the Swedish court, and we shall then speedily obtain these highly-valued pledges of your regard. You may in future make use of the same mode of conveyance; and I shall, by the same channel, send to the illustrious Count de Tessin, any thing that I know you want. I now send you a few roots of the *Orchis* tribe, recently gathered, as a slight token of my remembrance.

My brother, who has been examining the country of Peru, is not yet returned. May God send him back in safety! The works of Plumier, which I was desirous of publishing, have not yet appeared, nor can they be given to the publick till they are properly arranged, on the principles of natural classification; especially as they require an extension of our natural system; for which purpose the characters of these American plants are now undergoing revision. A friend of mine, resident in the French West Indian islands, has devoted himself entirely to this object. He follows your principles, having, at the first sight of your system, become a Linnæan, though originally a Tournefortian. He has already sent over descriptions of a few genera, drawn up in your manner, but, according to his promise, I expect a much greater number.

The Peruvian Bark has five stamens. The flower is seated upon the embryo, or fruit. It belongs to the same order as *Coffea*, *Randia*, *Nux vomica*, and perhaps *Cephalanthus*, the capsule of which last has two cells, with a single seed in each. To the same order belong various species of *Periclymenum*, enumerated in Plumier's catalogue, as also *Morinda*, his *Roioc*, &c.

I have ascertained the flowers of *Pilularia*, and the whole of its fructification, last summer. I have given its history in the Memoires of the Royal Academy of Sciences. This year I mean to follow it up with the *Lemma* * of Theophrastus, which, in natural characters, comes near *Pilularia*, but differs in some particulars, so that I retain both genera. As soon as the figures are engraved, I will send you them, with an explanation of the various parts therein delineated.

I am now well acquainted with the *Draco arbor* of Clusius, which is neither a *Cordyline*, nor a species of Palm, as is generally believed, and as indeed many botanists have suspected. It constitutes, in my opinion, a new genus, which I call *Draconthema* †; and I mean to read, at our next meeting, a paper on the subject, stating the grounds for the establishment of this genus. It ranges in the same natural order as *Asparagus*, *Convallaria*, *Tamus*, *Smilax*, &c.

Your distinguished Mæcenas, Mr. Cliffort, has been so good as to send me, in due time, the *Hortus Cliffortianus*. I have returned my best thanks for so great a favour. But I was already too much in your debt, on many accounts. I can only wish that it were in my power to render you more service. I am ashamed to have allowed so much time to pass without writing. I must beg your indul-

* *Lemma*, *Juss. Gen.* 16. *Marsilea* of Linnæus.

† *Dracæna*, *Linn. Mant.* 9. *Juss. Gen.* 40.

gence in favour of one who holds you in the highest esteem, and who can never cease to love and value you. Believe me your most obedient and entirely devoted friend, BERNARD DE JUSSIEU.

My brother desires his compliments; as well as our friends the good Father La Serre, Aubriet, the widow Vaillant, and Mademoiselle Basseporte.

MY DEAREST FRIEND, Paris, Feb. 15, 1742.

I received your welcome letter, and have several times been desirous of answering it, but have as often been hindered by various affairs. Pardon my past neglect, though I have permitted some opportunities, of testifying my regard for you, to pass by. I have been occupied in various journeys. All last autumn I was wandering on the sea coast of Normandy. I have met with many novelties, among which you will be surprized to find some additions to the animal kingdom. I mean, however, before I make my discoveries publick, to examine into the matter more fully.

I learn, with the sincerest pleasure, of your being appointed Professor of Botany at Upsal. You may now devote yourself entirely to the service of Flora, and lay open more completely the path you have pointed out, so as at length to bring to perfection a natural method of classification, which is what all lovers of botany wish and expect *. I know of nothing new here except an essay on the natural history of Cayenne †, and a catalogue of officinal plants. These little works will be conveyed to you by the surgeon of Count de Tessin, when he returns home. I shall also add a *Fasciculus* of medical questions, of the faculty of Paris. I have not yet received what you last sent me, but I return you many thanks for your repeated kindness. I beg leave to offer you, as a testimony of my gratitude, a few exotic seeds. May God preserve you long in safety! Believe me your most devoted,

BERNARD DE JUSSIEU.

My brother sends his compliments; and all our good friends, whom you have known at Paris, offer you their congratulations.

Paris, April 7, 1745.

Your letter by Baron de Scheffer was highly welcome. I learn from it that you do not forget me, and that the seeds, sent some time since, proved acceptable. Please to accept another small packet, consisting chiefly of what Mr. Bæck told me you wanted, before he returned to his own country.

* In the face of this testimony we trust it will hardly be asserted, in future, that Linnæus owed his ideas of natural orders to the excellent writer of the above letter.

† *Essai sur l'histoire naturelle de la France equinoxiale, &c. par Pierre Barrere. Paris*, 1741, 8vo. See vol. 1. 183.

This packet will soon be conveyed to you by Mr. Cleberg. But I am preparing another, which will be delivered to you by that distinguished young nobleman the Count de Spar, who brought me two dissertations, published under your auspices; one on *Ficus*, the other on *Betula nana*. Accept my best thanks for them. If you know of any thing newly published in botany or natural history, do not forget me. These things are my delight as well as yours. I have been much struck with the *Peloria*, nor does it seem easy to decide whether it be a metamorphosis, transmutation, or offspring of the common *Linaria*. By sowing of the seeds we shall know whether it will prove a monstrous production. I have sometimes noticed something of a *lusus naturæ* in the flowers of this *Linaria*, but nothing of so regular a shape. The nectary indeed was multiplied, having two, three, or four spurs; but I never met with so regular a figure in the limb as your plate represents. Besides, the alteration I saw was not in all the flowers of the same spike. I shall most thankfully receive the seeds of this plant which you offer me. With regard to the opinion expressed in your letter, relating to the *Lithophyta*, and the marine plants, as they are called, it wonderfully supports the ideas I have adopted. Your observations give great weight and effect to mine. I shall anxiously look for your promised dissertation on the Corals of the Baltic. As to my arrangement of the *Vermes* and *Zoophyta*, it is rather an essay than a perfect classification. I shall indeed be proud if you think it worth publishing in your Transactions. Perhaps I shall be indebted to your partiality for this, as well as for many other instances of the same. Among the chief of these is your having proposed and elected me a member of your Academy. Being known to you alone, I cannot but acknowledge as your act the honour which the other learned members of your body have done me. Be pleased to present my due respects to all of them. You may be sure I shall never forget this favour. Farewell! — do not withdraw your regard from Yours, &c.

My brother, and Father La Serre, send their compliments. D'Isnard and Aubriet are no more; but the widow Vaillant is living, as well as Mademoiselle Basseporte, who draws flowers and plants with great skill. Monsieur Clairvaut, an eminent mathematician, who met you in the *Jardin du Roi*, and talked Swedish with you, presents his best respects and good wishes.

Paris, May 1, 1745.

I would not omit this opportunity of expressing once more my regard for you; and I hope you will receive favourably the few seeds which accompany this present letter. You may be sure I will neglect no means of communicating any thing likely to be agreeable or useful. I know of nothing new in natural history. War keeps the Muses silent. I look for your dissertation *de Coralliis Balthicis,* and any botanical publication that may have appeared in your country. My brother desires to be remembered; and Father La Serre, the former companion of our journey, embraces you with all his heart. Adieu! do not forget me.

MY EXCELLENT FRIEND, Paris, Jan. 6, 1746.

I beg leave to solicit your kindness in favour of Mr. Cleberg, who has just heard of the death of the Greek Professor in the University of Upsal, and is desirous of supplying his place. He is preparing to return home, and, as you know him well, he hopes for your vote. I beseech you to give your support to this worthy man, and shall gladly acknowledge it as a favour done to myself, for which you will ever find me grateful. Pray contrive to have his name mentioned, as a candidate, to the king. I shall confide to him, at his departure, a small parcel of choice seeds, as a pledge of my regard. Farewell! &c.

Paris, May 7, 1746.

I now send you some seeds which you desired to have, and which I hope will speedily reach you. Mr. Cleberg, who is so much indebted to your favour, willingly takes charge of them. The services you have rendered this worthy man I feel as a true testimony of your attachment to me, which I shall ever be happy to return as far as possible. Let me know the title of the work you have recently published *, that I may recommend it to my pupils, who every year accompany me in botanical excursions, over the country round Paris. I shall have the more satisfaction in doing so, as I have no doubt that this book, so long wished for, will throw new light on natural history, and I have always pleasure in talking of you. I know also, by my own experience, how much benefit is to be obtained by studying according to your principles. Your publisher might therefore send a number of copies, a hundred for instance, to Paris, and I trust they would, most of them, if not all, be speedily disposed of. A bookseller of this town is now writing a letter to your bookseller, in answer to your proposals and those of Dr. Bæck. He is a most worthy man, distinguished for his honesty. I have received, at different times, your dissertations on *Ficus*, *Betula nana*, *Peloria*, *Corallia Balthica*, &c.

The *Ximenia*, *Baobab*, *Guiabara*, *Simarouba*, and *Lucuma* †, have never yet blossomed with us. The *Simarouba* is certainly no *Euonymus*, nor of the same natural order. *Fagara* and *Citharexylon* have shown flowers. The male of the former has five stamens; but Dr. Bæck has drawn up the cha-

* This appears to have been the first edition of the *Flora Suecica*. Jussieu had caused to be published at Paris, in 1743, an edition of the *Genera Plantarum* of Linnæus.

† *Achras mammosa*, *Linn.* *Baobab* is *Adansonia*; *Simarouba* is referred to *Quassia*. We cannot trace *Guiabara*.

racter of this tree. The female began to put forth flowers last year. I will send you its characters another time, as well as a description of all the parts of fructification of *Citharexylon*, except the fruit, which has hitherto not ripened. *Glans unguentaria* seems to me widely different from *Bonduc*, nor can I discover any thing akin to this tree.

The following Siberian plants, of which you have seeds, are wanting in my collection. I follow Amman's Index. *Cardiaca n.* 62. *Papaver* 81. *Sedum* 93. *Pentaphylloides* 116. *Lilium* 139. *Lupinaster* 143. *Vicia* 147. *Astragalus* 166. *Lathyroides* 151. *Melilotus* 158. *Delphinium* 174, also 175. *Absinthium* 193. *Lactuca* 211. *Aspalathus* 282, 283, 284. *Blitum* 239. *Amethystina* 70. *Ruyschiana* 64. *Lophanthus, Anandria*, &c. I shall be extremely obliged to you to send me any of these by means of some friend. Let me know of a fit opportunity to send you what you want out of our garden, a catalogue of which our friend Bæck takes with him. My brother salutes you, and the botanical surgeon La Serre embraces you with both arms. The damsel, who paints flowers so well, does not forget you. She hopes to hear of your welfare. Adieu, do not fail to esteem your old friend, &c.

Paris, Jan. 30, 1749.

Your botanical son, not mine, Mr. Missa, is set out in full sail, at your wish and instigation. From the day he began to study botany, he has followed you with the greatest devotion, and has been anxious only to imbibe your principles. I am glad that he has so happily attained the object he had in view, and I return you my grateful acknowledgments for the assistance you have rendered him, though without my request.

This young votary to Flora is gone for Africa. He is moreover well grounded and instructed in other branches of natural history; and we may expect from him an ample harvest, well accompanied with remarks. He proposes to make definitions of both plants and animals, after your method. He will send home dried specimens, and ripe seeds, as well as insects, to which he has very successfully paid attention in my company, during the last summer and autumn, following your *Fauna;* in which, however, we have found many things to correct, to alter, or to establish on a firmer foundation. Whatever novelties he may send from this burning country, you shall be sure, in the very first instance, to partake of, which I shall gladly take charge of.

The seeds herein enclosed are very fresh. Whatever else you may please to point out, I shall be happy to procure, and to send by the first opportunity, now the blessing of peace favours our intercourse. I have not yet had an opportunity of seeing your *Flora Zeylanica, Hortus Upsaliensis, Materia Medica,* the new edition of your *Systema,* or some other publications of yours. You say that Mr. Missa will transmit them to me; if so they will be long in coming. Perhaps you wish to fix a very severe curb upon my impatience, that I may write, and fulfil your wishes the sooner. I do, indeed, confess my past faults, with regret, and I trust therefore you will restore me fully to your former friendship.

But what am I talking about? Here are new wonders displayed to the scientific world! In the end of the last century, that laborious and faithful investigator of the secrets of nature, Leewenhoek, aided by the best microscopes, observed innumerable little bodies, *in semine masculino,* swimming about, and endowed with vital motion. These he anxiously sought, and detected, in the analogous fluid of every different kind of animal that he could procure. He gave to these bodies the name of Spermatic *Animalcula.* Bodies exactly similar in shape and size, agitated by similar movements, swimming about, like fishes in the sea, have, by the help of the microscope, been found and demonstrated, *in spermate fœmineo,* by M. de Buffon, member of the *Academie des Sciences,* superintendant of the Parisian garden, well known by his disquisitions in natural philosophy, and still more by his knowledge and publications in natural history. Having established this fact with regard to viviparous animals, he turned his attention to the oviparous tribe, such as birds, from whence still greater wonders came to light, respecting the times of breeding, and their indications; all which, however, terminate in a complete overthrow of Leewenhoek's theory, and involve the whole matter in more uncertainty than ever.

Please to accept the kind regards of my beloved brother, as well as of your faithful friend La Serre, and the worthy Mademoiselle Basseporte, who is very proud of the title you give her, of your second wife. I heartily wish health to yourself and all your family, rejoicing with my whole heart in the continuance of your friendship.

Many thanks for the fine specimen of the Norway Moss *, and the seeds of *(Axyris) Ceratoides*, welcome pledges of your kindness towards me.

Paris, Feb. 19, 1751.

In the course of last year, I obtained a very numerous supply of seeds, a considerable portion of which were sent from Peru, by my dear brother, who alas! is never to return from that country. Another most abundant harvest came from Africa, sent by M. Adanson, whom I mentioned to you formerly on his departure for Senegal †. A third collection, of no trivial importance, has been made near Pekin, the capital of China, by Father d'Incarville, a jesuit missionary. Out of all these I have selected such as were in the best condition, or such as have been afforded by plants, raised from the original stock, in the Royal garden. I wish this choicest share of our treasures may be agreeable to you, as a proof of our gratitude and respect.

* *Splachnum rubrum.*

† In conversation probably.

I have never observed Vianelli's worms shining, or rather sparkling, by night in sea-water. The author's figure shows them to be extremely minute *Scolopendræ*, or what you denominate *Nereides*, which another Italian, named Griselini, has since examined, and, by the help of a microscope, delineated still more accurately. Some philosophers have attributed to these little animals, the light which is often so widely diffused through the sea; but from the recent enquiries and experiments of a young physician at Paris, named Le Roy, it has been proved to originate from another cause, which is a phosphoric matter, mixed with the sea-water. This opinion appears to be supported by the shining of the waves in hot weather, and the luminous appearance of the whole surface of the sea, so frequent at night, as well as the sparkling tracks of ships, the radiant waves which surround them in their course, the shining of the wet oars when lifted out of the water, the sea-water itself, when poured out of a bucket in the dark, seeming like liquid light, and the shining appearance, here and there, of linen, when rubbed between the hands, after being sprinkled with sea-water, even though no sign of worms, or any sort of insect, could be perceived in the water used. He has read, before the Academy of Sciences, a very curious treatise on this subject.

I am not acquainted with Donati's History of the Adriatic. Pray inform me of the date and place of its publication, and the title of the work, so that I may be able to give orders for a copy; and I shall then let you know whether the worms of the Coral be correctly represented.

The spike of flowers, which some Swedish gentleman took to you from Paris, is an American *Phytolacca*, long cultivated in the garden. Its stamens are 30 or more, its styles five. Hence the plant belongs to *Polyandria Pentagynia*, being remarkable, as a species, for differing so much from the established character. You saw a dried specimen in my collection, and called it *Phytolacca caule arboreo* *. Plumier neither described, nor even met with this species, though a native of the hot parts of America; but it comes from the Spanish territories.

Chamæpericlymenum canadense, or *Baccifera mariana*, &c. *Petiv. Mus.* 363 †, is distinct in genus from *Lonicera*, nor has it any relationship to *Linnæa*. It is not far remote from *Coffea* as to natural order, but differs in generic characters; and is likewise allied to the family of *Galium*, *Rubia*, *Cruciunella*, &c. It flowered some time since in the *jardin du Roi*, and soon after perished. I have applied to a friend, who is one of the king's council, and a physician, at Quebec in New France, to procure me the live plant, as well as ripe fruit, and some dried specimens, all which I hope to receive by the end of autumn, and will immediately forward them to you. My specimens want flowers;

* It appears to be what Linnæus afterwards named *Ph. dioica*.

† *Mitchella repens*. *Linn. Sp. Pl.* 161.

and the figure which M. Aubriet has drawn, remains, as yet, imperfect.

Your mule plant, raised in the Upsal garden, is truly a remarkable production; and if it succeeds, so as to afford a fertile progeny, it will be a new, and altogether surprising, object in nature, in which, by the concourse of two different plants, the procreation of one distinct from both will be made manifest. Such a wonder is altogether unheard of, and perhaps is to be met with in the vegetable kingdom alone *.

The discoveries, remarks, and descriptions of your pupil Hasselquist, are worthy of all admiration. What he relates concerning the flowers of the *Sycomorus* is curious. You had so well instructed him, that he perceived many things which escape common eyes. You promise to publish various things worthy of notice, in the Upsal Transactions, which I beg and entreat you not to delay.

Kalm, another of your disciples, has promised you a valuable return from America, in the specific of the native Indians for the *lues venerea*. So noble a medicine, confirmed by repeated experiments and observations, proving a safe and ready cure of that disease, is much to be desired. It is

* This evidently alludes to a *Veronica*, named *spuria*, described in the *Amoen. Acad.* *v.* 3. 35. *t.* 2, and preserved in the Linnæan herbarium. Linnæus believed it to have been produced from *V. maritima* impregnated by *Verbena officinalis*. He has however recorded, by a manuscript note, in his copy of the above book, that it produced no seed.

to be wished that the same salutary effects may be experienced here, from what is so much wanted.

I will take care to note down precisely the days on which the leaves of certain trees come out of their buds, especially those of our native trees, as you request and recommend; also the days on which barley is sown and reaped in the ensuing season. If you wish me to do any thing else, do but write, and you will find me always most ready to obey your orders. I have never seen the essential character of *Cardiaca (Leonurus)*, consisting in minute snow-white globules, scattered over both sides of the anthers. As soon as any of the species of this genus, described by you, begin to flower, I shall treat myself with this spectacle.

The *Floræ Parisiensis Prodromus* shows Dalibard's carelessness; for he has erred more than once, even in what he has copied from you. This indeed is no wonder, as he is little versed in botany. He would consult nobody, lest he should lose a part of the glory, with which he hoped to shine forth in the spoils of yourself and others.

You may find the manner of preventing the evaporation of spirits of wine described by M. Daubenton, in the preface to vol. 30 of the Natural History of the Royal Museum, as well as by de Reaumur in the volume of the Memoires of the Academy, which is now printing, and will soon appear. All the experiments and details of several successive years, in pursuit of this object, are given separately by these authors; and you will learn the art in question better from them, than from any short account I could give you.

The Dissertations of Schiera * are not yet to be had of our booksellers, nor are they likely soon to get hither. The numerous and various things which you have sent me, are come safe to hand, as the *Philosophia Botanica*, *Acta Upsaliensia*, and many things besides. *Poa vivipara*, and the dwarf Bramble (probably *Rubus arcticus)*, of which you favoured me with both plants and seeds, are both thriving. I acknowledge my great obligations to you for many favours, that I can scarcely hope adequately to return.

I long to read what you have written about the flowers of the *Calamistrum (Isoetes)*, as also to see the figure given in your *Iter Scanicum*. I have learned from you the cause of a fatal disease in horses, as well as that of the blight in barley. You are very kind in thus communicating to me all your discoveries, as soon as made, which is a most valuable proof of your regard.

I am well pleased with your new genera of *Axyris* and *Napæa*. *Linaria scoparia (Chenopodium Scoparia)* has hermaphrodite flowers. Your Lapland plants, brought by Mr. Montin, were very acceptable. Pray let him know that I thankfully acknowledge this favour.

Missa has returned ungrateful. How he behaved

* *J. M. Schiera, Dissertationes duæ, de plantarum sexu, fecundatione, systemate sexuali, &c.* printed at Milan in 1750. They are mentioned by Haller, *Bibl. Bot.* v. 2. 395, but not in the rich Catalogue of the Banksian library.

in Sweden or Holland, I know not. I have only seen the *Prospectus* of a new *Materia Medica*, published by him at Amsterdam; but it is ridiculous. He came twice to see me, but said not a word of his voyage. He boasts of being entirely devoted to the practice of physic, and, like an upstart, avoids all his former friends and companions. There was not a word spoken of you. I am told he affects to praise Haller, Bæck, Burmann, Van Royen, and other physicians, whom he talks of as his intimate acquaintances.

If you have any thing new to send me in future, you may entrust it to Dr. Bæck, or Mr. Salvius, either of whom will forward it to M. Delaisement the apothecary, or to M. Dangerville at Rouen, by which means I shall be sure to receive any thing safe. I wish Salvius would send some copies of the *Systema Naturæ* to Paris for sale, as his edition is preferable to the Leipsic one. My brother desires his compliments. May you long live, the glory of our science! M. la Serre and Mademoiselle Basseport are always happy to hear of you. Do not forget me. May our mutual regard never know any diminution! Farewell!

The amiable writer of these letters died in 1777, aged 78. We have found nothing from him of later date than the above.

LINNÆUS TO ALBERT HALLER*, M. D. GÖTTINGEN.

[Latin.]

From Mr. Cliffort's Museum (Hartecamp, near Leyden),
April 3, 1737.

Three years have now elapsed since I first became acquainted with your celebrated name in the *Commercia literaria* of Nuremberg. The description there given of the *Androsace* pleased me so much, that I wished for nothing more than your acquaintance. When I first came to Holland, in 1735, I found you were engaged with Gesner in writing a *Synopsis* of Swiss Plants. When I returned, scarcely half a year afterwards, I heard, with the greatest pleasure, that you had undertaken the publick Professorship of Botany; which appointment I heartily wish you may long happily fulfil. Being at the Leyden garden a few days ago, I saw your inaugural dissertation, on the Methodical Study of Botany without a Preceptor, which I quickly perused, and with so much pleasure, that I earnestly beg you will favour me with a copy; in return for which I will send you any book, in my power, that you may want. If you wish for any of my little publications, I will readily send them, if you inform me how they may best be sent, whether

* This eminent physician and botanist was at this time settled at Göttingen. He subsequently enjoyed the highest honours and reputation in his native country of Berne, where he died Dec. 12, 1777, aged 68. This and the following letters of Linnæus are translated from the *Epistolæ ad Hallerum*, vols. I. II. and III.

by the publick conveyance or otherwise. If any of the exotic plants of the *Hortus Cliffortianus* would be acceptable, I promise to supply you before any body else.

I perceive you are intent upon the establishment of natural classes. I wish you would complete this work, and give it to the publick. I also have long laboured at this object, though as yet incompetent; but I think I shall be able to get together more fragments than many other people can, though much remains behind, and I doubt whether I shall ever complete them. I find the class of *Bacciferæ* has certainly no existence. I am ready to agree with you that the stamens and pistils lead to no natural system; having adopted a method founded thereon as a substitute, to excite curious observers to examine these parts of the fructification, hitherto reckoned so trifling and unimportant; for an alphabetical arrangement was always intolerable to me. Besides, an attention to the organs in question may have its use, though not altogether for the purpose of natural classes. You attribute eight stamens to the *Anblatum*. I never found that plant but once, in 1728, but I remember to have noticed four only. Pray tell me if I am wrong *.

My *Critica Botanica*, in octavo, is in the press, and the *Hortus Cliffortianus* in folio. My *Genera* were finished early in the present year. If you wish for a copy, I will take care that you shall have one.

* "Certainly they are but four." *Haller*. The plant spoken of is *Lathræa Squamaria*.

The *Flora Lapponica* was finished this week. I have dedicated it to the learned members of the Royal Society of Sweden. I have fallen into some errors, with the common herd of botanists, in this book, respecting C. Bauhin's synonyms, which I might have avoided, had I beforehand seen your dissertation. I have honoured an African plant with your name in the *Hortus Cliffortianus*, which I hope will not displease you, and which cannot now be easily revoked, having long been printed. A friend of mine received, about a year ago, from Mr. Gesner, a plant by the name of "*Stæhelina* of Haller." I wish to know where it is described, and what works you have hitherto published, not having met with any. I perceive Ludwig's Definitions of Genera are in the press. He follows the method of Rivinus. I have not yet seen the long-promised *Hortus Petropolitanus* of Siegesbeck. Martyn's fifth Decade is just come to hand. I lament the premature fate of Micheli. The two remaining parts of Seba's *Thesaurus* are not likely to appear, as his heirs are quarrelling for his purse, and care more about money than the fame of the deceased. When I was at Oxford, Dillenius was finishing the *Phytopinax* of Sherard, of which he had then entirely completed the fourth part. Burmann's *Thesaurus Zeylanicus* was published in the beginning of the year. He has undertaken the publication of the highly splendid work of Rumphius, called the *Plantæ Amboinenses*, which I wish he may finish. By my recommendation a young German physician

(Bartsch) is appointed physician in ordinary to the West India Company at Surinam, from whom, if from any one, I hope we are likely to obtain all the seeds and plants of that country.

If you are so kind as to answer this, pray direct to me at Mr. George Clifford's, Amsterdam, and I shall be sure to receive it. Wishing you all possible success, I recommend myself to your favour, and remain, &c.

The *Musa* has this day begun to flower. It flowered last season. If you have not my little tract on the *Musa Cliffortiana*, published last year, I will take care that you shall be furnished with it.

I must add a few words, having, since I finished my letter, received one from Dr. Gronovius, inclosing one from J. R. Iselius to J. F. Gronovius, dated Bâsle, March 10th, in which, amongst other things, are these words: "They say the celebrated Professor Haller, at Göttingen, has it in contemplation to write against this new method of Linnæus; I know not whether such an undertaking would be of any service to himself," &c. I know not what foundation there is for this report, having received the letter within an hour. Meanwhile, I beg leave to offer a few remarks to you. If the report be false, I have no doubt of your excusing what I have to say.

1. I must declare, that I am anxious to avoid, if possible, all anger or controversy with you; my wish is rather to act in conjunction with you; I should detest being your adversary, and, as far as

possible, I will avoid it. May there be peace in our days!

2. I have always, from the time I first heard your name, held you in the highest estimation; nor am I conscious of ever having shown a contrary disposition. Why then should you provoke me to a dispute? Let me know if I seem to have offended, and I will omit nothing to satisfy you. I ask but for peace.

3. If my harmless sexual system be the only cause of offence, I cannot but protest against so much injustice. I have never spoken of that as a natural method; on the contrary, in my *Systema, p.* 8, *sect.* 12, I have said, "No natural botanical system has yet been constructed, though one or two may be more so than others; nor do I contend that this system is by any means natural. Probably I may, on a future occasion, propose some fragments of such an one, &c. Meanwhile, till that is discovered, artificial systems are indispensable." And in the preface to my *Genera Plantarum, sect.* 9: "I do not deny that a natural method is preferable, not only to my system, but to all that have been invented. . . . . . But in the mean time artificial classification must serve as a succedaneum. Therefore, if you establish a natural method, I shall admit it.

4. If you detect any mistakes of mine, I rely on your superior knowledge to excuse them; for who has ever avoided errors in the wide-extended field of Nature? Who is furnished with a sufficient stock of observations? I shall be thankful for your

friendly corrections. I have done what I could of myself; but a lofty tree does not attain its full stature by the first storm that bursts forth.

5. I have been acquainted with most botanists of distinction, who have all given me their encouragement; nor has any one of them thwarted my insatiable desire of natural knowledge. Will you be more severe than any body else? You appear, by your dissertation, too noble to triumph over the ignorance of others.

6. You may, with great advantage, and without injury to me, display your profound learning, and intimate knowledge of the works of Nature, so as to acquire the thanks of all the learned world. Do but turn over the writings of botanists in general, and you will see, by their earlier performances, how they are puffed up at first with their own consequence, and scarcely able to keep from assaulting others; of which I myself have perhaps been guilty, which I greatly regret, having now learned better. But when these same people have passed a few years in the field of battle, they become so mild, candid, modest, and civil to every body, that not a word of offence escapes them. This chiefly leads me to doubt the truth of the report in question, for I know your reputation has already been long established.

7. It seems wonderful to me that I should have excited so much of your displeasure; for I cannot but think there is no work of any author more in unison with my ideas than this essay of yours.

8. I, and perhaps I alone, have acquired what I know entirely by the rules you have laid down, of studying without a master. I am still but a learner; and you must pardon me if I am not yet become learned. If knowledge is to be acquired by your mode, the hope of it, at least, still serves to illuminate my path.

9. I doubt indeed whether you, or any other professor, can enter into controversy with propriety. Professors and teachers should, above all things, acquire the confidence and respect of their hearers. If they appear in the light of students, how much of human imperfection must appear, and what a depreciation of their dignity! What man was ever so learned and wise, who, in correcting others, did not now and then show he wanted correction himself? Something always sticks to him. We have lately seen an instance of this in a most distinguished professor, the ornament of his university, who, having long indulged himself in attacks upon schoolmasters, has at last got so severe a castigation from one of this tribe, that it is doubtful whether he can ever recover his ground, and certain that he cannot entirely. A very wise physician has declared, that he would rather give up physic, and the practice of it, altogether, than enter into publick controversy.

10. Look over the whole body of controversial writers, and point out one of them who has received any thanks for what he has done in this way. Matthiolus would have been the great man of his day,

if he had not meddled with such matters. Who is gratified by "the mad Cornarus," or "the flayed fox"? * What good have Ray and Rivinus done with their quarrels? Dillenius still laments that he took up arms against Rivinus; nor has the victory he gained added any thing to his fame. Did not Threlkeld give him much more just cause of offence? But he was now grown wiser, and would not take up the gauntlet. Vaillant, at one time a most excellent observer, attempted to cut his way with authority through the armies of Tournefort; has he not met with his deserts? and would he not have risen much higher, had he left Tournefort unmolested?

11. I dread all controversies, as, whether conqueror or conquered, I can never escape disgrace. Who ever fought without some wound, or some injurious consequence? Time is too precious, and can be far better employed by me, as well as by you. I am too young to take up arms, which if once taken, cannot be laid aside till the war is concluded, which may last our lives. And after all, the serious contentions of our time may, fifty years hence, seem to our successors no better than a puppet-show. I should be less ashamed to receive admonition from you, than you from me.

Behold then your enemy, submissively seeking your friendship; which if you grant him, you will

* Titles bestowed on each other, in their controversial pamphlets, by Fuchsius and Cornarus. They followed those who ought to have set a better example.

be more certain of securing a friend, than of stirring up an adversary. I know you to be of a more generous nature than to level your attacks at one who has not offended, unless any enemies of mine should raise doubts in your mind against me. If, after all, I cannot obtain that peace, which, by every argument and supplication, I seek of you, I hope you will at least be so generous as to send me whatever you may print on the subject, and I will take care to convey my answers to you.

If the news I have heard be without foundation *, I earnestly beg of you to forgive me for the trouble I now give you.

In the hopes of your answer, I remain, &c.

HALLER TO LINNÆUS. [Latin.]

MY DEAR FRIEND, Göttingen, April 14, 1737.

I address you, though a stranger, by this title, that you may, in the first instance, perceive how empty was the report, and how false the intelligence, which Iselius sent to Gronovius. I certainly have no acquaintance with Iselius. He may have heard, from my friend Stæhelin, that I do not entirely agree with you as to the practicability of your system, founded on the sexes of plants. But it never came into my mind to enter into any contro-

* "The report was false, nor did it ever enter my mind to disturb a young man, of so much merit in the science of botany, in the commencement of his fame and fortune." *Haller.*

versy with you, or any one else, on the subject. Real lovers of science are necessarily united, in bonds of friendship, with those by whom science is advanced. Thus I was already much attached to you before you ever thought of me. Nor can any imperfections, in a meritorious character, interfere with my well-earned respect. We are all naturally liable to such. All therefore that you have, in your letter, so liberally and judiciously expressed, will only serve the purpose, and that no undesirable one, of giving me an insight into your character; who, even to a supposed adversary, showed yourself modest, ingenuous, commanding your temper, in spite of just provocation; and therefore, to a friend, you will doubtless prove most amiable and excellent. You shall, on my part, always find me anxious to render you any service; more especially as I am convinced that, in assisting you, I shall benefit all the botanical world.

Having said thus much, you will readily perceive that whenever, in our future correspondence, from which I anticipate so much pleasure, I may express any opinions different from yours, it will be in the secrecy of friendship, and not for publick exposure. So, in my recent pamphlet, I have treated of your system in such a manner, as neither to detract from your reputation, nor to condemn your sentiments. I have merely indicated some difficulties, which, in the application of your method to practice, might hereafter be obviated. Undoubtedly, the *Bacciferæ*, like the *Unisiliquæ* of Boerhaave, is untenable as a natural class. The *Tetrastemones, Pentastemones, Hexastemones,* &c. as well as the *Diangiæ, Triangiæ, Polyangiæ, Polycarpæ, Monopetalæ, Polypetalæ* of all kinds, run into them both. They cannot be separated on account of the tribe of Solanums, all connected by a certain peculiar flavour, and by medical qualities, though differing in seed-vessels, flowers, and organs of impregnation. This consideration prevents me from rejecting the class in question. But I shall take another opportunity of writing on this subject. The above is sufficient in reply to your remark, as I am just now very busy in finishing the description of an anatomical monster.

It is true that I have been labouring, with my friend Gesner, at a *Synopsis* of Swiss Plants. We have had the assistance of Stæhelin, a close investigator of the minute tribes. I have now undertaken the last revision of the whole, and am deeply immersed in the *Fungi,* whose characters and synonyms I find altogether inextricable. I have got through the Mosses, but not without many doubts. I am preparing indeed, not a mere *Synopsis,* but a great work, containing original characters and descriptions, with synonyms, of 3000 native plants, and a critical history of the whole, disposed after a system of my own. Dillenius's *Phytopinax* would greatly assist me. I wish we were likely to have it soon.

I now send 13 copies of my dissertation to Frankfort, from whence any Dutch merchant of your acquaintance may obtain them, by applying to James Renier of that place. Six of these are for Boerhaave, Albinus, Gronovius, Gaubius, Burmann, and Van Royen; the rest are at your disposal.

Anblatum (Lathræa) has indeed four very large *apices* (anthers), whose stalks, surrounding the capsule, are inserted into them like arrows into their heads. I know not how I came to reckon them as eight, which is contrary to my notes, except that my mind, at the time I wrote, was overset by the death of my beloved wife, which may easily account for any such oversights. Your *Genera*, *Systema Naturæ*, *Musa*, *Flora*, *Critica*, and *Hortus Cliffortianus*, will all in their turns prove most welcome. I will take care to transmit you the value of each, by a merchant whose name, I think, is Reuss. Be so good as to send the books to the abovenamed Renier of Hamburgh, who will at the same time transmit the money, or any thing else, from me.—Lapland Mosses will be highly acceptable, and any foreign plants of your own, or from Mr. Cliffort. I offer you, in return, plenty of Alpine plants, with any remarks of mine which may be worth your notice, or any thing else in my power.

That the organs of impregnation become more accurately scrutinized, descriptions of plants more precisely drawn up, and founded on their most important characters, is what I, like all lovers of truth, shall rejoice to see. I have myself followed that path which alone leads to solid information, and which is far more easy than the tuition of a preceptor. Go on, therefore, in this path, and enrich us with your discoveries. I thank you for your favour with respect to the *Halleria;* an honour I do not deserve, and which is never becoming, except when conferred on the great leaders of botany.

Your literary news is very acceptable. Our mutual friend Ludwig has adopted a method more popular in Germany than satisfactory to himself. It is certainly one of the most decided *non-naturals,* and founded exclusively on one part, which is not one of the most important. I know nothing of Siegesbeck's work. I am very sorry to hear of the unexpected death of Micheli. He was very diligent, though not a man of letters; and he has detected the most minute distinctions, among a vast number of species. He has forestalled me in a great many of these, among the *Jungermanniæ,* the *Lichenes,* and the near relations of the *Fungi.* He has excelled every body in these tribes. Yet there are some things in his work which I either cannot see or cannot approve.

I should be very greatly obliged to you if you would purchase for me, from the booksellers, Burmann's *Thesaurus,* as well as some older books that are occasionally to be met with in Holland. I could, in my turn, furnish you with Hercynian plants, and others that might be worth your having; as also minerals, if I knew whether you would wish for any. I mean, after midsummer, to undertake a journey towards the Bructer and other neighbouring mountains.

I write in haste, being occupied with other matters. I will mention some of my doubts another time, and shall now only add the titles of some books in my library which I could spare you, if wanted. These are, *Caroli Avantii notæ ad cœnam fieri, Patavice*, 1649; *Severini lapis fungimappa, ibid.*; the best edition of *Matthiolus*, printed at Venice in 1565; the life of Gesner, with some figures by Wolfius; the second volume of Feuillée, bound, with the plants, Paris, 1725; Zannichelli dell' Ippocastano, Venet. 1733; *Gesneri Index Plantarum quadrilinguis, Tigur.* 1542; Bruckman *de Koszodrewina*, and *de Arbore Limbowe*, the *Pinaster* of Micheli; *Costæus de naturâ stirpium, Turin*, 1578; Neumann on *Opium*, Cloves, Wine, Tea, Coffee, &c.; *R. J. Camerarius de Cervariâ nigrâ, et Pini conis*; and various other German dissertations, &c. Also all the treatises upon *Herbaria*, the method of making them, how to take impressions from the fresh plant, &c. These seem to have escaped you, in consequence of the great number of more important things.

What I wish to declare chiefly to you, and to you alone, is, that I have always had a high regard for you, and now, that I know your mind, I shall be still more anxious to cultivate your friendship.

Farewell, &c.

LINNÆUS TO HALLER.

From Mr. Cliffort's garden, May 1, 1737.

Welcome, most welcome, was your letter, professing so much friendship and partiality! I will take care not to be ungrateful. I rejoice, with all my heart, that the rumour was unfounded, for indeed you and Dillenius are the only people I would not wish to have for adversaries. You have read the volume of Nature as I have. I care not for those, however learned, if they be only so from books. I wish I had any means of serving you; you may command me, and every thing in my power.

I sent my *Characteres* and *Flora*, through the hands of the Waesbergs, the last day of my stay at Amsterdam, and have no doubt of your having received them. You must know this *Flora Lapponica* is not written for the learned, but for the unlearned. You were the Mercury who came from heaven to untie the knot of the *Gentianæ*, and certainly far more happily than I could ever have expected from any mortal hand. You have laid me and every body else under obligations. I wish you could do as much with the *Salices*.

Mr. Cliffort is not at all conversant with Mosses, nor am I, nor any person in Holland. We are all devoted to the love of exotic plants, especially those from America. I do not profess to be even a *tyro* in Mosses, the study of which requires a man's whole attention; and this I have not yet been able to bestow. Holland produces very few of this tribe, in which Sweden abounds. I entreat you, by all that is sacred, to tell me any thing you know with certainty relative to the fecundation of Mosses. I perceive in them a diversity of sex; for instance, in *Polytrichum*, where one plant bears a capsule, another terminates in a star. I can scarcely doubt that the dust and the capsules of *Lycopodium* are flowers; but the structure of these capsules is widely different from *Polytrichum*. The *Marchantia*, or *Lichen polymorphos* of Dillenius, has a twofold fructification; one sessile, with orbicular, flat, upright seeds in a cup, which I am certain is the fruit and seeds; the other a compound inverted flower, supported by a stalk, which must consequently be the male. If I could but ascertain any thing certain about *Polytrichum*, it would determine many other points; indeed as much as I want. You write that the powder of Mosses is their seed, if I understand you rightly; have you made any experiments to prove this *? This powder, seen under a microscope, exactly agrees with the dust of the anthers in other plants.

If you have an established garden, I beg you will take advantage of an offer I am authorized by Mr. Cliffort to make you, of plants from his collection. He is better able to supply you than any other person I know of. No return will be expected, except you happen to have a few fossils, which are now his chief passion.

* "Stæhelin has." *Haller*. This doctrine is now established by Hedwig.

I am most ready to supply you with my little publications, which I wish may prove worthy of your perusal; but I hope you do not take me for such a blockhead as to take money for them from one of the first of botanists. I have all of them, except the *Systema Trium Regnorum Naturæ* (the first edition, in folio tables), which is sold by Gronovius. Burmann's *Thesaurus Zeylanicus* sells for the horrible price of 15 or 16 florins; and I am sure you would not like it.

The *Musa*, now flowering in our garden here, is twice as large as that of last year, and bears a fructification of double the size of the former.

The *Benzoe* * is now in bloom with us, bearing a calyx, or involucrum, like that of *Cornus;* but the flowers are, in other respects, those of a *Laurus.*

I have heard you have published something upon Alpine plants, but I know not when or where, nor the title of the book. When are we likely to see your observations, or your great work?

I return you my best thanks for the 13 copies of your dissertation, and shall write this day to James Renier, to send them, by any merchant of Amsterdam, to Mr. Reuss or Mr. Cliffort, who are both friends.

Nobody in Holland cares about genera, except myself, a little man among the prophets. Gronovius, Burmann, and Van Royen are only anxious for dried specimens; yet Van Royen begins, in

* *Laurus Benzoin. Linn. Sp. Pl.* 530.

some degree, to examine genera. Dr. Boerhaave loves nothing but trees, and is more solicitous about their varieties than species. Albinus is entirely engaged with anatomy. There is nobody in England who understands or thinks about genera except Dillenius. Rand is satisfied if he can get one synonym for a plant, and Miller if he can procure American plants, living or dried. Martyn is a good sort of man, but not very scrupulous about science. In Sweden the only botanist is Olaus Celsius, principal professor of divinity, who loves plants without thinking of their genera, and is an assiduous collector of Mosses. Rudbeck is very infirm.

If you know any thing about the *Liquidambar*, which some have referred to *Acer*, I shall be obliged to you for information. I have the fruit, but never saw the flower. I suspect the *Palmaria* of Muntingius is the same tree in blossom.

If among your dried specimens you have flowers of the *Celtis*, of the *Tribuloides* of Tournefort, or of his *Aphyllanthes*, I should be thankful for a single one, to see the stamens and pistils. I once saw the *Salicornia* of Tournefort in bloom, and found but one stamen in every instance. Whether the pistil is on the same plant or a separate one I know not. If you have any information on this point, I request you will communicate it.

By means of the *Capnorchis* of Boerhaave, I have lately traced a close affinity between *Hypecoum* and

Epimedium, through the *Fumariæ*; beware therefore of separating them in a natural arrangement.

In like manner, *Papaver*, *Argemone*, *Glaucium*, *Chelidonium*, *Bocconia*, *Sanguinaria*, *Anapodophyllum*, and *Christophoriana*, belong to one class; but it is difficult to give a character that will include them all.

So also *Mirabilis*, *Plumbago*, *Verbascum*, *Hyoscyamus*, *Nicotiana*, *Belladonna*, *Mandragora*, *Solanum*, *Capsicum*, and *Alkekengi*, constitute a class; and *Ahouai*, *Cerbera*, *Rauvolfia*, *Vinca*, *Nerium*, *Cynanchum*, *Stapelia*, *Asclepias*, *Periploca*, *Ceropegia*, *Apocynum*, *Plumeria*, *Cameraria*, and *Tabernæmontana*, another.

My *Tournefortia*, though its fruit is a berry, absolutely belongs to the *Asperifoliæ*; as does my *Prasium* to the *Verticillatæ* of Ray, though furnished with a berry likewise.

I pass my harmless time in preparing the *Hortus Cliffortianus*, and am now at the *Syngenesia*, where I am struck with the supreme sagacity of Vaillant *. In the work I am about, I reduce varieties to their respective species, and apply specific definitions, more correct, as I hope, than former ones.

In my *Critica*, sections 210 to 335 of the *Fundamenta Botanica* are explained, and proved by examples. I there undertake to demonstrate that one third of the generic names are bad, particularly

* See the very different sentiments of Dillenius, in his letter to Linnæus, p. 93 of this volume.

those of Vaillant ending in *oides*, and those of Heister in *astrum*. I prohibit all specific names; but time will determine *.

Many thanks for the mention of several authors' works; but among these, in what form did Zannichelli publish on the *Ippocastano?* What author wrote on *Herbaria viva?* Was it Camerarius? What is the *Myriophyllum pelagicum*, on which Zannichelli wrote a whole book, and to what genus does it belong? I have never been able to meet with Zannichelli's work on the Plants growing about Venice. Does it contain any thing curious, or any good remarks?

My *Halleria* is the *Caprifolium afrum, folio pruni levitèr serrato, flore ruberrimo, baccâ nigrâ.* Its generic character is as follows:

Cal. Perianth of one leaf, permanent, flat, spreading, short, cut half way down into three very blunt segments, the uppermost twice as broad as the rest.

Cor. of one petal, ringent; globose at the base; reflexed and tumid in the throat; with a small, oblique, four-cleft border; whose uppermost segment is longest, obtuse, emarginate; the lateral ones shorter, broader, and still more blunt; the lowermost very short, and acute; all of them bent forwards.

* Most happily for the popularity of his system, and the convenience of botanists, Linnæus himself subsequently determined this, against his first decision, and established those trivial (or specific) names, which have been of the greatest benefit to all natural history.

STAM. Filaments four, bristle-shaped, straight, inserted into the tube of the corolla, the two nearest ones shortest. Anthers roundish, of two lobes.

PIST. Germen ovate, within the calyx, terminating in an awl-shaped* style, longer than the stamens. Stigma simple.

PER. Berry roundish, of two cells.

SEEDS. Generally solitary.

I can add no more, as I am now obliged to set out for Leyden. Pray excuse haste. Take care of yourself, and cherish your regard for me.

HALLER TO LINNÆUS.

May 24, 1737.

I have duly received yours. Chatelain has, I trust, presented you with my *Programma.* Not merely two species of *Globularia,* and a specimen of *Aphyllanthes*, such as I have, shall be sent you, but I hope to furnish you with about 200 alpine plants, by way of Frankfort, from whence you may procure them. The things mentioned in your last are not yet arrived at Frankfort. Cannot you induce your bookseller to take, on proper terms of exchange, about 100 copies of the *Programma,* by which means it will become more known, and I shall be benefited.

I may possibly this summer publish a history of

* *Subalatum* in the original seems a mistake for *subulatum.*

the class *Diangiæ*, by way of a specimen of my *Synopsis;* or it may be only of *Pedicularis* and *Saxifraga.*

All that you can send me will be highly acceptable. The *Salices* are difficult to settle, especially the common ones. I have seen none in Caspar Bauhin's herbarium; they have perished. The alpine ones are numerous, but not difficult. He affords us little help as to Mosses. In Micheli's *Clathroides*, I have examined the elegant structure of the capsule, which has a lid, like that of a Moss. When the lid falls off, spiral threads are protruded, of a purple, yellow, or other colour, discharging masses of seeds, like little perforated rings. The structure of some Mosses is similar, but I am not sufficiently acquainted with their flowers, as the *Pinax* (or *Synopsis* of Swiss Plants) occupies almost all my spare time.

I have without doubt seen at many different times, the seeds of *Marchantia*, *Hepatica*, *Lunularia*, *Marsilea*, and *Muscoides*. The flowers of these plants appear to consist merely of innumerable minute anthers. In *Equisetum* I have obtained a very pleasing view of the four-legged spermatic bodies, agitated by a spontaneous elastic motion; but what is your opinion of the seeds of *Equisetum* *?

I thankfully acknowledge Mr. Cliffort's kind in-

* Hedwig first suggested these four legs, or appendages, to be the stamens, accompanying the germen. Hence the theory of some botanists, that the elastic ring of genuine ferns is their anther, or anthers, acquires a degree of confirmation.

tentions as to my garden, as well as my obligation to you. I have preserved some of Heister's plants, in another garden.

I have published nothing upon alpine plants, except an account of a journey to the Alps in 1731, in a periodical work called the *Tempe Helvetica*, vol. 1. sect. 4.

I thank you for having thought of Burmann's book. I am, however, most anxious for works that have a connexion with my intended *Synopsis*. If you can obtain for me any of these, that I want, you will assist me materially, and I will contrive to repay you by Mr. Reuss, who is in correspondence with my father-in-law. I want, for instance, Columna, for which I would give as much as 40 florins; I want also both the Floras of Holland, as well as Cæsalpinus, and other things. I am sorry to hear your account of Van Royen, Burmann, Rand, Miller, and Martyn. If every body were to do thus, Botany would be at an end. Do you but go on as you have begun, in the path of true science, and never trouble your head about adversaries. Whatever means I may have in my power, shall be entirely devoted to the furtherance of your praiseworthy objects. My *Iter Hercynicum* is just ready, as well as some minerals. I shall send them to you. I am aware of the affinity of *Papaver* and the other genera you mention. They nearly agree with *Tetrapetalæ (Cruciformes)*, but differ in their numerous stamens, and deciduous calyx and petals. *Mirabilis* and the following are *Pentastemones, monopetalæ*, &c.

I am not content with the order of *Bacciferæ*, and think of striking it out, as I have the *Unisiliquæ*.

What say you? do you approve Vaillant's distinction of genera in the following instances, between *Filago* and *Elichrysum* *, *Hieracium* and *Hieracioides* †, *Rhaponticum* and *Rhaponticoïdes* ‡, *Jaceæ*, as one of the class of *Cinarocephalæ*, and the *Discoideæ*?

Why do names in *oides* so much displease you? Why may we not call a *Rapunculus* with two cells *Rapunculoides*, and a *Lychnis* with three cells *Lychnoidea*?

I have Zannichelli on the horse-chesnut, printed at Venice in 1732, consisting of two leaves and one plate. The authors who treat of herbariums § are Simon Paulli, Kniphof on his own work, Ehrhard of Memmingen on a little work of his own, Geyer in his *Thargelus Apollini sacer*. I am not acquainted with the *Myriophyllum pelägicum* ||, which is a plant of Clusius. I have Zannichelli, which is not a work answerable to its handsome appearance. The book contains some new figures, but not faithful ones. The descriptions are bad, nor is there

* *Gnaphalium*. *Linn. Gen.* 419.

† *Crepis*. *Linn. Gen.* 403.

‡ Both referred by Linnæus to his multifarious genus *Centaurea*, as well as the *Jacea*, in which last the presence or absence of neuter florets, in the *radius*, proves not even a specific distinction.

§ Or rather, chiefly, of impressions taken from plants.

|| No plant, but a Coralline, *Sertularia Myriophyllum*. *Linn. Syst. Nat.* v. 1. 1309.

any thing new, except perhaps a species of marsh *Ketmia* *. There are but few plants on the whole, and about three of the *Lichen* or *Fungus* tribe. He compiles the medical qualities from other people. Farewell, I have nothing new or important to tell you.

LINNÆUS TO HALLER.

From Mr. Cliffort's museum,
June 8, 1737, in haste.

You overwhelm me with so many good offices, that I know not which way to turn myself, and I must hasten to make some return, lest I should die too much in your debt. I received the Dissertations, *Programma*, &c. by M. Chatelain, and distributed all the former, for which I am charged with due thanks from every body concerned. Albinus, who spent a whole day with me here, above a week since, tells me he is well acquainted with you. Dr. Gaubius says the same. If what I sent to the care of Messrs. Waesberg, by the publick waggon, is not come to hand, pray let me know, and I will replace the parcel. I have written to them on the subject, as Ludwig, to whom I sent at the same time, has not received his packet.

You must know that there are no greater rogues in the world than booksellers. They will not take charge of any thing, unless you are content to wait for payment till all is sold. Those of Holland do

* *Hibiscus pentacarpos. Linn. Sp. Pl.* 981.

not much like botanical works; a clear proof that neither the students, nor other people, esteem them. I have enquired at several shops, but have not, as yet, met with any one willing to deal in such matters.

I wish I were master of your system. I would readily communicate any observations likely to be worthy of your notice. Do you comprehend under the *Diangiæ*, both *Pedicularis* and *Saxifraga?* Do you not separate plants with two cells to their fruit, from those that have two distinct capsules? Surely you ought!

Your *Diangiæ*, I think, will consist of all the genera enumerated under my *Didynamia Angiospermia.* These compose a natural class, nor ought any of them to be thrown out, though I am still a little doubtful about *Selago* and *Limosella.* But the following require to be added to them; *Veronica, Pæderota* (now *Wulfenia), Clandestina (Lathræa), Pinguicula, Lentibularia (Utricularia), Verbena, Collinsonia,* perhaps *Morina;* whether *Protea, Dipsacus, Scabiosa,* and *Globularia,* I know not; certainly *Coris, Belladonna, Mandragora, Nicotiana, Hyoscyamus,* my *Lonicera, Diervilla, Mirabilis,* and *Ruellia.* Please to consider whether these are sufficiently distinguishable from the *Verticillatæ;* think how many characters the latter have, in common with the genera just enumerated, in the calyx, stamens, corolla, pistils, in the form, situation, and structure of which, they differ from all other plants in the world; do they accord with your *Diangiæ?* Then, as to the consideration whether any one mark of difference, should be allowed to separate what are united by so many characters; consider the *Verbenæ*, and other things, which combine these classes. I speak to you as to a botanist who is only in search of natural classes. Are the *Valerianæ* to be separated from these? do you well consider the *Valerianellæ?* I would fain know what you do with the *Salicariæ*, whether you refer them to the *Cassiæ*, the *Bauhiniæ*, and the *Siliquastra?*

Whilst you are making good use of your eyes, I confess myself to be blind. Your account of the *Clathroides* * delights me. I lament that I have not been able to examine very minute productions. I am also much pleased with what you say of the elastic powder of *Equisetum.* I know nothing about the imperfect tribes of plants, and must confess my ignorance whether what I see is seed, or dust of the anthers. The capsule of *Equisetum* bursts exactly like an anther. The two kinds of flowers in *Taxus*, described by Tournefort, I have not seen. How there can be any impregnation without an egg, I cannot understand; consequently I do not comprehend the theory of Dillenius (in Mosses), of farina of anthers, but no pistil. The powder in *Marchantia* does not appear to me to consist of anthers, as all anthers burst; neither can I ever believe it to be of the nature of an egg; but it is rather of an impregnating nature, as those orbicular bodies, found in the cups, most certainly germinate,

* Now comprehended under the *Trichia*, *Cribraria*, &c. of recent writers on *Fungi*.

performing the function of seeds. If these bodies therefore are really seeds, and the peltate disks, whether stalked or radical, be each a calyx bearing male flowers, the yellow matter they produce being the dust of anthers (pollen); if, I say, this be the case in *Lichen (Marchantia)*, may it not be so with the Mosses properly so called?

Burmann is still indisposed. Columna's book is not to be had here, and whenever it occurs at an auction, as last year, it fetches an enormous price. When I return to Sweden, I may perhaps buy you a copy. Cliffort himself has not this work. I have not been able to meet with Cæsalpinus. I am mnch pleased with this author, and those short descriptions of his, in which he differs from all others, and has always something peculiar.

You absolutely expunge the *Bacciferæ*. You see the calyx form a *bacca* in *Morus*, *Morocarpus (Blitum)*, and *Ephedra;* the annular receptacle in *Taxus* does the same; and the common receptacle in *Fragaria*, as also in *Rosa*. Mark well the dry berry of the Elm, Almond, and Walnut, and the pulpy cone of *Anona*, &c.

No man in his senses can, by any means, distinguish *Filago* from *Elichrysum (Gnaphalium.**)*. I am much in doubt concerning *Hieracium* and *Hieracioides (Crepis)*. If you separate *Taraxacum (Leontodon)*, *Hieracium*, and *Hypochœris*, you may do the same by *Hieracioides;* but if you join them all together, you will commit no absurdity.

* "Linnæus did subsequently distinguish them." *Haller.*

I comprehend under *Jacea* (now *Centaurea), Centaurium majus, Cyanus, Calcitrapa, Calcitrapoides, Rhaponticum, Rhaponticoides, Amberboi,* &c.; in short, all with a sterile (or rather neutral) radius; as the structure of the calyx affords no just limits, and the crown of the seeds gives the most fallacious differences. They ought not absolutely to be distinguished, as a class, from the *Carduus* tribe, for they approach that tribe so nearly as scarcely to be distinguishable from it. *Carduus galactites* of John Bauhin * has the flower of a *Jacea.* All genera of the *Cynarocephala* of Vaillant are good, and ought all, as natural genera, to go together, except *Xeranthemum,* which appears to me nearer the *Elichrysa* †.

Petasites, Cacalia, Elichrysum, Conyza, Eupatorium, Senecio, Absinthium, Artemisia, Santolina, Abrotanum, Gnaphalium, and *Tanacetum,* ought to be referred to the *Radiati,* from which however I remove *Carlina.*

"Why does the termination *oides* displease?" Because it is the asylum of ignorance. Botanists of the present day have scarcely introduced any new name but what ends in *oides.* Thus if I were shown a thousand new genera, I could name them all extempore, by the addition of *oides.* You must have noticed this evil among recent writers. I do not indeed approve any names which differ only in their termination, or in having as it were different tails,

* *Centaurea galactites. Linn. Sp. Pl.* 1300.

† Some Linnæan species of *Gnaphalium.*

or appendages. These nails drive one another out of the memory. Witness *Alsine*, *Alsinoides* of Ray, *Alsinella* of Dillenius, *Alsinastrum* of Vaillant, *Alsinastroides* of Kramer, *Alsinastriformis* of Plukenet, *Alsinanthemos* of Ray, and *Alsinanthemum* of Kramer. Can you keep all these distinctly in your head? I can neither recollect them, nor many others like them. Is it not an abuse?

You ask further, why I "object to *Rapunculoides* and *Lychnidea?*" Because natural genera ought to be kept entire, nor ought a difference in one part merely, to introduce a new genus. Every botanist who founds a new genus, should ascertain its essential character. Number of cells is not an infallible distinctive mark. You have yourself been aware of this in *Ruta*, *Euonymus*, *Moschatellina (Adoxa)*, *Orobanchoides*, &c. As you maintain natural classes, I have no doubt that you still more insist upon natural genera; for less injury is done to botany by classes, than by genera, that are not natural. If, then, genera be distinct, why should not their names be kept perfectly so likewise? Otherwise you might call *Cerasus*, *Prunoides*, and *Malus*, *Pyroides*. Have the antients done thus? Let us take an example from the animal kingdom. Suppose we were to call *Anas* (the Duck), *Anseroides* (a Goose-like bird), and *Cygnus* (the Swan), *Anserastrum* (a Bastard Goose); or imagine any other case, not so directly shocking to our prejudices. Most assuredly I could not help laughing, when I saw a certain botanist establish a genus by its name alone, calling it *Con-*

volvuloides (resembling a *Convolvulus)*, because it had an upright stem!

I shall send you my *Critica* in about a fortnight. Pray inform me, in the mean time, whether you have received my *Genera, Flora Lapponica, Musa*, &c. that I may replace them if lost. I know you are hostile to all change of names. I know you are about writing against this practice *. Botanists seem to me never to have touched upon nomenclature as a subject of study, and therefore this path of their science remains still unexplored. If you were to collect all the generic names which have been changed from the time of Tournefort to this present day, they would exceed a thousand, though insensibly introduced. What is the cause of all this innovation? I can perceive no other, than there having been no laws laid down, by which names could either be made or defended. Nothing is more certain than that the whole stock of specific names are erroneous. This our successors will perceive. If specific names require alteration, why may not false generic ones likewise be changed, at the same time? Those who come after us, in the free republic of Botany, will never subscribe to authorities sanctioned only by antiquity, if we retain such intractable names as *Monolasiocallenomenophyllum* and *Hypophyllocarpodendrum*; why therefore should we retain barbarous or mule names, or names distinguished only by their tails?

* "I never had any intention of writing controversy." *Haller.*

As far as I can perceive by your dissertation, you appear to me adverse to those botanists who obtrude upon us varieties for species. I wish you might agree with me on this subject, and that I could have a day's conversation with you! How glad should I be of your support in this matter! for I know all other botanists will be against me. If Micheli's species are genuine, and not, for the most part, varieties only, I could undoubtedly exhibit, any day, 5000 new species. The neglect of the philosophy of our science, has bewildered botanists in this visionary pursuit.

You refer *Trollius* to *Populago (Caltha)*. It ought to be associated with *Helleborus*. See the essential character in their nectaries. As *Adonis* differs from *Ranunculus* in those parts only, so does *Populago* from *Helleborus*. Be pleased to examine these plants.

Pray strike *Melianthus* out of my *Genera*, as the character I have there given is fallacious. I had nothing for examination but unexpanded flowers, which were destroyed by the winter. I hope soon to draw up a better description.

Accept my never-ceasing acknowledgments for so many rare plants which you promise me. I wish I had them now at hand. How many are the stamens and pistils of *Aphyllanthes?* How many are those of Tournefort's *Tribuloides* *? Are the

* *Trapa natans. Linn.*

two sexes separate in *Salicornia?* and how many are its stamens and pistils? Farewell!

HALLER TO LINNÆUS.

Gottingen, July 3, 1737.

Your elegant presents of the *Flora Lapponica* and *Genera Plantarum*, reached me in safety, a few days ago; but the *Musa* appears to have been omitted. As all your publications sell here at a very high rate, I beg the favour of you to procure me the *Hortus Cliffortianus.* I do not mean to encroach upon your liberality for this work. My brother-in-law, Mr. Wyss of Berne, is in correspondence with Mr. Reuss, and by this means I can contrive to indemnify you in any way you please. Your works are also forwarded, I believe, to Professor Ludwig. Have you seen his *Characteres?* He entirely follows our worthy old Boerhaave, my late preceptor, but not in botany, which at that time I neither liked nor studied.

How am I to distinguish the *Biloculares* from the *Bicapsulares?* Would you call them *Dicarpæ?* These, though entirely distinct, ought to lead to the beginning of the *Multisiliquæ.* But I mean, if I can, to follow a different track, which promises rather to accord with nature, than to establish classes, tracing out the affinities of plants, rather than the subdivisions of system, which frequently offer violence to natural genera. You seem always to think that I am aiming attacks at your system, in which you do me injustice. I merely mean to propose my own speculations, without offence to any one. I do not desire, if I were ever so able, to refute any body whatever, but of all mortals I least wish to oppose you.

I conceive that the *Capitatæ* are connected with the *Discoideæ*, by means of *Carlina* and *Xeranthemum*, and the *Discoideæ* with the *Radiati* through *Bidens*, &c. The *Radiatæ* are allied to the *Cichoraceæ*, and these to the *Scabiosæ*, which are next akin to *Verbena* and *Mentha*, and thus we come to the *Verticillatæ* with four naked seeds. These lead us to the rest of your class *Didynamia*, not excepting *Pinguicula*, and we gradually proceed, through *Polygala*, to the *Papilionaceæ*. But I have not made up the matter completely in my mind. I find in *Polygala* eight monadelphous stamens only. Such is the case with *Chamæbuxus;* whose undivided beard should seem to afford a generic distinction, compared with the fringed lip of *Polygala*. I am in doubt on the subject of *Belladonna*. We meet with a natural class in the *Solanum* family. Can it be separated from them?

In synoptical arrangements we ought to keep exactly to our characters, lest our definitions should not always answer to the objects defined. I therefore, in my system, separate those of the *Lychnis* tribe which have three cells, from the rest. But not so in natural orders. The system, or method, is intended to furnish the learner with unexceptionable characters; but in studying affinities, we seek out the hidden chain of nature. I therefore agree with you, as well as with your adversaries *.

I have not yet attended to the *Rosæ*. I should certainly leave the bulbous tribe with *Polygonatum* and Lily of the Valley; next to which follow *Arum* and its allies; then the *Orchis* family; and next *Orobanche* with *Hypopitys*. The *Umbelliferæ* constitute a natural class. So do nearly all the *Gymnopolyspermæ*, from which *Thalictrum* leads us to *Christophoriana (Actæa)*, *Glaucium*, *Papaver*, and even *Ulmaria*. So the genus *Sedum* will come near to *Saxifraga*, under which last you rightly include the *Geum* of some authors. *Salicaria* will, for many reasons, be found not widely separated from *Agrimonia*, any more than from *Chamænerium (Epilobium)*, because of the character they possess in common, of the stamens being double the number of the petals.

Whatever I have seen in *Marsilea*, *Marchantia*, and the Mosses, as well as in *Equisetum* and *Clathroides*, is all of the nature of pollen. The female flowers are conspicuous in *Marchantia*, *Marsilea*, *Lunularia (Marchantia cruciata)*, and a few Mosses,

* This idea was subsequently much better understood and developed by Linnæus, when he asserted the necessity of an artificial system for practical use, and of the study of natural orders for a philosophical knowledge of plants. The great fault of the french school is the confounding these two distinct objects.

but in the greater number they are obscure. The anthers in all these plants consist each of a thread, having spermatic globules (or pollen) attached to each side. The latter in *Clathroides*, and all neighbouring genera, are usually annular.

You cannot fail of some opportunity of meeting with Muntingius in Dutch, both the Floras of Hermann, Cæsalpinus *de Plantis*, and the inaugural dissertations of Feldmann, Van Royen, Gronovius, Breynius, &c. If you can procure any of these at a fair price, I shall be happy to do any thing for you, in my power, in return.

My opinion respecting the *Bacciferæ* is, to retain the narcotic tribe of *Solanum* only, excluding the rest.

The termination *oides* has certainly been carried too far. But in a *method* (or synoptical arrangement) this adjunct is not inconvenient, as it expresses more than a proper name, or any vague word of antiquity. It ought by no means to be admitted into a natural system; for all plants are of equal date; nor ought one to precede, or give a name to, another*. The practice in question has, indeed, been carried to a ridiculous extent in the

* Happily this scheme of having one set of names for common use, and another for the students of natural orders, has never been carried into effect. The nomenclature of Botany is sufficiently extensive without it. The ingenious writer, however, confirms the opinion of Linnæus, that natural orders must be studied independent of artificial or popular methods of arrangement.

tribe of *Alsine*. But at this time of day people are accustomed to establish genera before they are masters of their distinctions, or the power of their characters. Dr. Moering, the founder of a garden at Jever, has sent me his catalogue. He has made six new genera. He says the climbing species of *Fagopyrum (Polygonum)* have two petals, and a three-leaved calyx. *Asphodelus palustris* he gives as a new plant of Friesland *. He applies the name of *Hyacintho-butomus* to the umbellate hyacinth †. He separates *Arnica* from *Doronicum*, for the same reason that Vaillant referred it to *Solidago*.

May not *Convolvuloides*, or something like it, be tolerated in the natural orders, though certainly not in the methodical system? Your *Critica Botanica* will be highly welcome. I have just finished the *Diangiæ*.

I differ in a few points from your most elegant *Flora Lapponica*. *Saxifraga alba petræa* of Pona ‡ is not *Sedum tridactylites tectorum* (Bauh. Pin. 285). The leaves are often five-cleft, their lobes directed backward, a character scarcely observable in the latter. It is also larger, though an alpine plant; nor is it found in Switzerland. I have described it from C. Bauhin's Herbarium. *Veronica alpina frutescens* of Clusius; *V. petræa sempervirens;* and *V. alpina bellidis folio hirsuta*, are three distinct plants, which you seem to have combined,

* A good genus, called by him *Narthecium*. See *Fl. Brit.* 363. † Now *Agapanthus*.

‡ *Saxifraga petræa*. *Linn. Sp. Pl.* 578.

giving a figure of the second of these *. I cannot reconcile any of my *Pediculares* to yours, except the *flava (flammea, Linn.)* *Odontites* seems really distinct from *Euphrasia*; and *Teucrium alpinum inodorum* † is rather a *Melampyrum* than an *Euphrasia*. Having made my description of this plant in the winter, and having been deceived by authors, who attribute an undivided lower lip to *Melampyrum*, I took my plant for a new genus. I now take it for a new species of *Melampyrum*. The smallness of the flowers, in specimens not less tall than the common kind, and their deeper colour, distinguish it. Though I thus dispute with you in my trifling letters, I should in publick do so with great diffidence, or, what is better perhaps, not at all. May I hope for your *Silene, n.* 185 ‡? which perhaps is one of the new species of spurious *Lychnis*, in my *Synopsis*. Many other dried specimens, of exotic, and especially alpine, plants, as Willows, &c. would be highly acceptable, if your stock be not exhausted, as mine nearly is; for I have given away, to various people, more than ten bundles of dried plants.

In mentioning *Silene*, I must say the admission of the stamens and nectaries into their characters pleases me highly; though in a System formed after

* This is incorrect. The plant of Linnæus, which his plate represents, is *V. alpina*, *Sp. Pl.* 15, figured also by Haller in his great work, t. 15, f. 2: but not one of the above synonyms belongs to it, nor does Linnæus suppose so.

† *Bartsia alpina*. *Linn. Sp. Pl.* 839.

‡ *Lychnis alpina*. *Sp. Pl.* 626.

Nature I cannot allow of separating the bulbous tribe, the *Alsine* family, the Rosemary, &c. from their allies, on account of their three or two stamens. The *tubæ* (styles), however, appear to me to afford very ambiguous characters, as a simple style with three stigmas and a three-cleft one are hard to be distinguished, as is the latter from three distinct styles; and a cloven or hooked style (stigma), common in the *Didynamia* class, from a simple one. Yet these differences form the characters of whole classes (rather of orders). Whether the styles in *Rosa*, *Sambucus*, and some others, are one or several, may puzzle any body, especially a learner*.

I am not fond of new species. Several of my own, which I published with great satisfaction, I am now disposed to condemn, as two *Orchides*, two *Veronicæ*, and a *Saxifraga*, described in the *Commercium Litterarium*. The *Androsace* indeed is new; and I think I am possessed of another new one, different from your *Diapensia*, as well as from the hairy kind with many flowers †. The *Xeranthemum* is new ‡; and so, I think, is the *Melampyrum* with a small yellow flower §, and the *Astragalus* ||. Tournefort undoubtedly has an infinite

* A very ordinary degree of observation and common sense, in studying Nature herself, or in reading the rules laid down by Linnæus, will settle all these questions.

† *Androsace villosa*. *Linn. Sp. Pl.* 203.

‡ *X. inapertum*. *Willd. Sp. Pl. v.* 3. 1902.

§ *M. sylvaticum*. *Linn. Sp. Pl.* 843.

|| *A. uralensis*. *Linn. Sp. Pl.* 1071.

number of varieties for species, as well as more recent writers, Loesel, Helwing, Ruppius, &c. I think, for instance, that difference of size alone can never make a species; difference of colour very rarely, except it be yellow. An assemblage of characters, derived from different parts, affords easy specific distinctions. I will examine the *Trollius.* I did not advert to its nectaries. The flowers, as well as the seed-vessels, are like *Populago (Caltha).*

My specimens of *Aphyllanthes* are imperfect. I want *Tribuloides (Trapa)* and *Salicornia.* I have bestowed nearly all my attention upon our native Swiss plants; for I had to pursue, at the same time, anatomy and the practice of physick; unlike you, the consecrated priest of Flora. I have always cultivated botany, in spite of all obstacles, since the year 1728, when I accomplished a laborious journey of 200 leagues, through the Alps, on foot. I have since visited those mountains ten successive times. But I am near-sighted, which is a great inconvenience. I have laboured much at Mosses, and such plants. I hope to settle a good many doubtful matters, though many must remain undetermined. My family was always finding fault with my pursuit; but I do not repent. On the contrary, I regret that I did not devote more of my spare time to these things. I propose this winter, or the ensuing spring, to publish my *Synopsis,* without figures. It will be twice the size of your *Flora Lapponica,* and will contain from 2500 to 3000 plants*. Yet there are, unquestionably, many things I have never seen, in the Valtelline, the Rhætian alps, the Valais, and even about the little town of Avenche, whose neighbourhood is extremely fertile, and its climate scarcely inferior to that of Aquitain. All my journeys were necessarily hasty and clandestine. But I have intruded upon you too long with my gossiping.

Here are a few titles of books to add to your *Bibliotheca Botanica*. The figures of Zannichelli are 311, but occasionally bad and superficial.

Theophrasti sparsæ de Plantis sententiæ, per Cæsarem Adonum. Bonon. 1561, 4to.

Campi Spicilegio Botanico, overo dialogo di Baldassar e Michele Campi, nel quale si manifesta lo sconosciuto Cinnamomo degli antichi. Lucca. 1654. This I saw at Hanover.

Gesneri Historia Plantarum, 1541, with notes from Paul Ægineta, Theophrastus, Dioscorides, &c. I have.

Johannes Burckhard, de Methodo Plantarum non ad unam partem adstringenda, Epist. ad Leibnitz. Wolfenb. 1702.

D. Vignæ Animadversiones in libros de historia & de causis plantarum Theophrasti. Plsls. 1625, 4to.

* This most excellent work did actually appear in 1742, in a handsome folio volume, under the modest title of *Enumeratio Methodica Plantarum Helvetiæ*, with 24 beautiful plates. The second edition, or *Historia*, published in 1768, is infinitely less accurate.

Petri Castelli, an Smilax aspera sit Sarsaparilla Americana. Messan. 1652.

Leonh. Ursini, Rosa menstrua. Lips. 1662.

Pfautz, Descr. Graminis medici plenior. Ulm. 1656.

Schelhammer, Cat. Plant. horti domestici. Helmstad. 1683.

P. Borelli, Hortus simplicium. Paris. 1669, 8vo.

Theod. Schoon. Waare oeffening der planten. Gravenhage. 1692.

Martini Mylii, Hortus philosophicus. Goerliz. 1597.

Adieu!

LINNÆUS TO HALLER.

No date.

I return you my best thanks for your letter, and for your many new and curious communications. Mr. Renier writes, that your liberal present of plants will soon reach Amsterdam. I now send my *Musa Cliffortiana*, and *Critica*. Being unfortunately a stranger here, I am obliged to scrape every thing together in great haste, and confide the superintendance of the press to other people; no wonder if errors occur. You regard the matter only, not barbarisms of language. Conversing with Laplanders, Finmen, and Norwegians, for several years, has made me more barbarous than Micheli. I was, moreover, obliged to write my *Critica* in secret, and every thing in the greatest haste, having composed this work, the *Flora Lapponica*, and the *Hortus Cliffortianus*, within three quarters of a year. The latter is printing. Four hundred pages, in large folio, are struck off, so that it will come out before the end of the year. You may depend on having it, which I hereby promise; and I shall be more than repaid if you send me your *Synopsis*. I owe you much beyond what I am able to repay.

I have studied Professor Ludwig's Characters. His labour has been very great. I wish his classical authors may not have misled him. All Boerhaave's decisions are not infallible. I am every day enlarging or correcting my characters, which are no other than generic descriptions, differing from those of Ludwig and Tournefort, as a specific character of a plant from its full description. Both are necessary in botany.

My *Bicapsulares* and your *Tricarpœ* are perhaps the same. The *Semiflosculosi* of Tournefort and his *Flosculosi* are combined by the *Elephantopus*. The chain of connexion from the *Radiati* to the *Cichoracei* is difficult; and I fear there is a gap between these last and *Scabiosa*. I am perfectly aware that *Rosmarinus*, *Ziziphora*, *Salvia*, *Pinguicula*, &c. ought to be united with the *Didynamia*; but I was obliged to submit to laws, else no method could exist; as I have owned at my class *Didynamia*.

I erred respecting the number of stamens in *Polygala*, from examining the small European species only. Having now seen the African ones, I have acknowledged my mistake in *Hort. Cliff.* p.

353. *Linagrostis (Eriophorum)* has three stamens; *Stellaria (Callitriche)* one; *Linosyris* (perhaps *Thesium)* five; *Asarum* twelve; *Tithymalus (Euphorbia)* many. *Uva Ursi* has ten; *Persicaria (Polygonum)* never quite so many. *Fagopyrum* has nine. One *Tamarix* has five, the other ten. In *Mercurialis* there are nine or twelve, whether ever more I know not. *Xanthium* has five. *Fumaria officinalis* has three anthers to each stamen. If you find me wrong, pray let me know. I entreat you not to make a distinct genus of *Chamæbuxus.* You certainly would not, had you seen the specimens I have. This plant has certainly the little fringed lip (of *Polygala),* though more remote from the summit. If you do separate it, you will unquestionably alter your mind hereafter.

Why you should hesitate about *Belladonna,* I cannot see a single reason. If you know where to place *Nicotiana* and *Hyoscyamus, Belladonna* and *Mandragora* must go along with them; and they are so near akin, that you may rather hesitate whether to keep them as distinct genera from each other. Compare all the parts of fructification. I know not what can seem doubtful in *Solanum. Alkekengi, Solanum, Lycopersicum,* and *Capsicum,* can by no means be far separated. Are they not of the same natural genus? They are all bicapsular. Why then should they not go to the tribe of *Nicotiana?*

The *Bulbosæ* may go with *Convallaria,* but assuredly not with *Arum,* nor with the *Orchis* family.

Pray look over the species of *Arum* described by Plumier.

I am absolutely convinced that the *Palmæ*, *Canna*, *Costus*, *Kæmpferia*, *Maranta*, &c.; the *Orchis* family; with *Iris* and *Sisyrinchium*; form one class, to which belongs *Arum*; and even my *Stratiotes*, nearly related to the Palms. See *Musa Cliffortiana*.

Pray do not separate *Orobanche* from the *Personatæ*; you can find no reason for it in the structure of the flower. The family of *Hypopitys* I cannot discover.

Thalictrum has certainly nothing to do with the *Papaver* tribe, but is closely allied to *Clematis*. Examine all the species.

If you remove *Ulmaria* away from *Spiræa*, you will commit a solecism. I doubt whether they do not form one genus. See the *Barba Capræ* *, the *Filipendula*, and the species in Plukenet. You surely ought not to separate these from the *Icosandria Polygynia* family. *Agrimonia*, if I am right, has little to do with *Salicaria*. The latter has some relationship to *Azedarach*, *Bauhinia*, *Cassia*, &c.; but *Agrimonia* belongs to the same tribe as the *Icosandria Polygynia*, as well as *Alchemilla*, and *Percepier (Aphanes)*. Look how the styles are inserted at the base of the germens, and you will see the calyx does not become the fruit.

Have you ever seen powder discharged by the bursting of what you and Micheli term anthers, or

* *Spiræa Aruncus. Linn. Sp. Pl.* 702.

apices, in *Marsilea*, *Marchantia*, &c.? Did you ever see an anther which did not discharge powder? I judge your anthers to be the powder (pollen).

You may procure Muntingius; but what can you want with him? I am sure there is not a more worthless botanist. His figures are very neatly engraved, but merely to catch the eye, for I never saw any worse. Hermann's *Flora* never came out, but when half printed it was suppressed. I will take care to supply you with the treatises of Van Royen and Gronovius. I cannot but regret the paper wasted in the work of Muntingius. If you will have it, let me know. It sells dear enough, though most worthy of the fire.

I have desired all the booksellers to buy me Cæsalpinus, but it is not in the shops. I may by chance soon meet with it.

I could wish to see Moehring's *Hortus*. Where and when was it printed? He must be joking when he characterizes the climbing species of *Fagopyrum* as having a three-leaved calyx and two petals. I have, like him, separated the *Asphodelus palustris*, it being tricapsular, whereas the rest are trilocular; see *Anthericum* *. I have also distinguished his *Hyacintho-butomus*, what a horrible word! from *Hyacinthus*, joining it with the *H. tuberosus* (Tuberose), to which all its characters answer, by the name of *Polianthes* †.

* *Anthericum ossifragum*, *Linn.* now *Narthecium*; see p. 264.

† Surely this is a great botanical error. The plant in question is the *Agapanthus* of Solander in Ait. Hort. Kew.

"*Convolvuloides* differs from *Convolvulus* in having an erect stem." So you see how writers make new genera, because the original one had an incorrect name! I never saw any thing more fallacious in botany than this. You will find my opinion respecting it in the *Critica*, p. 34.

You delight me by saying the *Saxifraga alba petræa* is distinct from *S. tridactylites*. I perceived a difference, and was aware of the leaves being often five-cleft, but took it for a variety. I hope you can show, by sufficient marks, it is not a variety, that there may be no doubts or mistakes respecting this point in future. Does not the larger alpine plant likewise turn red with age?

I wish you would sufficiently describe those alpine *Veronicæ*, if you have all the three, and furnish us with marks of distinction. Surely two of them are varieties. Whether the third be so, I am doubtful *.

I can hardly as yet make *Odontites* distinct from *Euphrasia*, for I am acquainted with but one species of *Euphrasia* besides. They may, nevertheless, be sufficiently distinguished.

I already perceive *Teucrium alpinum inodorum* † to be distinct in genus, and widely different from *Euphrasia*, *Pedicularis*, and *Melampyrum*; but it

* The writer had here in view *V. alpina*, *bellidioides*, *fruticulosa*, and perhaps *saxatilis*, four as distinct species as any in nature; but he had competently examined the first only, scarcely seen, at this period, by any one else.

† *Bartsia alpina*. *Linn. Sp. Pl.* 839.

agrees in genus with *Horminum tenui coronopi folio virginianum* of Morison *, and with *Pedicularis maritima, folio oblongo serrato* (of Tournefort's Inst. 172), which I have mentioned in *Hort. Cliff. p.* 325 †, and is among the new genera I mean soon to publish.

You at length allow the little yellow *Melampyrum* to be a variety. I know both the plants under consideration perfectly. Perhaps that with a violet-coloured tuft may, likewise, not be a distinct species ‡.

It seems to me easy to ascertain the number of styles, except alone in the *Herniaria* and *Amaranthus* tribe; as I count the styles, if present, at their base; and if there are none, the number of stigmas determines the order. There is no difficulty in *Rosa*, if you cut the calyx through longitudinally; for the fruit or berry is not a pericarp, but a calyx become coloured and succulent. Consequently the germens themselves are all perfectly separate, as well as naked. This plant is absolutely gymnopolyspermous, the neck of the calyx being but little coloured. In *Sambucus* I count the stigmas, as the styles are wanting.

Is my *Diapensia* the same plant with that of the authors I quote? Their figures and descriptions are but obscure.

* *B. coccinea, Linn. Sp. Pl.* 839.

† *B. viscosa, ibid.* scarcely different from *Rhinanthus Trixago.*

‡ They are all three now justly made distinct.

In the *Hortus Cliffortianus* I have rejected some thousands of varieties, noting them, with Greek letters, under their respective species.

I should be glad to see the truly alpine plants marked in your *Synopsis* with an asterisk, or some such distinction, so as to perceive them at one view. An Alpine *Flora* of Europe is a work much to be desired.

If you possess a flower of Micheli's *Franca* *, pray tell me the number of its stamens. I fear they are but five, though Martyn says ten.

Your *Synopsis Helvetica*, if it alone contains about 3000 † plants, may take place of all other *Floræ*. You who have already prepared so fine a work, who possess so superior a collection, who excel most botanists, if not all, in acuteness of judgment, and are so eminent for observation, as well as for a correct acquaintance with plants and their synonyms; you, I trust, will not allow your work to be a mere list of Swiss plants. You may render it extremely valuable by additional information; such as an indication, under each species, of its essential difference from its congeners; for all our theories, all our descriptions and figures, tend to this object, of knowing the species of plants correctly, and distinguishing them by appropriate names. This, the great end in

* *Frankenia* of Linnæus. The stamens are six, double the number of the stigmas, as well as of the valves of the capsule, but having no numerical reference to the petals or calyx. They are said, however, to vary.

† "Only 2490." *Haller.*

view, however difficult to others, will be much easier to you.

Having fixed the species, you will reduce the varieties to their proper places under each, as I do not doubt your having the same opinion of them as I have. Have you observed what multitudes of varieties are put forth as species by Pontedera, Micheli, and others? If every minute difference, every trifling variation, is to establish a new species, why should I delay to exhibit ten thousand such species? and who cannot point out as many? I have always preferred taking two distinct species for one, reckoning them but varieties of each other, so long as I was doubtful of a clear and obvious mark of difference; rather than publishing any doubtful plant as a certain species.

Jussieu is my friend, and so is Dillenius. I had never any acquaintance with Vaillant. He was a man full of himself, ambitious of raising his own fame on the overthrow of his teacher, the honourable and excellent Tournefort. Vaillant was merely demonstrator in the Paris garden, and rude in literature. He set himself up against Jussieu, and once laughed Dillenius to scorn. He was poor, &c. All this is nothing to me. I wish to be a just and reasonable man, as well as a botanist. I confess I never yet read any writer who was more accurate than Vaillant, who made more discoveries in botany, who laboured harder, or reaped a more sparing reward. Is a man to be handed down to posterity either as a scoundrel, a madman, or the most stupid of all mortals, merely because he has pursued, honoured, and laboured to improve botany? Jussieu, as I am informed, has solemnly sworn hostility to the memory of Vaillant, during his own life; nor is Dillenius content with the numerous cavils with which he has insulted his *manes* in the *Hortus Elthamensis*. Admit that Vaillant has his faults in synonymy, or perhaps other respects. Who has ever been free from botanical errors? He is a wise man who can distinguish good from evil; and that general may be esteemed happy who conquers and disperses his enemies with the loss of half his own forces. Who is more meritorious in exotic plants, though not a systematick, than Plukenet? but who was ever more unprincipled, more of a heretic in botany, or a greater scandal to our science, than either Plukenet or Vaillant? If the authority of the *Hortus Elthamensis* is to be followed, I should have nothing to do with Vaillant, nor against him. But an honest man ought to do justice to every one's deserts. If you give due praise to Vaillant, posterity will be just to your memory. In this respect I care not for either a Jussieu or a Dillenius.

I entreat you to remark that *Xanthium* and *Ambrosia* bear amentaceous flowers.

The *Veronica* of *Fl. Lapp.*, sect. 7, is *V. alpina frutescens*, as I learned from Dillenius, and the Sherardian specimens, at Oxford *. Yet no part of

* Linnæus was here misled, this synonym belonging to *V. fruticulosa*, figured in Haller, t. 16. Willdenow gives a correct

mine is shrubby, except the lowest joint of the stem. Is this because of the colder climate of Lapland?

What is meant by *V. alpina, bellidis folio, hirsuta*, is no where better to be learned, as I think, than from Burser, who was an attentive botanist, and who records having sent that plant to Caspar Bauhin, which the latter confirms. What is preserved in Burser's Herbarium is precisely my plant, though somewhat more hairy *.

If there be three distinct species, I wish their characters were laid down clearly, without which nothing is certain.

On looking at my specimens, I cannot make *Saxifraga tridactylites* and your alpine one more than varieties. Their fructification indicates this, where there are more sources of discrimination than in all the rest of a plant. I shall be thankful for an explanation of these parts in the present case †.

Be so kind as to tell me whether you have found, on your Swiss Alps, the following plants of my *Flora Lapponica*: No. 13, 85, 88, 115, 164, 165, 166, 172, 174, 181, 207, 208, 231, 232, 236, 242, 243, 244, 245, 302, 303, 319, 342, 353, 368, 443.

view of these disputed species, now in general well understood, in his *Sp. Pl. v.* 1. 61—63.

* Here again Linnæus is mistaken. Burser's plant is *V. bellidioides*. Haller justly observes that Linnæus subsequently made three species of these plants, and correctly. The fourth, *V. saxatilis*, was published in his *Supplementum*, p. 83.

† Haller truly remarks that these plants are perfectly distinct, though he did not explain their characters.

What number of seeds and stamens has *Tozzia?* Have you ever observed the male and female flowers of *Viscum* on entirely separate plants? and whether those are to be termed petals in the female which are seated on the flower (germen)?

You justly complain of the want of characters in the history of animals, as in Ichthyology, &c. I am now employed in printing the posthumous works of my late friend Peter Artedi, in which, if I mistake not, you will see more perfection than can be expected in botany for an hundred years to come. He has established natural classes (orders), natural genera, complete characters, an universal index of synonyms, incomparable descriptions, and unexceptionable specific definitions.

If you are satisfied that I, whom you never saw, love and esteem you with all my heart, you will not take it amiss that I make a few remarks on your dissertation; a work certainly of great labour, from the perusal of which alone I have conceived a higher idea of you than of any other botanist of my acquaintance, unless I should compare you to Dillenius. I consider you as a man worthy of all commendation; and you may be assured that whatever I say shall be known to no mortal but ourselves.

Page 7 *. You are here justly aware, that when the System of Ray was spoken of as perfectly natural, all botanists must have been blind, unless, like Dillenius, they hoped for a professorship, or were

* *A. Halleri, &c. de Methodico Studio Botanices absque Præceptore, Dissertatio Inauguralis. Gotting.* 1736, 4to.

compelled, by the authority of the English, to give to Ray supreme honours. What was he? Undoubtedly an indefatigable man in collecting, describing, &c.; but in the knowledge of generic principles, less than nothing, and altogether deficient in the examination of flowers. I beg of you to compare the first edition of his *Methodus* with the second and third, where he has learned to take every thing from Tournefort. I know not why the discoveries of Cæsalpinus have escaped all observation, whilst every thing has stupidly been ascribed to Ray. Cæsalpinus appears great to me, inasmuch as he was the father of systematic botanists. Morison was vain and puffed up with conceit, like .... Yet he cannot be sufficiently praised for having revived system, which was half expiring. If you look over Tournefort's genera, you will readily allow how much he owes to Morison, full as much as the latter was indebted to Cæsalpinus, though Tournefort himself was a faithful investigator. All that is good in Morison is taken from Cæsalpinus, from whose guidance he wanders in pursuit of the chain of natural affinities, rather than of characters. You will, doubtless, profit by his error, to take a more judicious course *.

* Linnæus could not but perceive, by Haller's preceding letters, that he was in danger of falling into the same mistake. He was himself already aware that the study of natural affinities and the construction of a practical system were distinct, if not incompatible, objects.

You have been a severe critic of Rivinus, but I shall think you very great if you do not fall under the same censure. If you assume rules, you ought to follow them, else you can accomplish no system. If you lay down natural classes, you ought to conform to them. How much circumspection becomes necessary! If you infringe any of their limits, you commit a fault. If you can combine them all, and proceed in safety through them, you will be more fortunate than any other man. But if you go in search of affinities, independent of rules or combinations, what do you produce? not the promised system. Would you follow the traces of Morison?

Are the outermost flowers of the *Opulus* irregular*? and what then is your definition of regularity?

Why is any one wrong in assigning a solitary seed to *Sanguisorba*, and four seeds to *Tormentilla* †? The latter may be as correct as eight or sixteen.

Why may not trees with an umbilicated fruit be associated with such as are otherwise? I fear you do not well understand what is meant by this term. Do you place *Rubus* in a different class from *Rosa?* Are you not rather incautious here?

I should like to see your characters of *Orchis* and its allies, not taken from the spur. Do but turn over the American species in Plukenet, Petiver, and

* "They are." *Haller.*

† "There are more." *Haller.*

Ray; and in the first place the European ones in Vaillant, &c.

Page 8. You object to Tournefort's system. I allow that he has his errors, and in great number; yet no system more natural than his has been seen before or since. I confess that many of his classes are entirely arbitrary. I wish you may be able to reduce them all to natural classes. Tournefort's *Labiatæ, Cruciformes, Liliacei, Umbellati, Papilionacei,* and *Compositi,* are all good *, with a few slight additions or removals.

Page 9. Pontedera struck upon rocks which Tournefort avoided, for Pontedera was less solicitous of following nature. He undoubtedly created great confusion in the order of *Compositi;* but in many instances he was useful, though less so than Vaillant. Pontedera was almost the only philosophical botanist, though I cannot every where subscribe to his theory.

You say "Knaut's method was overthrown by Dillenius." Why is this? There surely was never a more unjust judge than Dillenius, in his apology for various methods. I wonder he was never answered. He certainly deserved it. He wrote learnedly indeed, and was therefore worthy of a learned reply.

As to Magnol, I regret that we have not more artificial methods, founded on various parts. If we had, they might easily help us to decide in which

* "But they are not Tournefort's own." *Haller.*

classes any particular part of the fructification was of importance, and in which it was not.

With regard to Vaillant, I have never yet met with any body more sagacious in genera than he was, and I am daily sensible of this. Possibly he may have given characters where you say there are none. And what if he has not correctly ascertained the plants of Bauhin? Can this be always done with accuracy from imperfect descriptions, without figures, and without particular information? If I should publish an absurd definition, are all to be reckoned unlearned who may not understand me?

In your judgment of the systems of Boerhaave and Ray I agree, that their classes are faulty in the point in question; but are there any preferable to them?

I wonder at your conjunction of *Lychnis* and *Rapunculus!*

That trees ought not to be separated from herbs, is evident upon inspection; for in what does an herb differ from a tree?

Page 10. Are there not herbs with an umbilicated fruit and others not so, as well as trees?

Are not *Cachrys* and *Simpla nobla* * arborescent, though of the tribe with two naked seeds?

Is not *Tournefortia*, though a tree, one of the *Asperifoliæ?*

Are there not shrubs without end among the *Labiatæ*, or plants with four naked seeds?

* *Phyllis nobla. Linn. Sp. Pl.* 335.

How many arboreous species of *Conyza* have we from Africa and America, as well as other very large trees with compound flowers?

What do you understand by the double calyx of *Chenopodium*, or a calyx of the fruit different from the common calyx*?

Have you traced any distinctive characters among the cups of the Lichens? or do you derive any from the quantity of their leaves?

Page 11. I wonder that, after having seen Plumier's book on Ferns, you can look for their characters in the shape of the leaves.

Certainly I never saw your genera, nor those of Dillenius, before my own were written. It is curious to observe how we, nevertheless, have adopted the same opinion in various places. So it happens with respect to the flower of the *Unifolium*, my account of which occurs in a page of the *Flora Lapponica*, printed in April last year.

Page 13. You cite Vaillant and Dillenius whenever you wish to give an example of any faults. Who has always avoided errors? happy is he who commits the smallest number! I would not, if I could, pick out the faults of good authors. There are easier modes of correcting others.

Page 16. You assign to the *Uva Ursi* a four-cleft calyx; and you elsewhere say it has eight stamens. The *Uva Ursi* of *Fl. Lapp.* t. 6, f. 3, known

* Haller apparently means the *Utriculus* of some recent authors, a fine close membrane covering the seed in *Chenopodium*.

to me from my childhood, has always ten stamens, with a five-cleft calyx*. Have you any other species under this name?

You object to a definition of parts, founded on their use, or physiology; and you reprehend Pontedera on this account. I doubt whether you could, in zoological subjects, define the nose, otherwise than by its use, so as to render your definition applicable to every kind of animal, as birds, fishes, insects, &c. The task would surely be difficult.

This very day your kind present of plants is come to hand, for which I cannot sufficiently thank you, but I will do all I can to show my gratitude. Among them all the *Aphyllanthes* has given me most pleasure, as your specimen enables me to supply what was wanting in my character of that genus.

Anonis alpina humilior, radice amplâ†, seems a species of *Trifolium*, not of *Anonis*, though there are many seeds in each pod!

Androsace alpina angustifolia glabra, flore singulari‡, appears to be wanting, which I much regret. I wish I could see but a flower, to know certainly whether it be my *Diapensia* §.

Salix alpina, alni folio rotundo glabro is *Fl. Lapp. n.* 355 ||.

* "Right," says Haller in a note. He had evidently confounded the *Arbutus* in question with *Vaccinium Vitis Idæa.*

† *Trifolium alpinum. Linn. Sp. Pl.* 1080.

‡ *Androsace lactea. Linn. Sp. Pl.* 204.

§ "They are widely different." *Haller.*

|| *S. herbacea. Linn. Sp. Pl.* 1445.

Alchemillæ affinis alpina * seems a species of *Alsinella*, or rather *Arenaria*.

The *Thora* † I never saw before, and I rejoice to have now seen its nectaries, so as to ascertain that point.

Cruciata alpina latifolia lævis, in my garden is only a male plant. What you have sent has stamens and pistils in the same flower ‡. Have you ever seen the male? Is it a different species from the *Rubia quadrifolia vel latifolia lævis* of C. Bauhin, or *Rubia quadrifolia italica hirsuta* of J. Bauhin?

Polygaloides and *Chamæbuxus* are but varieties of the same species §.

Acini pulchra species || is a species of *Teucrium*, and comes very near the *Marum Cortusi*. The upper lip of the corolla is wanting.

Salix alpina rotundifolia incana, is *Fl. Lapp. n.* 359, t. 7, f. 1, 2 *(S. reticulata)*.

Astragalus alpinus, foliis viciæ, ramosus, &c. is *Fl. Lapp. n.* 267 *(A. alpinus)*. I perceive, by another specimen, the *A. Onobrychis dictus* is sufficiently different in appearance. Whether these two

* "*Cherleria.*" *Haller.*

† *Ranunculus Thora. Linn. Sp. Pl.* 775.

‡ "It is always androgynous." *Haller.* The plant is *Asperula taurina* of Linnæus.

§ "The same plant." *Haller.*

|| *Thymus alpinus. Linn. Sp. Pl.* 826. Haller rightly says, "the upper lip is not wanting."

are really distinct species*, you can determine with certainty, having gathered them wild. If I have erred, it was by the persuasion of Dillenius.

Veronica alpina bugulæ folio, calyce hirsuto, is *Fl. Lapp. n.* 7, *t.* 9, *f.* 4.†

Uva Ursi. I now see what your *Uva Ursi* is, and you have rightly determined the number of stamens, as well as the segments of calyx, and of corolla; but the plant is my *Vaccinium* ‡, *Fl. Lapp. n.* 144, in which the calyx becomes a fruit with many seeds. In the *Uva Ursi*, however, as Tournefort likewise observes, the pistil becomes the fruit. I have given its history as an *Arbutus (Uva Ursi)*, *Fl. Lapp. n.* 123, *t.* 4, *f.* 3. The stamens of this are ten.

These remarks of mine ought to be written over again, but I really have not time. A friend excuses every thing. They were put down at various times, and in various words.

* "Widely different." *Haller*. The latter is *A. montanus. Linn. Sp. Pl.* 1070.

† *V. alpina. Linn. Sp. Pl.* 15.

‡ "Right. My error arose from C. Bauhin's having described the leaves as dotted." *Haller.*

HALLER TO LINNÆUS.

Göttingen, Sept. 12, 1737.

Your letter, with the *Critica* 'and *Musa,* arrived safe, and I send you in return my botanical essays on *Pedicularis* and *Veronica.* Your letter, as well as these publications, is somewhat adverse to me; but I am sure that all your remarks proceed from good will, and I have no objection to improve by them.

I must premise, that, immediately after my first coming to settle here, the loss of my most amiable and accomplished wife so overwhelmed me, that any faults which may have crept into the inaugural dissertation I was then obliged to compose may be entitled to indulgence.

As to *Stellaria (Callitriche),* my observation is confirmed by that of Vaillant. About *Linagrostis* I am uncertain. About *Linosyris* he is wrong; my notes mention five stamens. In *Persicaria* there are six or seven, never ten, as you truly remark. In *Fumaria* you may say there are either six or two.

Why do you take up the cause of Rivinus? His method is intolerable and impossible, on account of the extreme difficulty of drawing the line between a regular and irregular flower, the great uncertainties attending the petals in many iustances, and other things too well known to you. You seem to differ from me rather from system than real opinion. Has *Sanguisorba* but one seed? This is contrary to me and every author. The number of seeds in *Tormentilla* are uncertain; the petals are likewise so, and this plant is very erroneously removed from those with five. Our alpine *Tormentilla,* whose root is very large, has often five petals, and usually more than five seeds. An umbilicated fruit is elegantly distinguished from one which is not so; nor is that character doubtful, as you know better than I. The *Orchides* have two stamens, concealed in a hood. Their tube (style) is none, or very obsolete; their flower unequal, &c. In *Helleborine* and its allies the hoods of the stamens are very conspicuous; whereas in your *Neottia,* and some related to it, they nestle, as it were, in a little spongy substance.

Wherever Tournefort has a natural class, he almost always deserts his own system, and bears witness against himself. There is no certain foundation in his method, neither with respect to flower or fruit; for he joins, in the same classes, very different flowers, as well as fruits. He characterizes his classes by forms of the flower, which are variable as well as indefinite. Although I, who want time and opportunity, may not do better, yet Tournefort and Rivinus are not the less in fault.

You admire Magnol! Surely you have never read him! I do not see how he can be defended. He, in fact, hardly takes the calyx into consideration at all, though he makes his principal distinction between naked flowers and those furnished with a calyx. All his interior distribution, about the flower, fruit, &c. you must allow to be as bad as can be.

When the synonyms of Bauhin are not supported by sufficient citations, they ought to be omitted, or quoted with a mark of doubt; not taken out of those of Tournefort, and worse applied. The characters of capitate, discoid, and dipsaceous flowers, in Vaillant, really amount to nothing; but the reader is warily satisfied with a character that is, indeed, true, but not distinctive.

By Boerhaave's rules, strictly followed, the *Lychnides* divide into the *Monangiœ* and *Triangiœ;* the *Rapunculi* into *Diangiœ* and *Triangiœ;* as the *Alsines* ought to be separated from the *Alsinellœ,* and *Alsinastrum tripetalum* from the *pentapetalum,* according to the system of Ray. You, who seem to take so much pleasure in criticizing me, might well talk of the patience botanists ought to have with each other!

I agree with you about herbs and trees; rather unwillingly, but there are no limits between them.

Lichens are divisible into sections, not genera, as follow: 1. Some bear cups, or cylindrical horns, without shields. 2. Others bear leaves and shields. 3. Some have leaves without shields. 4. Others shields without leaves. 5. There are some consisting of a mere expansion, without either.

I have Plumier's works; nor am I inattentive to the leaves of Ferns, which in those plants is of principal moment. Yet I do not think it absurd to look for assistance from more recondite characters. You have principles of your own.

Concerning the *Unifolium*, you cannot suspect me of copying you, when I drew up the character of this plant in 1731; as I never, before the month of June 1737, saw a line of your writing, except the characters of *Statice* and *Limonium*, sent by Gronovius to Gesner, and by the latter to me. Nothing vexes me more than to be suspected of plagiarism.

The dotted leaves, attributed to the *Uva Ursi*, misled me. I long for nothing more than specimens of your *Uva Ursi*, your *Andromedæ*, and your own *Linnæa*. I will put down the names of the Lapland plants I wish to have, if you have duplicates remaining.

When I reject definitions founded on the uses of the respective parts, I mean those doubtful uses which Pontedera has attributed to the petals and stamens. It is easy enough to give a definition of the nose, by means of the pituitous tunic, and the structure of the *ossa turbinata*, without including the use of this organ, though evident enough, and undisputed. You censure me, in defending Pontedera.

I was wrong about *Opulus*. I meant to say sterile flowers.

To return to the alpine plants. I now add my *Diapensia*, which was omitted. The *Alchemillæ affinis (Cherleria)* never shows any petals. I have no memorandum about the sex of the *Cruciata*. *Polygaloides* is Johren's name for Tournefort's

Chamæbuxus. I might perhaps send you two specimens by mistake. The *Acinos pulchro flore* has a vaulted upper lip, and is no *Teucrium*. The two *Astragali* differ but little. One has rounder leaves, and variegated flowers; the other is a smaller plant, with sharper leaves, and purplish flowers*. But I will enquire after them in their wild situations. That there are four or five of these Alpine *Veronicæ*, I hope my treatise will convince you.

That most beautiful work, the *Hortus Cliffortianus*, will be but ill compensated by my *Synopsis*. I have finished the *Polyangiæ*, being now about the *Siliquosæ*, a difficult tribe; and I want the *Eruca genevensis* †, as well as some others.

Is all the *Solanum* tribe, in your opinion, bicapsular?

The *Orchis* family differ but little, in natural character, from the bulbous plants; but have systematic distinctions in the stamens, tube (style), and ovary (germen). *Arum* differs from both, and is apetalous. Whither must it go?

I will send you the *Hypopithys*. Dillenius's character is excellent, and I do not alter it in any point. The plant is related to *Orobanche*, though different in flower and fruit.

Agrimonia and *Salicaria* are like each other in flower, calyx, stamens, and even fruit.

You enquire how the anthers of *Marsilea* burst. I have seen them a thousand times. When they

* The first is *A. alpinus*, and second *montanus*.

† Perhaps no other than *Sisymbrium tenuifolium*. *Davall, MS.*

send up the globe from the sheath, it is green. When ripe, it bursts into four segments, and in the centre is a quantity of a yellow woolly substance; akin to the Mosses.

I am in no haste about Muntingius, but can hardly do without Cæsalpinus. The *Hortus* of Moehring was printed at Oldenburgh, anno 1737 (1736), in octavo. His *Asphodeliris,* as he informs me by letter, is the *Phalangium palustris, iridis folio* *. His *Convolvuloides,* or upright *Convolvulus,* is an elegant section, or relation, of the genus *Convolvulus.* I have vowed against making any new genera. In *Pedicularis* you will see four characters. I have not ventured to fix the limits of the genus. I have retained my *Staehelinia* †, which, though distinct in species from *Melampyrum,* is not a genus.

I would not be positive about *Saxifraga alba petræa,* but it is remarkably taller than the *tridactylites,* though, being of alpine growth, it should, if a variety, be of a more humble size. The flower and other parts are much larger.

You will judge whether to adopt my name *Staehelinia,* which you would oblige me by doing, to commemorate an able, though over modest, man ‡. You will see by my *Synopsis* how much he has done.

* *Narthecium ossifragum,* as heretofore mentioned.

† *Bartsia* of Linnæus, who has a different *Stæhelina.*

‡ Linnæus would not change the name he had long before printed, for this indubitable genus, but he honoured Stæhlin with a most elegant one of the Composite family.

You do not allow plants to be distinguished by their place of growth. According to what rule then shall I distinguish them? Shall it be by the Thermometer and Barometer? This would afford no reason why our alpine plants are meadow plants with you. I will try, nevertheless. Thus, for instance, almost all the species of *Pedicularis* are alpine, never mountain plants, or found on the Jura. I believe I have the *Franka (Frankenia)*. It sometimes occurs.

My *Synopsis* will certainly contain not much less than 3000 species. I will endeavour to indicate specific differences; but this cannot always be concisely done. You know I have not yet worked at the trees, *Uva Ursi*, &c. When I put the last hand to my undertaking, I hope to correct many errors. There are many things I want, and much assistance. You have it in your power to supply me with numerous alpine plants. Dillenius has a great many more. I do, however, make some progress, though I can hardly find any spare time. I have finished the *Fungi, Lichenes, Musci, Diangiæ, Triangiæ,* and *Polyangiæ* entirely. In these I have altered nothing. In the course of a year's labour I shall correct the remainder. I shall range the varieties, marked by figures, under each species. Whenever I meet with not merely one part, but many, and those in the flower, differing from others, in various respects, I make distinct species; I have, for instance, two *Cristæ Galli*, and two yellow *Melampyra*, &c.

You blame me for pointing out the blunders of Dillenius and Vaillant; while you criticise thousands, and change all their names. If I have not ventured to open my mouth on the subject of Mosses, is that a sign of ill will? Have I not praised Dillenius? or have I done any wrong to Vaillant? In a word, might I not be allowed to indicate my having observed what had escaped the eyes of so many persons?

Tozzia is very closely allied to *Ageratum (Erinus, Linn.)* and belongs to the *Didynamia*, but, according to Micheli, the capsule is of one cell only. I will examine this.

Among your Lapland plants, No. 85 *, 88 †, 231 and 232 ‡, are all, as it seems to me, natives of our Alps. No. 236 § certainly is so. It grows at the lake of Morats, where Burser gathered it, and where I noticed it in 1728; but it seems not distinct from *Ranunculus palustris (R. Flammula)*. No. 244 ||, 303 **, 353 ††, at least Scheuchzer's plant, and 443 ‡‡, are certainly found on the Swiss Alps. Concerning the remainder of your list (see p. 279), I cannot inform you. Some of them are undoubtedly not Swiss plants.

Adieu! Proceed with your labours in the advancement of Botany; and may your regard for me go on increasing! No difference of opinion ought

* *Campanula uniflora.* † *Diapensia lapponica.*
‡ *Ranunculus lapponicus* and *nivalis.* § *R. reptans.*
|| *Pedicularis flammea.* ** *Tussilago frigida.*
†† *Salix myrsinites.* ‡‡ *Lichen croceus.*

ever to interfere with that. How highly I esteem you will soon appear in our periodical work, published here. I will notice your *Critica Botanica* another time. I have a few books of travels at your service, in return for *Hort. Cliff.*

LINNÆUS TO HALLER.

Amsterdam, Oct. 8, 1737.

I received your letter this very evening. Tomorrow I leave this place. I must be in Sweden before the end of two months, but Mr. Cliffort has kept me till now. I wish I may be able to get away from Leyden, where my friends wish me to make some stay; that I may not lose too much time, I will wait upon you at Gottingen, if convenient. I wish to learn something from you about Mosses; which favour I hope you will not deny me.

You vex me by seeming to think I wrote with any intention of hostility to you. I call God to witness that I never held any botanist in higher estimation, honour, and affection, than yourself! Do not therefore think ill of me. You appeared to me of a frank character; and I hoped you would receive from me, as I from you, whatever an unrestrained pen, in hours of haste and business, might throw out.

I have made out a list of the names of such of your genera as differ from mine; not for the sake of criticism, but of information, that I might hereafter improve.

I am sorry you take unfavourably what I wrote concerning the *Unifolium*, by which I never meant to express, that either of us had borrowed from the other. I wrote as I did because Dillenius, whilst I was with him, so absolutely denied the accuracy of my view of the matter; while you, at the same time, observed the very same thing. Do not think amiss of me, for I am not maliciously disposed; and if I were so to every body else, I certainly should not to you.

If the plant you inclosed be truly the *Androsace*, which I made a synonym of my *Diapensia*, I must allow my mistake to have been very great, and ought to beg pardon, requesting you to correct me with as gentle a hand as possible. Your plant is clearly an *Androsace;* and my *Diapensia*, which I send you herein, proves hitherto a nondescript.

The *Solanum* genus is bilocular, though often variable, as *Lycopersicon*, but still bilocular in a certain manner. There is no membrane separating the cells, but only pulp; there is, however, a double kidney-shaped receptacle of the seeds, which grows convex, tumid, and is gradually enlarged.

Agrimonia you surely cannot separate from the plants composing the *Icosandria Polygynia*. Do but consider its parts, and their analogy, and you will perceive its calyx to be coloured, as in Roses.

I am sorry you apply to yourself what I said on the subject of Vaillant, as I have hardly seen a word of yours against him ; but I have much regretted to observe the posthumous hostility of Dillenius.

I had no doubts concerning the anther of *Marsilea* as you understand it, but only of the explanation in Micheli.

Time will not permit me to write much more at present; only I entreat you to banish from your mind all that offended you in my last. Pray continue your accustomed candour, and you shall never have any cause of anger from me. When we meet, I hope to secure your regard and good opinion. When I am absent, even my enemies cannot but declare how highly I value you. I deeply indeed lament that I should have done any thing to wound your generous spirit. I deplore my fault, and beg your pardon. This I hope will satisfy you, and that you will be as much my friend as ever.

My *Corollarium* is printed, and is left, with the *Methodus Sexualis*, in the hands of Mr. Reuss, who will forward them to you. You shall also have the *Hortus Cliffortianus*, as soon as ever the plates, now wanting, are engraved. I should like to have your opinion of my distinctions of the different kinds of leaves, given in that work.

The well-founded merits of Staehelin have not escaped my memory, without my finding a genus for him, at your request.

The *Hura* is now in flower at Mr. Cliffort's.

I should have thought the *Orchis* family more akin to *Canna*, *Kœmpferia*, *Maranta*, and *Amomum*, than to the *Liliaceæ*; and the latter to

Musa; Musa to the Palms. The *Orchideæ* are not trilocular. Examine the situation of the anther in *Canna*, and of the petals in *Kæmpferia.* No one will deny that these two last are very nearly related.

Whatever you send me will be highly acceptable, but I beseech you not to keep any reckoning between us. I know how much I am in your debt, though I can send no alpine Lapland plants till I get home. My publications are all of a small size, such as suit an exile or a traveller, who must carry all his property about him. I have lately met with the females of two species of *Cliffortia*, a female *Cluytia*, and various other genera. I hear Siegesbeck's *Hortus Petropolitanus* and Gerber's *Flora Moschoviensis* are come out, but am sorry to say I have never yet seen them.

Asphodeliris then proves to be my *Anthericum*, of which I have given a third species in the *Hortus Cliffortianus.*

Gronovius will, doubtless, soon publish the plants sent by Clayton from Virginia, unless he considers too long about the matter. Bartsch has lost his life at Surinam. If he had lived, he would have given a most accurate *Flora* of that country; and I am sure he would have sent me plants from thence, as I procured him his appointment. Van Royen has already begun to compose a new *Hortus Lugduno-Batavus*, on a systematical plan.

If you have any superfluous specimens of Minerals, they will be very acceptable to Mr. Cliffort, but you need not deprive yourself of any of the rarer sorts. When you happen to obtain any duplicates, you may send him what are least valuable. He already owes you a copy of his book, for the many valuable specimens last sent, most of which are deposited in his museum ; and if he does not give it you, I will.

In my description of the *Uva Ursi*, the leaves are said to be opposite, which should be alternate.

I now send you a few alpine Lapland plants. You shall have more when I get to Sweden. Adieu !

HALLER TO LINNÆUS.

Gottingen, Oct. 13, 1737.

Do not suppose our friendship is impaired by any of the heedless little bickerings which have fallen from you or me. I love you as a man born to advance the science of botany, and as my kind and beneficent friend. Let us forget all differences of opinion. Do but contrive to let me see you safe in Gottingen very soon, and accept, without reserve, of my accommodation, such as it is. We will look over my plants together; and we may confer together upon many points, that can only be settled when the specimens are before us.

Your *Diapensia* appears different from all my *Androsaces.* What I sent you is really an *Androsace,* frequent on our Alps, and on mount Jura. It has a sweet smell.

I would not have you part with Cæsalpinus on my account, unless I were sure you could easily replace it. My friend Staehelin has a copy, which he will readily lend me. His drawings of *Fungi*, *Lichenes*, and *Musci*, are now here, and I am at present employed in comparing them with my own as well as yours.

I am much gratified by all you have sent me. The plants I sent were intended for you, not for Mr. Cliffort. They consisted of what I had then by me. I wish you to do what you will with the minerals and stones, of which I have not many, though some beautiful ones, prepared for you and your friend.

Lapland plants will at all times be highly acceptable.

The writings of Siegesbeck and Gerber, mentioned in your last, I have never seen; but only the *Primitiæ* of the former. This is truly a work sufficiently full of novelties. He calls your *Linnæa* by the name of *Obolaria*, and your *Andromeda* by that of *Ledoides*, and so forth.

I understand Father Charlevoix's History of Japan treats of plants, but I have not seen it. I have nothing new to inform you of. I am now having the ground of my garden ploughed very deeply for sowing in the spring. Pray beg a supply of seeds for me from Van Royen, Burmann, and Cliffort. They will be highly acceptable. I have been deep in the study of *Fungi*, and not without success. I flatter myself I shall be able to fix some characters in this tribe, as in the *Lichen-agaricus*, *Fungoidastrum*, &c. of which latter I have many species. I think of publishing them in a kind of dissertation, like that I have just finished on *Pedicularis.* I would not send you this, nor my treatise on the *Veronicæ*, wishing to present them to you in person. You will see by them my manner of treating the history of plants, and of studying their synonyms, &c. I propose to extract from the larger work, now almost concluded, a methodical *Synopsis*, which shall be published next year. I am still extremely desirous of contriving an arrangement approaching to a natural one, or founded on the affinities of the plants, at least of the few families of which I have to treat. Artificial classifications are perhaps most useful to a learner, but they are less suitable for local Floras, in which when plants that are akin are placed together, they illustrate and define each other; as is the case with the Grasses, Bulbous plants, and others. My Swiss plants will prove not much under 3000; and if I did not exclude varieties, they would exceed that number; particularly by means of the *Fungi*, of which I admit none that I have not, by repeated observation, ascertained to be distinct.

Moering has not even yet had a sight of your works, as I learn by a recent letter from him. The generality of German purchasers will hardly buy your publications, at the enormous price fixed upon them by the booksellers, which is two rix dollars of 99 creutzers. Nor do they easily submit to changes in botanical nomenclature. The colour of flowers is, begging your pardon, of great moment with them. Indeed, among the *Pediculares,* many species are not readily to be known but by this means. Perhaps it may be had recourse to in the characters, when it is known to be constant, otherwise it may mark a variety.

I am now entirely devoted to Anatomy, upon which I lecture for three hours every day. I snatch a little time to work at the *Siliquosæ Tetrapetalæ,* or *Tetradynamia,* among which there is great confusion of genera, as well as of species. Some of the most vulgar kinds, as *Sinapi rapi folio*,* I can scarcely recognize, for I have neglected to collect specimens of culinary plants especially. From this tribe I shall proceed to the *Papilionaceæ;* in treating of which I lay no stress on the foliage. All the species of *Lotus* and *Coronilla,* as well as the *Hedysara,* with one, three, or many leaflets, are very closely related to each other. *Trifolium* becomes an enormous genus, especially if I am not to remove from it the *Trifoliastra.* The determination of the synonyms and species is attended with prodigious labour, as much among the common as the rare ones, or more so. It is the same with the *Orchis* family, the *Salices,* in elucidating which I hope for much help from you, and the *Carices.* What has detained me till now is my want of confidence in determining genera; and likewise the want of a well-furnished garden. But when my history of

* *Sinapis nigra. Linn.*

the Swiss plants is completed, and my garden in order, I shall carefully peruse your publications, which are undoubtedly most accurate, and which I confess afford a conclusive answer to those who dispute the paramount utility of your system. Still I do not yet so much despair of that philosopher's stone, a natural method, as to be willing to adopt any artificial one.

I hope you will soon come to us; meanwhile farewell, and believe me, &c.

Göttingen, Dec. 21, 1737.

Though long expected, you are not arrived. I have collected together, for your examination, paintings of *Fungi*, Mosses, Lichens, all the plants within my reach, as well as stones and metals. With these I invite you, and I shall endeavour to find you out, wherever you may be.

Your parcel has been delivered to me, in which I find much reason to acknowledge your kindness, in bestowing upon me unmerited praise, and claiming for me, as it were, the publick favour, in your splendid work. May I, as well as my friends and disciples, retain this in grateful remembrance! I hope to find some opportunity of showing myself one of the chief of your friends and admirers.

Amongst your plants, which were highly acceptable, your contending so positively that the *Uva*

Ursi is a box-leaved plant, with which the epithet of *folia carnosa punctata* does not well agree, makes it highly probable that what you call *Uva Ursi* is a different plant from your *Vitis Idæa sempervirens* *. I have not my books now at hand, being deeply engaged in anatomy, but so it seems to me. Your *Pedicularis* † comes nearest to mine with a beaked flower ‡, but is nevertheless different. Your *Geum* seems to differ from ours in having round petals, instead of long and narrow ones §.

I have forwarded to Mr. Cliffort's care the copies of my treatises on the *Pediculares*, and the alpine *Veronicæ*, destined for you. I should have sent them before, had I not expected you here. In both of them you will see my manner of treating such subjects, in a more diffuse way than is proper for a *Compendium*. I have written upon several other plants in this manner, and propose to do so by all that are natives of Switzerland. I have, this autumn, met with many species of *Lichen-agaricus (Sphæria)*, of various forms; but in every one there are globules full of pollen (rather seeds), ranged under some sort of bark. To these belong many of the *Lichenes exusti* of Micheli. I have

* These plants are now well known to every body, as *Arbutus Uva ursi*, and *Vaccinium Vitis Idæa*.

† *P. lapponica*, *Linn. Sp. Pl.* 847.

‡ *P. rostrata*, *ibid.* 845.

§ This remark answers to no *Geum*, properly so called. Perhaps Haller adverts to the old *Geum (Saxifraga)*, and speaks of *S. stellaris*.

caused several *Fungi*, near 150, to be delineated in their proper colours, and I shall go on with this undertaking. I have noticed, in many of them, the seminal pollen, of a black, blue, or other colour; but never any vestiges of the flowers as described by Micheli. Whenever I have time, I will examine these plants microscopically.

With regard to the plants of your class *Tetradynamia*, I fear their genera are very difficult to determine, and their species confused. All the *Siliquosa* differ but slightly from one another, as *Sisymbrium*, *Turritis*, *Eruca*, *Leucojum*, *Hesperis*, &c. I will examine them by your distinctions.

I have not procured Cæsalpinus, but I can borrow it. The work of à Turre, bought for me in Italy, is of little use. Whether the travels of Shaw or Kæmpfer are yet published, I know not. They must both be highly interesting. I asked Mr. Cliffort for seeds for my garden, but he has sent me ores, fossils, and dissertations. Pray recommend me to your friend when you have an opportunity. I have built a stove, which I wish to furnish in the spring. The anatomical theatre is also finished, and there have been many dissections in it this winter.

If you fix your residence in Sweden, pray collect some of your Mosses, and north-country plants. You shall have Lichens, Mosses, and Hercynian plants, if you please, in return.

Pray give my compliments to my old friend Rosen, though I am not personally known to him.

He has now attained a very advantageous appointment. I was particularly acquainted with a surgeon named Ribbius, the son of your king's physician. If I have an opportunity, I will send something to our Society. Farewell! and do not deprive me of your regard.

LINNÆUS TO HALLER.

Leyden, Jan. 3, 1738.

I shall remain here till the 3d of February. If, meanwhile, your work is sent me, I will, if you please, take care to have the different types appointed for the classes, generic and specific names, and synonyms. If I can be of use to you in any other way, I am entirely at your service.

Though the *Hortus Cliffortianus* has long been printed, it is not yet published, owing to the tardiness of the engravers.

Dr. Boerhaave told me lately that a second volume of Micheli's is now printing *. I long to see the flowers of Ferns.

Professor Amman writes me word, that he has laid before the Petersburgh Society a treatise on such Ferns as have some leaves bearing flowers and others not, in the same species †. He hopes to

* "This never came out." *Haller*. The unpublished plates, in the hands of the writer of this, contain Corals, Corallines, *Fuci*, *Spongiæ*, and some *Zosteræ*, but no Ferns.

† "*Struthiopteris*." *Haller*.

have his work, on the rarer plants of Siberia and Tartary, ready in the course of next summer.

Monsieur de la Croix relates that Reaumur has undertaken to write upon fossils. I fear he will not shine so much in these as in insects.

If you know any thing of the *Holosteum* of Dillenius, pray inform me. I certainly never saw an *Alsine* with three-cleft petals *. How many stamens has it? are they ten? and how many pistils?

So also of Tournefort's *Tribuloides (Trapa, Linn.)*. Has this four petals? what calyx, and how many stamens and pistils has it? I have in vain sought for the flowers of this plant in every herbarium to be seen in England, Holland, or Sweden †.

When I leave this place, I shall, as far as I can tell at present, go directly for Göttingen, that I may put myself under your tuition in Mosses. I wish you may not find me a troublesome guest.

I have written, in my *Systema*, that all botanists acknowledge the fundamental importance of the fructification, except perhaps Heister alone, whom I have mentioned as a botanist. It appeared to me that this eminent man did not admit the above principle, as he founded his orders and genera on the leaves. He, being displeased at my remark,

* *Holosteum* is, however, well distinguished by its jagged, or notched, petals; see *Fl. Brit.* Dr. Wallich has recently sent a new species from India.

† All the parts in question are clearly and faithfully represented in the exquisite wooden cut in Camerarius's *Epitome*, p. 715.

wrote to Siegesbeck that he meant to take his part, having recommended him to his present appointment at Petersburgh; and, as I hear, this good man has accordingly published a critical dissertation, in which he severely takes me to task, though he has written me many civil letters. So my old friend Ludwig announces, in his last letter, that he is about to write something controversial against me. I shall be easily conquered, as I already lay down my arms, and have no intention of defending myself *.

When I come to you, I hope to tell you more about these and other matters, and to afford you some amusement.

Artedi's work on Fishes, hitherto certainly unrivalled, is printing here. You will be astonished at his index of synonyms. He has genera and their characters, classes, species with their names and definitions, &c.

I am printing the Systems of Cæsalpinus, Morison, Ray, Christopher Knaut, Boerhaave, which last is just struck off, Rivinus, Ruppius, Ludwig, Christian Knaut, Tournefort, Magnol, my own on the organs of impregnation and on the calyx; as well as those of Vaillant and Pontedera on the Compound flowers, of Scheuchzer and Ray on Grasses, and of Dillenius on Mosses and Fungi,

* Heister and Ludwig did write, and possibly some of their pupils might read, but Linnæus had no occasion to reply. Heister, amongst other things equally to the purpose, complains he could not find the stamens in an *Aloe,* and therefore the Linnæan system was useless.

with a chronological index of genera, &c.; that, by knowing any name, I may see what each genus is called by all systematic writers, and that learners may readily turn to each method of arrangement, and if one is defective, consult another. I every where subjoin the synonyms to each genus.

Far be it from me ever to uphold artificial arrangements, as if they were in any measure comparable to natural ones. I wish we knew more about natural classes. I write in haste, and must request your indulgence. Farewell! May you long live and flourish, as the ornament and arbiter of Botany!

HALLER TO LINNÆUS.

Göttingen, Jan. 9, 1738.

It was very kind in you, my dear Linnæus, to engage Wishoff to let me have his types to print my work. The form of it will be that of your *Flora Lapponica*, but its size somewhat larger. As I intend to give an entire history of the *Fungi*, it cannot go to press before November. By that time three fourths of the whole will be completed. There will be only the trees and grasses to finish. In the mean time the plates may be preparing, in case Wishoff means to have any. I am anxiously looking for your arrival, for you will be a most welcome visitor; though I fear much of the honour you now heap upon me will be found, on a near approach, to be far above the standard of my merits.

I regret that the pursuit of anatomy prevents my devoting my attention entirely to botany. During the present holidays I have finished the *Cruciatæ*, or *Tetradynamiæ*, to which I have been giving my attention for eleven years past. But you, who have done so many other things, shall not be deprived of your just praise. What obscurity is there, every where, in settling the genera of the most common plants! You need not care for the censure of Siegesbeck or Ludwig, though it might as well be prevented. I will write to Heister and Ludwig, for it does not become them to depreciate the merits of those who are devoted to the elucidation of nature.

I will lay by some stones and ores, of which you shall take what you please, and I will give the remainder to Mr. Cliffort, to whom I have written. I will also either keep till you come, or put into my parcel to Mr. Wishoff, my essays on *Veronicæ* and *Pediculares*. I never saw the *Tribuloides (Trapa)*. *Holosteum* is here very common. It has undivided petals, with a few marginal notches or teeth, like an *Alcea;* five stamens, and three wavy styles. The bluntly conical, somewhat cylindrical, seed-vessel opens at the top into five parts. The seeds are attached by short stalks to an obtusely triangular receptacle. You may depend on the calyx having five leaves. Farewell, my much-desired guest! let me see you soon. Remember me to Gaubius, Van Royen, and other friends. If you can scrape together any of the books I have already mentioned, I shall be greatly obliged to you; and when you are here I can the more easily repay you.

LINNÆUS TO HALLER.

Leyden, Jan. 23, 1738.

You, my distinguished friend, have already given me so many testimonies of your high regard, that I am not conscious of being more deeply indebted to any of my friends or patrons. I wish it may ever be in my power to show, by my actions, how sensible I am of your favour. I should be a stock or a stone were I not affected by these pledges of your friendship. Your regard for me, though personally a stranger, is altogether paternal; and I promise, with all my heart, that mine for you shall be filial.

Dr. Ludwig, whose displeasure could be appeased by you only, has written to me, with a promise of friendship. Pray take no trouble about Heister, nor prostitute your consequence in defending what is unworthy and trifling in itself. Besides, your remedy would be too late, for Siegesbeck's essay has long since been sent to the press. I will answer it when it comes to hand, but without any agitation of mind. He shall never provoke an angry word from me, though he has poured out thousands on my devoted head.

My mouth waters at your two dissertations, especially that on *Pedicularis*, whose species I never could well understand. I long to see them; but perhaps you had better not send them, lest they should not arrive till after my departure, and thus be lost. It will give me more pleasure to receive them from your own hands. I wish my company may not prove troublesome to you. I wish to become acquainted with Mosses; and I wish to see a man so distinguished, and so kind to me as you have proved yourself.

I have seen, but have not yet obtained a copy of, Shaw's Specimen of African Plants, collected in Barbary, Egypt, and Arabia, printed a few days since, in 12 folio pages, in which there is a short catalogue of 667 plants, written probably by Dillenius; but it is too concise, there being only one name to each.

Dr. Burmann has begun to publish his African Plants in Decades. I have seen the plates of the first part, but am sorry to say there is nothing new, and the figures he has given, drawn by I know not who in Africa, are bad, and the plants published already by Plukenet and Petiver.

I have at length received from London a specimen of that American *Polygala*, for which a certain empyric received a thousand pounds sterling from the English American Society; this man having found its root a specific for I know not what disorder.

The plant has many characters in common with the *Penæa* of Plumier. Gronovius, no doubt, will describe it amongst his Virginian plants. I wish he had not so long hesitated to publish his book.

He is too timid. You would scarcely believe how many of the vegetable productions of Virginia are the same as our European ones. There are Alps in the country of New York, for the snow remains all summer long on the mountains there. I am now giving instructions to a medical student here, who is a native of that country, and will return thither in the course of a year, that he may visit those mountains, and let me know whether the same alpine plants are found there as in Europe *.

My last letters from M. de Jussieu mention that his brother, long since supposed to be lost, is well. If he returns in safety, we shall have many good things from him. Bartsch has been about a month at Surinam. He is indebted to me for his appointment, as well as for what he knows of botany; and he is one of the most diligent of men. I expect much from him. He is by birth a German †.

Both Van Royen and Burmann are possessed of coloured drawings of your *Halleria*. If the latter does not soon publish it, Van Royen will. Two species of *Halleria* ‡ are described in the *Hortus Lugduno-Batavus*.

* The editor has lately received several of these, as *Menziesia cærulea, Azalea procumbens,* and some new species, from the White mountains of New Hampshire, gathered by Professor Peck, Dr. Bigelow, and Mr. Francis Boott.

† Jussieu died in Peru, and Bartsch survived but a short time at Surinam, which Linnæus very pathetically laments, under the genus *Bartsia,* in *Fl. Suec.*

‡ These were subsequently judged to be only varieties.

I have no more news to communicate. Van Royen and Gaubius send you their best wishes.

HALLER TO LINNÆUS.

Göttingen, Feb. 11, 1738.

Having a copper-plate to send to Mr. Wishoff, I would not let it go without some token of my friendship. I therefore send you a few of my dissertations, some of which you will please to give, as before, to our friend Van Royen. I shall communicate to you many things when I see you. By these you will see the nature of my present undertaking, though they are by no means in a finished state. I have, in fact, scarcely any spare time. I have been dissecting ten subjects, and am obliged to prepare for speedy publication some anatomical dissertations, with a description of a monster which I have been examining. In my *Synopsis Stirpium* I am got to the *Papilionaceæ*, among which *Trifolium* is one difficult genus, and *Vicia* another.

You may justly laugh at Siegesbeck, and all such people. It is really unworthy of learned men to use such quarrelsome language. I have written to Heister, but he is particularly angry with you for speaking of him as not a systematic writer.

Among the *Pediculares* you have sent me, one is very near mine with a beaked white flower *, but

* *Pedicularis tuberosa. Linn. Sp. Pl.* 847.

differs in the leaves, habit, and other respects, as I perceive by the specimen for which I am indebted to you. What think you of my objection, that Caspar Bauhin speaks of the *Uva Ursi* as having *dotted* fleshy leaves ; and that this is rather suitable to a species of *Vaccinium* than to a plant whose leaves are "by no means dotted," as you say of your *Uva Ursi*?

I am possessed of but few Mosses, and every system is deficient as to the genera of this tribe, as well as their species, more particularly among the *Hypna*. Dillenius has also many more species of *Bryum* than I have. I have plenty of *Polytricha*. Pray bring as many with you as you can conveniently.

I send you a View of your Botanick Characters, written last July, but not taken by the director of our periodical work into his third *fasciculus*, so it must appear hereafter. You will see my intention. It was an irksome task to accommodate the German language to matters never mentioned in it before.

All botanical and literary intelligence is welcome to me. If you can pick up any thing for me, I shall not only repay you in money, but be very much obliged to you. Cæsalpinus has been purchased for me out of Micheli's library. Do not omit Boerhaave's oration on resigning the charge of the garden, which I have not got. Pray urge Wishoff to send me the catalogue which is wanting. I value Hernandez on account of what Columna has introduced. As to Columna's book, there is nothing

I wish for more. It was sold by Vander Aa for 50 florins, a price I would now gladly pay.

I would not have you trouble Burmann on my account. I have long been in possession of his *Thesaurus Zeylanicus*, which indeed is not very necessary to my line of study.

We are now all undisturbed by the Muses, those Muses to whose service you are devoted. I have no more to say, but to entreat you to come hither as soon as possible, where you will find yourself most truly welcome. Compliments to Van Royen and Gaubius.

Göttingen, March 14, 1738.

Though I have not much to say, I would not send a parcel to Wishoff without a word to you. I am every day expecting you. Having done with anatomy, I can now devote myself to you and Flora. I am arranging the *Papilionaceœ*. The *Viciœ* and some of the *Trifolia* stop my progress. Mr. Cliffort has not answered my letter; nor has Van Royen sent any seeds. Pray come soon. Ludwig writes that he is not undertaking any thing in opposition to you. Siegesbeck is an enemy to both of you, and has published something of a caustic nature against our Leipsic friend. The latter (Ludwig) has sent me his *Aphorismi Botanici*, an elegant compendium of the structure of plants, but he has made great use of your principles. The book is but just published. I hope to send Mr. Huber to the Alps, at my own expense, for the purpose of making a collection of specimens, to assist me with regard to my *Synopsis Stirpium Helveticarum*, as I find myself every day deficient in some little plant or other, to my great mortification.

I have sent no minerals to Mr. Cliffort, as he has not answered my letter. You shall select what you please, for him and yourself. Be so kind as to try to persuade Wishoff to have figures engraved of some of my new plants. The cost will be little, and the plants are truly elegant *. I would have none but such as are new, or extremely obscure in their history. Figures please amateurs, and prevent a book from seeming dry and tedious. Adieu! may you soon be a welcome visitor at Göttingen!

LINNÆUS TO HALLER.

Hartecamp, March, 1738.

I have duly received your most agreeable letter of the 11th of February; but all my plans have been interrupted by a severe illness, which has kept me in bed these six weeks. Within a few days my health has begun to amend a little, so I have been able to quit Leyden, and am come to Mr. Cliffort's garden, in order to breathe a freer air, and recover my strength and spirits.

* This great and long-promised work of Haller was, at length, printed at Göttingen, in 1742, with 24 fine plates, in folio.

During my confinement I have perused, with the greatest pleasure, your dissertations on *Veronica* and *Pedicularis;* in both which you have, as usual, displayed your eminent abilities. I wish we had many things written with the same attention and learning. I begin to get an insight into your system. I shall study it more and more, and follow the same path, as far as I am able.

I have received from a friend Professor Siegesbeck's *Verioris Botanosophiæ Specimen,* with his *Epicrisis* on my writings. This author has been very hard upon me. I wish he had written these things when I was first about publishing. I might have learned when young, what I am forced to learn at a more advanced age, to abstain from writing, to observe others, and to hold my tongue. What a fool have I been, to waste so much time, to spend my days and nights in a study which yields no better fruit, and makes me the laughing-stock of all the world! His arguments are nothing; but his book is filled with exclamations, such as I never before met with. Whether I answer him or keep silence, my reputation must suffer. He cannot understand argument.

He denies the sexes of plants. He charges my system with indelicacy; and yet I have not written more about the polygamy of plants, than Swammerdam has about bees. He laughs at my characters, and calls upon all the world to say if any body understands them. I am said to be ignorant of scientific terms. He judges me by the principles of Rivinus, and hundreds of the vilest scribblers.

Inasmuch as this man humbles me, so do you, whose learning and sense have been made sufficiently evident, exalt me. It distresses me to read the commendations you are pleased to heap upon so unworthy an object. I wish there might ever be any reason to expect that I could evince my gratitude and regard for you. I hope life will be granted me, to give some proof of my not being quite unworthy.

The *Uva ursi* of C. Bauhin I am not acquainted with. Tournefort's is certainly the same as mine. The leaves of this often become wrinkled by drying, and then seem as if they were dotted; but I know nothing certain about Bauhin's plant.

As soon as I recover my health, perhaps towards the end of April, I shall go to Paris, where I shall have an opportunity, never perhaps to be had again, of inspecting the collections of Tournefort, Plumier, Surian, &c. But I can devote a fortnight only to Paris; after which I must return, by Cassel and Göttingen, towards my own country; and then I may perhaps be a more welcome guest to you in June, or the end of May. I wish also to see the library of d'Isnard, if he be living, and Reaumur's insects. The Jussieus have received many rarities from Peru, by means of their brother.

Dr. Boerhaave labours under an asthma, I fear a vomica; but, as yet, he goes through all the duties of his situation. He cannot indeed ascend the steps of the schools.

When I had written thus far, your delightful letter of the 14th of March reached me. The *Viciæ* are very troublesome. There are many varieties among them, difficult to determine. The *Trifolia* are not all easily known, and the generic character is excessively bad. It is certainly not to be founded on the shape of the legume, or number of the seeds.

Mr. Cliffort does not intend writing till he can send you his *Hortus*, which I hope he will be able to do in a fortnight, or three weeks at longest.

I will be with you, if I live and be well, on or about the first of June. I am obliged to you for helping me out of a quarrel with Dr. Ludwig.

If you have any doubts concerning species, which can be settled by Tournefort's herbarium, or if there be any other service which I can render you at Paris, pray let me know. To convey many books indeed, in so long a journey, would be very difficult, as I am already encumbered with much luggage.

If you send a botanist to the Alps at your own expense, you will rank foremost among the promoters and restorers of botany, labouring yourself to excess, as you do.

James Sherard is dead, without leaving any thing to Dillenius, which, as he died without children, I wonder at *.

* This is sufficiently accounted for by a letter of Dillenius to Dr. Richardson; see p. 154 of the present volume.

HALLER TO LINNÆUS.

Göttingen, April 7, 1738.

I am happy to hear the fever, which so long afflicted you, has gradually left you. I fear your too great ardour in the pursuit of our delightful study may have been unfavourable to your recovery. Pray spare yourself. The fruit you gather may be somewhat later, but it will be the sweeter. I wish you a prosperous journey to Paris. The commissions I could give you there would more than fill this sheet. There are numberless points on which I could wish you to consult Tournefort's herbarium; nor do I know where to select my questions. While you are making your examinations respecting Lapland plants, you will make discoveries for me. I shall enrich myself with your remarks, during your stay here.

The seeds, as well as your letter, are sent to Van Royen, and you have procured me his friendship. Why should you care for Siegesbeck, who has also, as I hear, written against Ludwig? Were slanderers ever wanting, or will they ever be wanting, to calumniate all who distinguish themselves by their discoveries or their abilities? Are you destitute of those who are more just to your merits? Or did you ever hope to conciliate the approbation of all these Siegesbecks into the bargain? Do but proceed boldly, and strive to adorn still more the science in which you have already acquired so much true glory.

The *Uva ursi* of C. Bauhin, like mine, has closely crowded points, not merely wrinkles, all over the leaves. But we shall settle this matter when we meet. I never offered 50 florins for Hernandez, but for Columna. I have Cæsalpinus by me, as well as Turre, which you pointed out as extremely rare. It is a thick folio volume, which I bought very cheap.

If you discover any curious books at Paris, I shall be obliged to you to mention where they are to be had to Louis Philip Guerin, bookseller, who may send them to M. Gachet at Berne. The latter has a list of the books I have long been in search of.

The *Hortus Cliffortianus* will be one of the most splendid of works, and in giving an account of it I shall do justice to you and your Mæcenas. Boerhaave's illness is sad news. This man certainly deserves, not merely for his knowledge, but for his most amiable disposition, that not only his pupils, but all mankind, should rejoice at his happiness, and grieve at his afflictions.

I will prepare some figures, according to Mr. Wishoff's desire. Staehelin has very beautiful drawings of the *Trifolium capite folioso* *, and the *Lenticula aquatica quadrifolia* †, with the whole progress of its growth. I shall take care to have drawings made of the new species of *Pedicularis*, *Astragalus*, &c. which must be done from my

* *T. repens*, with a leafy head, figured in *Ambros. Phyt.* 541.

† *Marsilea quadrifoliata. Linn. Sp. Pl.* 1563.

dried specimens; but the characters which Mr. Huber may delineate in his journey to the Alps, can be added to the plates hereafter.

I am now among the Ferns. I have but few, and want some of the Swiss ones, as the *ramosa mollior*, and *non ramosa mollior*. What are all the *Dryopterides*, and *Filiculæ fontanæ?* I fear many of them are but vain distinctions. After these I shall proceed to the *Orchideæ*.

My health is in a feeble state, and I am very anxious about one of my boys, so that my studies are much interrupted. I hope things will be better when you come, and by that time nearly half my collection of plants will be so far arranged that you may quickly look them over. I am preparing my plates. I could find above 200 plants that are not already figured. My garden is not yet walled in, nor prepared for use. My house is built in the garden, so I shall live in the midst of flowers, and soothe the remainder of my life with this most innocent pursuit.

My system is a nullity. I labour, nevertheless, at settling the species of plants. The characters of too many are not to the purpose, and would require to be examined over and over again. The herbarium of Scheuchzer, which I have ordered to be bought after the recent death of John Scheuchzer, will conduce, in no small degree, to enrich my *Synopsis*. He saw much himself, and received many things from his pupils. His brother collected a vast number of grasses. I wish the herbarium of John Bauhin could be found; or that I had more opportunity to make use of that of Caspar. M. Huber has been to the Rhætian Alps, and the Valteline, a district of Switzerland, with the climate of Italy; and from hence he has furnished great additions to my catalogue.

I am obliged to conclude, being ill at ease in mind and body. I sincerely wish you health, and hope soon to have the happiness of seeing you under my roof.

LINNÆUS TO HALLER.

Rouen, June 22, 1738.

I have at last been at Paris, and passed a month there. I saw no small number of exotic plants, and had an opportunity of examining almost all Vaillant's *Orchideæ* in flower, at Fontainebleau. The *Orchis* resembling a spider, and those which imitate a bee and a fly, are only varieties, as I learned from seeing them in flower. The character of *Crambe* consists in the filaments of the stamens, as in *Prunella,* the four larger of the six filaments being forked. All the species of *Rapistrum* are not, therefore, to be referred to this genus of *Crambe.*

I have seen the herbariums of Surian, Tournefort, and the Jussieus, all beautiful. I found it difficult, from these collections, to distinguish varieties, and refer them to their genuine species; but still more so to make out specific characters. I have seen abundance of botanical libraries, both publick and private, so that I have materials for another edition of my *Bibliotheca*, being acquainted with double the number of books that are in the former one.

The elder Jussieu is occupied in the practice of physick, nor will he stir a step from Tournefort. He knows species well, but makes them too numerous. The younger Jussieu, Bernard, is a good botanist enough. He begins to observe with attention, and he taught me to distinguish you from other botanists.

What the fates have long promised me in Holland, I was obliged to leave that country without altogether obtaining. I am therefore under the necessity of giving up my visit to your country, which I have so long wished to see. I must go by sea, as being the shortest way, and get home as soon as I can. But when I can there get a little money together, I mean to travel again; and if a few more years be granted me, I will see Germany. I ardently long to visit your mines, the insects of Frisch, Hebenstreit's shells, and you and your plants.

A German here has lately informed me, that you are not in good health, that you labour under a calculous cough, and that you have a design of returning to Switzerland. If this be the case, I am most truly concerned. Such is the reward of all our labours!

If you favour me with any more letters, send them to Stockholm, and I shall be sure to receive them. May God grant you happiness! I hope long to be remembered by you, as I shall ever remain your most devoted admirer. Again and again farewell!

HALLER TO LINNÆUS.

Göttingen, Nov. 24, 1738.

I received your letter from Rouen, and have long looked for one from Stockholm, but in vain. Wherefore, that I may not be wanting to myself, I would not omit writing, any longer, on the subject of our common pursuits, though uncertain whether my letter may reach you.

Since that time I have visited the Hercynian forest, and have given a narrative of my journey, with a figure of an entirely new kind of *Sisymbrium* *. I have subsequently completed my history of the *Gymnomonospermæ*, so that now the following orders only are wanting to finish my *Synopsis: Gramina, Arbores, Gymnodispermæ, Gymnotetraspermæ, Gymnopolyspermæ*, and *Monangiæ;* all which I mean to get ready for the press in the summer of 1739. Ten plates are engraved, but there will be fifty in all, of the natural size. The work is to be in quarto †, and will be published here by Vandenboek, as Wishoff puts dif-

* *Arabis Halleri. Linn. Sp. Pl.* 929.

† It appeared in folio, with 24 plates, containing 43 plants, besides numerous Mosses, Lichens, and minute *Fungi.* The Swiss *Orchideæ*, complete, were added to the second edition, published in 1768.

ficulties in the way, and does not answer my letters, though I have written to him often. The *Fungi* prove excessively troublesome. I have 300 of this tribe, drawn in their natural colours, after having employed much attention in getting rid of spurious species, caused by disease, or by different periods of growth. I have met with a number of Agarics, entirely of a fungous and tender substance; but I wish so to dispose them that the *Agarico-fungi* should be referred to the *Fungi;* the *Agarico-suilli* to the *Suilli (Boleti);* the *Agarico-polypori* to the *Polypori;* and the *Agarici plani* to the *Fungoidastrum* and *Fungoides.*

I am now engaged in Anatomy, and in publishing Boerhaave's Commentary on his *Institutiones Medicæ,* which is my chief labour at present. I have repented of this undertaking, ever since I was aware of its extent; for to a man whose time is otherwise so much occupied, it is almost unsurmountable. The whole of the anatomical part requires correction throughout, and to have proper references to authors, whom he has cited vaguely. His materials, however, shall be kept sacred, with merely so much polishing as the author himself would have practised.

Our friend Huber is very busy in finishing his dissertation, describing his Rhætian journey, undertaken at our joint expense. In it are described and delineated many fine plants, which I either had never seen, or did not know to be indigenous. They are full 400. He is preparing for another much longer journey, from which much is to be expected. Meanwhile I shall prepare and plant my garden.

I have sent some fossils to Mr. Cliffort, and he writes to desire more. I have not yet received the *Hortus*, which vexes me. I hear it is only to be had as a present from the editor. Burmann, meanwhile, of whose *Stirpes Africanæ* I have given an account in our Diary, has for some time improved upon the original merit of his publication. Van Royen has had above an hundred mosses from me, but has not made any return of seeds, as I earnestly requested he would. He has plenty at his command, as well as various kinds of dried specimens, and particularly maritime plants found on the coast of Holland. But my illustrious friend Dr. de Hugo has sent me Hercynian plants, and Malabar grasses. I hope, if it please God, next summer to make a journey to the Pennine Alps, and the Valais, that I may put an end to my long botanical undertaking. The latter is a very hot country, and has never been well examined, except by Burser and myself; nor do I doubt that many things still remain to be gathered.

The *Orchis muscam referens* is unquestionably distinct from the Bee Orchis, whether the variety of the latter with a rounded lip be distinct or not from the variety which has a point from the middle of its lip *.

* They are all three unquestionably distinct, as well as many more species, whose synonyms Linnæus confounded under his *Ophrys insectifera*.

Farewell, my dear Linnæus! may you enjoy your health and your botanical pursuits, with every advantage for the prosecution of your labours! My studies and engagements, of a different kind, draw me unavoidably aside; but my inclination always leads me to the charms of Flora. To Botany I wish to devote my leisure, my old age, and my fortune, in collecting drawings, plants, and books. May you, from whom Flora expects more than from any other mortal, make the most of your advantages, and one day or other return to a more genial climate! If at any time my native country should invite me, or I can ever, as I hope, return to it, I have fixed upon you, if the situation be worth your having, to inherit my garden, and my honours, such as they are. I have spoken on this subject to those in whose hands all these concerns are placed.

As soon as I hear from you, I will tell you all the news I can, for I shall be happy to resume our agreeable correspondence. Adieu! do not forget me.

LINNÆUS TO HALLER.

Stockholm, Sept. 12th, finished the 15th, 1739.

Your letter, whose value to me is beyond estimation, though dated Nov. 14, 1738, did not reach me till the 12th of August of the present year, when I received it from the minister of the German church at this place. Of the cause of its delay I am ignorant.

A thousand times have I invoked the honoured shade of Hermann! How well did he deserve the compliment of having all the fountains in the royal gardens play on his arrival, if we consider his liberal conduct towards Tournefort! Hermann had previously offered to resign the botanical professorship (at Leyden) in his favour, intending himself to seek some other situation, during Tournefort's life. But what shall I say of you, who have conceived so strong an affection for a stranger as to invite him to accept your professorial appointment, your honours, and your garden! A man could scarcely do this for his brother, or a father for an only son. I can only say, in one word, I have had a numerous acquaintance among my fellow-creatures, and many have been kindly attached to me; but no one has ever made me so bountiful an offer as yourself. I would express my thanks if possible, but cannot find words for the purpose. Your memory shall be engraved on my heart whilst I live, and shall be cherished by those who come after me*.

* It would be ungrateful in the editor, while perpetuating the remembrance of these generous actions, to omit recording one of the same kind, in which he was deeply concerned. The venerable Professor Martyn at Cambridge, incapacitated by his great age from fulfilling the duties of his appointment, was ardently desirous that the writer of this should either be invested with the professorship, on his resignation, or undertake its functions as a deputy. This proposal was sanctioned by the highest authorities, both in and out of the University; by all who had the interests of science and literature at heart; and by many from personal regard for the Professor or his friend.

I cannot give an answer; but as you have placed yourself in the light of a father, and me of a son, I will lay before you a sort of history of my life, up to the present time.

In the year 1730 I taught botany in the garden at Upsal. Our common friend Dr. Rosen returned thither the same year. I, then a student of medicine, was Professor Rudbeck's deputy in botany, as Rosen was in anatomy, he being likewise the *adjunctus*, or coadjutor, in medicine. In 1732 I went to Lapland, and returned; after which I read lectures on botany and metallurgy for a whole year. I then quitted Upsal, and, as Providence ordained, went into Dalecarlia. Having accomplished my journey, I returned to Fahlun, the principal town of that province. Here I lectured on mineralogy, and followed the practice of physick. I stayed a month at Fahlun, where I was received with universal kindness. A physician named Moræus resided there, who was esteemed rich by the common people. Indeed he was one of the richest persons in that very poor country. With regard to learning, he might undoubtedly claim the first rank among

But the most sordid interests being brought into competition with the above laudable views, the most futile pretences were made to defeat them; and it did not become an independent man to embroil himself in local parties and intrigues, of whose existence he had previously no conception. Much of the history of this affair is before the publick in two pamphlets, entitled, *Considerations respecting Cambridge*, and *A Defence of the Church and Universities of England against such injurious Advocates as Professor Monk*, &c.

the medical men of Sweden. I have heard him say, a thousand times, that there was no line of life less eligible than the practice of physick. Nevertheless, he was much attached to me. I found myself frequently a welcome visitor under his roof. He had a handsome daughter, besides a younger one, the former of whom was courted, but in vain, by a gentleman of rank and title. I was struck when I first saw her, and felt my heart assailed by new sensations and anxieties. I loved her, and she at length, won by my attentions, listened to my proposals, and returned my passion. I became an accepted lover. I addressed myself to her father, avowing, not without much confusion, my total want of fortune. He was favourable on some accounts, but had many objections. He approved of me, but not of my circumstances; and desired that things might remain as they were for three years, after which he would tell me his determination. Having arranged my affairs, and made the necessary preparations for a journey, I quitted my native country with 36 gold ducats in my pocket. I immediately took my medical degree (at Hardewick in Holland), but was not in circumstances to return home with much comfort. I remained, as you know, in Holland. In the mean time my most intimate friend B . . . regularly forwarded the letters of my mistress by the post. She continued faithful. In the course of last year, 1738, which I passed at Dr. Van Royen's, with the approbation of the young lady, though it was the fourth year of my absence, and her father had required but three, B . . . thought he had himself made considerable progress in her favour. By my recommendation he was made a Professor; and he took upon him to persuade my betrothed that I should never return to my own country. He courted her assiduously, and was very near obtaining her, had it not been for another friend, who laid open his treachery. He has since paid dearly for his conduct, by innumerable misfortunes.

At last I came back, but still destitute of a maintenance. The young lady was partial to me, and not to him. I settled at Stockholm, the laughing-stock of every body on account of my botany. No one cared how many sleepless nights and toilsome hours I had passed, as all, with one voice, declared that Siegesbeck had annihilated me. There was nobody who would put even a servant under my care. I was obliged to live as I could, in virtuous poverty. By very slow degrees I began to acquire some practice. But now my adverse fate took a sudden turn, and, after so long a succession of cloudy prospects, the sun broke out upon me. I emerged from my obscurity, obtained access to the great, and every unfavourable presage vanished. No invalid could now recover without my assistance. I began to get money, and was busy in attendance on the sick, from four in the morning till late in the evening; nor were my nights uninterrupted by the calls of my patients. Aha! said I, Esculapius is the giver of all good things; Flora bestows nothing upon me but Siegesbecks! I took my leave of

Flora; condemned my too numerous observations, a thousand times over, to eternal oblivion; and swore never to give any answer to Siegesbeck.

Soon afterwards I was appointed first physician to the navy. The magistracy immediately conferred upon me the regius professorship, that I might teach botany in the seat of government at Stockholm, with the addition of an annual stipend. Then my fondness for plants revived. I was also enabled to present myself to the bride to whom I had been for five years engaged, and was honourably received as her husband. My father-in-law, rather fond of his money, proved not very liberal to his son-in-law; but I can do without it, and those who come after me will enjoy it.

Just now, both the medical professorships are likely to become vacant. Professors Rudbeck and Roberg, both advanced in age, are about offering their resignation. If this takes place, probably Mr. Rosen may succeed Roberg, and I may obtain Rudbeck's appointment. But if I do not, I am content to live and die at Stockholm, nor shall I oppose the pretensions of any competitor. If, therefore, I should not obtain the botanical professorship at Upsal, and you, at the end of three months, should invite me, I would come, if I may bring my little wife with me. Otherwise, if there be any chance of my ever seeing you at Hamburgh, for that reason alone I would go thither, though I live here at a great distance. My regard for you makes me wish to know you personally, to see and talk to you, before I die. Farewell! may you long continue to be the load-star of our science!

P. S. Sept. 15th. If you possibly can, pray let me have your treatises upon *Veronicæ* and *Pediculares*, but above all, your *Synopsis* of Swiss plants.

If you favour me with an answer, it must be franked to Hamburgh, in a cover to the Royal Society of Sciences at Upsal.

How glad should I be to see the account of your Hercynian journey! I hear your *Synopsis* is to be printed at Leyden. I can easily have it from thence, and wish it may soon be sent. I perceive the Commentaries on Boerhaave's lectures are printed, and have long enquired for them from Holland, but, as yet, in vain, though I hope soon to have them.

It rejoices me to hear that Mr. Huber's tour is finished, and I wish it may speedily be published. You are thus rendering more service to science than any other person.

I earnestly entreat you to write me word whether or not Cliffort has sent you his *Hortus*. I have mentioned it to him a thousand times. If you have it not, I will forward it by the first vessel to Hamburgh; only say where you would have it left, as Cliffort has given me ten copies.

I can scarcely wonder at Van Royen's not having written to you. You are probably aware of what he has gone through this year, and what he has had to do respecting the *Prodromus* of his *Flora*. He will write to you. He is well acquainted with you through me.

I have never yet seen the ***Primitiæ Floræ Petropolitanæ*** of Siegesbeck, nor Ludwig's botanical aphorisms. How can I procure them?

With all my heart do I rejoice, that what the French told me, concerning your health, was false. May you long enjoy life and health, inasmuch as I would devote my life to your service!

I am here buried in obscurity. If I get to Upsal, I shall give up the practice of physick, and devote my whole attention to plants. If you have seeds, or any thing else, to send me, they will be welcome.

HALLER TO LINNÆUS.

Göttingen, Jan. 11, 1739.

Your letter so strongly expressive of your affection, and so kindly making me a participator in your happiness, I value extremely. May you long enjoy your domestic felicity! I continue in the same intentions as to my garden. I shall probably remain here but a few years, nor could I ever give up my appointments to a more worthy successor. It is very probable that I may have occasion to go to Hamburgh. There is in that place a copy of Kerkring's *Thesaurus*. If this should come to be sold, it will be bought for publick use by our University. It will be delightful to confirm, by personal intercourse, that friendship which we have hitherto cultivated, like spiritual beings, by intelligence only. Pray tell me the name of some merchant at Hamburgh or Lubeck, to whom I may send, for you, all my before-mentioned publications, and with them my *Synopsis*, not yet finished, of which 13 plates only are, as yet, engraved. Boerhaave's first and second volumes are published. Of the former there is a second edition, enlarged. Huber's Tour shall be published, provided I can raise the money by next autumn. He has certainly collected many plants, as well as rare ones, but no new species. His harvest in this journey amounts to 400.

Mr. Cliffort has sent me his *Hortus*, for which I am obliged to you. There is no reason you should deprive yourself of a copy of this most beautiful book, which you can well bestow on some other friend. Van Royen has sent me his *Flora*. It is entirely your own. I have not yet seen the *Flora Virginica*. Siegesbeck is done for. The plants of Amman are rare, and sufficiently well described and figured. Some of them seem nearly related to those of my country. The work of Breynius has not fallen in my way, nor the posthumous publications of Plumier. Dillenius writes word that he is about publishing his *Sylva Muscorum*, of which 36 plates are already executed, so that we may hope soon to have all our difficulties, in this tribe, removed. Targioni continues to threaten us with the publication of Micheli's second volume, for which purpose he certainly has money enough. No doubt Micheli has great merit; but his work contains many trifling things, and too many varieties, especially among the

Fungi. Of this family my drawings now amount to 300, or more; but I think they continue to increase. I have, with peculiar pleasure, observed an Agaric, which confirms your observation, being of a hard, and truly Agaric-like, substance, but in other respects a true *Fungus*; as also another doubtful kind, growing on wood. Staehelin has lately made some new discoveries. I cannot attend to botany in winter, because of anatomy. Formerly I used to study it, in my own country, where the winters are often very mild. The Mosses flourish in January and February, *Marsilea* in March and April, the *Pezizæ* in December. Your Mosses, and every thing that you send, always give me great pleasure. Farewell! and may you ever love me as you are used to do!

In England lately, a lady, of the name of Blackwell, has published a Herbal, consisting of 500 copper plates, exhibiting original figures of plants, for the most part very common.

Göttingen, Sept. 26, 1739.

You appear not to have received a letter I wrote you some months ago, and I therefore trouble you with this, to give you a concise account of my journey to the Alps, and a few other matters. Know, therefore, that in June and July last I visited various districts of the Jura, and of more alpine regions, not without success. I discovered a new kind of *Astragalus*, with pointed silky leaves, and a large blue flower *, different from my beautiful Swiss one †. I gathered the *Knawel alpinum* ‡, which bears, at each leaf of its calyx, a yellow crescent-shaped petal, or nectary. The stamens are ten. A sort of *Mnium*, whose hollow fruit-stalk gradually dilates itself into a cup §. I saw the *Tozzia*, which you have rightly referred to your *Didynamia*; and *Tribuloides (Trapa)*, which is no less certainly tetandrous. I also met with *Limodorum austriacum*, *Salvinia*, and *Dipsacus laciniatus*, which is distinct from the common one. On the whole, I saw about 300 of the more rare plants, of which I mean to publish a catalogue and history. From this journey of mine, and Professor Huber's two expeditions, about 60 species will be added to my work. I am now turning my attention to it, and am busy with the *Didynamia Gymnospermia*; among which I am particularly pleased with many of your alterations, though in a few instances, as must sometimes happen, we differ. The *Umbelliferæ*, *Gymnopolyspermæ*, Grasses and Trees remain to be done. These last I shall mix with other plants, after the opinion of the great Jungius and yourself. Indeed I meet with many very striking affinities between

* *A. uralensis.* *Linn. Sp. Pl.* 1071.

† *Hedysarum obscurum.* *Linn. Sp. Pl.* 1057.

‡ *Cherleria.*

§ Could this be *Blasia*, now proved by Professor Hooker to be a *Jungermannia?* See his admirable Monograph, t. 82, and *Engl. Bot. t.* 1328.—See also Linnæus's letter, p. 349.

trees and herbs, as between *Fagus* and *Xanthium*, *Taxus* and *Equisetum*. I wish to arrange my *Synopsis* according to a natural method as much as possible. Eleven plates are engraved, six of which are occupied by plants of the Thistle kind, particularly some *Cirsia*, either new or not well understood. I have sown my garden this year, whence I have had opportunities of observing the characters of numerous plants. All those from Surinam have perished, owing to the ignorance of my gardener. I advance in my knowledge of *Fungi*, of which I have collected much above 300. But many of this tribe prove mere varieties, as I shall endeavour to point out.

What is new in the botanic world you know better than I, such as Plumier's posthumous figures, Burmann's ten *Decades*, and the Russian plants of Amman. Pray let me know what you, the most indefatigable of men, are now doing. I pass away my time in anatomical and botanical pursuits. I have opened my new theatre; I have commenced the publication of the Commentaries on Boerhaave's *Institutiones;* I have been writing on the *Allantois*, *Corpus luteum*, and vessels of the heart, as well as an account of my Hercynian journey. In the course of the year 1740, if it please God, I hope to finish my *Flora*, which will doubtless comprehend above 2700 plants, omitting varieties of colour. I have made almost a complete collection of synonyms to the whole. I have acquired several rare books, such as Columna. I have also been employed in overlooking my draughtsman, who is undoubtedly an excellent one. I shall endeavour so to conduct myself as to preserve your good opinion and regard. Farewell!

Göttingen, Aug. 25, 1740.

Though I have long been without an answer to my letters, I am rather willing to attribute your silence to some unfortunate accident, or to your numerous occupations, than to any neglect, or failure in your esteem for me. I therefore repeat my entreaties for your correspondence, and send you two of my Itineraries, in which, if I now and then differ from you, it is only in matters of very little moment. I have sent to press the beginning of my *Synopsis* of Swiss plants, where the *Fungi* are described, a mutable and treacherous tribe. Since I began to print, I have met with a few more species, and particularly a beautiful red kind of *Trichosphæra (Trichia?)* with a stalk; its cortical hairs disposed in elegant convolutions, among which the seminal dust is lodged; also a soft though woody sponge, which seems to be in a young state; and a species of the *Coralloides.* This autumn, if God grants me strength, I wish to complete the nomenclature of the *Musci* and *Lichenes,* upon which the admirable work of Dillenius is now printing. I shall give figures of about 20 species of this family. Almost all my plates are ready.

With respect to Grasses, which I have recently been looking over, your genera please me very much. The worthy Scheuchzer certainly cannot be followed, on account of his varieties without number made into species; his scattering the Loliaceous grasses into four or more different classes; his spurious and ill-defined tribe of the *heteromalla;* and his too numerous divisions, founded on density, laxity, beauty, and such uncertain characters. His descriptions themselves, being so very diffuse, are altogether useless, for they want specific differences, which, with all the pains I can take, I often cannot elicit from them. The greatest difficulty attends the correct establishment of genera, throughout the whole vegetable kingdom, which may well be the case, as it requires a perfect knowledge of the characters of all the species in the world in one person's mind. You have undoubtedly set us an excellent example. I cannot presume to enter on such a task in my *Flora*, nor have I before me a sufficient number of characters. My aim will be to establish the species of plants, so that we may be a little more certain which are truly distinct and which are not.

You, who have access to the maritime and mossy regions of the North, may assist me essentially, provided your various business will permit, in sending me dried specimens of many little plants, as also seeds of various wild species. You shall find me not ungrateful, nor unwilling to make a similar return.

I have long ago offered you my minerals. Whenever you desire it, I will send you my whole collection, which is tolerably rich in the productions of Hercynia and Saxony.

I know not whether you look to me for botanical news, but I have none to communicate. I have not yet even seen Gronovius's *Flora Virginica.* The Germans care for nothing but money, nor have they much of it. They have no gardens, that of Heister excepted, and you know how that is managed. My own collection of plants is small. I am not assisted by the publick, nor can I put myself to any expense. I preserve an account of the characters of whatever I cultivate.

I wish we could have an European *Flora* written upon your principles. As to an universal System, it seems hardly to be hoped for, except from some man to whom every botanist would communicate his whole stock of observations, and all his dried specimens. Dillenius has great advantages of this kind. I wish he might accomplish something before he dies! The French, so rich in materials, do nothing. Why is this? Much has been said about the posthumous works of Plumier; but on enquiry I find nothing has yet been done with them. We are led to expect from Italy something more of Micheli's, who was an industrious man, though a prodigious multiplier of varieties, and not absolutely correct in his elegant characters. In *Hepatica, Marchantia,* the Mosses, &c. he frequently mistakes buds *(gemmæ)* for capsules of seeds.

I am not able to add more at present. A fatal child-birth deprived me, about two months since, of my wife, who was endeared to me by her manners, her accomplishments, and her connexions. God still reserves me for whatever may be his pleasure.

Adieu! May you long live happy with your *Moræa,* and enjoy deserved fame! But may the Supreme Governor of all things teach you, as well as me, that there is nothing in this uncertain state which can shield us against the terrors of an approaching and inevitable eternity; fame, riches, and the dearest attachments, are of no avail; nor any thing else but the Divine favour.

LINNÆUS TO HALLER.

Stockholm, Sept. 15, 1740.

Professor Rudbeck, at Upsal, having died this year, I was desirous of succeeding him. My whole summer has been devoted, though in vain, to this object, or I should certainly have paid you a visit. It would have been suitable to my wishes to have removed to Upsal, where I should entirely have given up the practice of medicine, and devoted myself exclusively to botany; but fate has decreed otherwise.

At last I have got your Travels in Hercynia, as well as in Switzerland, and your Lectures upon Boerhaave. They are all so tempting, that I know not which to read first. They are now in the bookbinder's hands; but I will soon write you a long letter. I doubtless learn much from you at all times. I long to have your *Synopsis* published. As soon as I have seen that, and Dillenius's book on Mosses, I shall send my *Species Plantarum* to the press. I regret that I have not yet been able to obtain your works on the *Pediculares* and *Veronicæ*. Have you published any thing more upon botany?

I lament the fate of Joseph de Jussieu, who in returning from Peru on board a ship which attacked an English vessel, was killed by a cannon-shot, as my friends at Montpellier inform me *.

Your *Cherleria* is a very pretty genus. I wish you could inclose a bit of it in a letter, that I might investigate the flowers. As far as I understand from your description, this plant seems of an intermediate character between *Sedum* and the *Lychnides* †. The nectaries evince a manifest affinity to the *Sedum* family; but you represent the fruit, or capsule, as of a simple form; not three distinct capsules, but one, with three styles. I long to see it.

I wonder you did not observe in the *Tribuloides (Trapa)* those other four stamens without anthers, or may be nectaries. I know not, by dry specimens, what they are. I have seen the *Limodorum* ‡, but not the *Salvinia* §. *Blasia* grows very plenti-

* This was not true. Haller justly records, that he remained in Peru, where he died.

† It is altogether and exclusively one of the order of *Caryophylleæ*.

‡ *Orchis abortiva. Linn. Sp. Pl.* 1336.

§ *Marsilea natans. Ibid.* 1562.

fully here. Are Plumier's posthumous works come out? I have Burmann's *Decades*, and Amman's Russian Plants.

During the summer I have been searching for insects, not finding any new plants. I have dedicated my new edition of the *Fundamenta Botanica* to Dillenius, yourself, Van Royen, Gronovius, the Jussieus, Burmann, and Amman, placing your names in this order, which seems to me the just one, you being all the most eminent botanists of your time.

I have been describing the Gold fish of the Chinese, the Alpine Sparrow * of Lapland, the Norway Mice † of Wormius, and the *Picus tridactylus* of Dalecarlia. I have also published my *Systema Naturæ* in an octavo form, where many things are corrected.

What I have collected on the subject of Dietetics, you shall one day see. I have been for ten years intent upon it.

You esteem the *Melampyrum* with a small gaping corolla to be absolutely distinct from the common species. I wish you would raise it from seed ‡.

Pray send me also, if possible, in a letter, your *Pyrola rotundifolia, staminibus pistillisque erectis* §.

* *Emberiza nivalis. Linn. Syst. Nat. v.* I. 308.

† *Mus Lemmus. Ibid.* 80.

‡ "They are different." *Haller.* Undoubtedly so, as Linnæus makes them in *Sp. Pl.* 843.

§ "I never saw it." *Haller.*

I am sorry I find it so difficult to send you any thing. If I live, I will procure for you Rudbeck's *Campi Elysii*, which I am certain you have never seen. This work is full of figures. Scarcely ten copies are extant in the whole world. I have but one.

If I had an opportunity, I would send you Browallius's *Examen Epicriseos Siegesbeckii*. I have never been able to meet with the *Primitiæ* of the latter. I would give a gold ducat for this little book, if it came in my way.

As soon as I get your Travels from the bookbinder, I will run them over, and freely point out whatever seems to me doubtful. The more errors of my own that you can point out, the more I shall be obliged to you. By such means I may be enabled to correct all that is wrong before I die, for no one can amend his own works in the grave.

It is certain that I cannot, as yet, understand the *Equisetum*, or whether the supposed seeds are really such, or dust of the stamens. Their configuration is, unquestionably, very like the spike of the Yew, *Taxus*. Have these seeds been sown, and found to vegetate? If they are seed-vessels, where are the stamens? If they are stamens, where are the seeds? Pray instruct me on these points.

Blasia, as you know, has an erect abrupt tube, beautifully represented in Micheli, seated on a pervious capsule, from whence numerous seeds are discharged, whenever the tube is filled with water, or with nocturnal dews. These seeds are visible to the naked eye. I have seen likewise those tubercles within the substance of the leaf, which Micheli takes to be capsules, but I have no seeds from them. Be so good as to send me the *Salvinia* in a letter.

Farewell! &c.

HALLER TO LINNÆUS.

Göttingen, Sept. 19, 1740.

Your letter, after so long a silence, proved extremely welcome, and would have been indeed perfectly satisfactory, but for the unwelcome intelligence that you are disappointed of the situation you have so long aspired to, and so well deserved. This loss, however, will, I trust, be soon repaired by the accession of some other honour*, though botanical distinctions are among the most rare. I was formerly in correspondence with Dr. Rosen, and was associated with him in the promotion of a society, which at that time flourished; but there was never a word of botany passed between us.

I wish you would read my little publications, and then give me your opinion upon them. The Lectures on Boerhaave were composed too hastily, scarcely more than four winter months having been bestowed on each volume, and those perpetually interrupted by anatomical pursuits. I have already dissected eleven bodies, since this winter began. My account of my Journey is also somewhat unfinished; but Greece was lost by delay. All my collection of *Cryptogamia* is at Leyden. In spring I

* Linnæus soon after obtained the professorship in question.

shall undertake the apetalous classes; and, by God's assistance, all the work will be finished before next winter. I have also got through half the third volume of the Lectures.

No copies of the *Veronicæ* remain; but if I meet with a conveyance, you shall have the *Pediculares*. Inclosed is a little branch of *Cherleria*. It is akin to the *Alsine fugax* *, in habit, and structure of the tube (pistil), but differs in its quinary number of parts, and in the dry petals, to which the stamens are attached.

I will send the *Salvinia* with the books. I never saw either *Blasia* or *Valisneria*. Nothing is now said about publishing the remains of Plumier.

Your memorial of me in your *Fundamenta* is very kind. I have received that work. The species of *Melampyrum*, as you are aware, do not grow readily from seed; but I will make the attempt. Both the species in question grow together, so the soil cannot cause any difference. But there are several plants with a larger flower than ordinary, in whose leaves there is no alteration; and these I mean to treat upon, such as the various kinds of *Thymus*, *Calamintha*, &c.

Pyrola has indubitably no erect stamens; they are always irregular in posture. The style indeed is perfectly straight. No plant is more common here.

Huber's Tour will never be published. He has given up botany, by which nothing is to be gotten. I never had a sight of Rudbeck's book. No acqui-

* *Sagina procumbens. Linn. Sp. Pl.* 185.

sition could be more desirable. Gleditsch also has written against the contemptible Siegesbeck, but he does not abound with observations. Browallius will be very acceptable.

The *Equisetum* in which I have observed the elastic filaments, is that common field species, whose flowering stalks have no leaves*. The capsular part of the fructification I take to be the male, amongst which the females are intermixed, as in Ferns.

I know scarcely any botanical news. You have seen a specimen of Dillenius's History of Mosses. The Germans publish nothing, except a miserable description of *Phellandrium*, and another of *Piper* by Heister, which last I have not read. Neither have I seen Seguier's *Bibliotheca Botanica*. The thirteenth volume of the *Phytanthozaiconographia*† is come out, and is somewhat better than the preceding ones, but still bad, and unworthy of its pompous appearance.

I have prodigiously increased my figures of *Fungi*, and have established some new genera, distinguishing the tribes of *Embolus*, *Spœrocephalus*, and *Agaricus*. Four plates only remain to be engraved, in which are near 30 new Mosses, much better drawn than the former ones, with magnified portions annexed. I have not more than four plants besides to engrave. I have drawn up their characters in my garden, but have had no time to re-examine them. I hope to raise about a thousand plants this year, and that my garden will go on

* *E. arvense*. *Linn. Sp. Pl.* 1516.

† By Weinmann.

perpetually increasing. My kind friend Van Royen has promised many additions. If I am able, I propose again to visit the forests of Hercynia, chiefly in pursuit of Mosses and seeds. The place where I am is but a barren field for botany, excepting *Fungi*, which are very plentiful. I have detected a very curious elastic motion in the common sessile *Peziza*, of a dirty white hue. The whole plant contracts spontaneously, and discharges a powder upwards, with a sort of hissing noise. This doubtless is the seed. The *Clathroides* are also elastic, and their dust is expelled spontaneously, though more slowly. I have discovered no flowers of *Fungi*.

This I believe is all I have to say at present. Wishing you and your lady every felicity, I remain, &c.

LINNÆUS TO HALLER.

Upsal, July 26, 1742.

The longer, my valued friend, I am deprived of your letters, the more earnestly I wish for them. I am anxious to hear of your continuance in health, engaged as you are in so very laborious a life.

I have found in this country a species of *Sedum* *, not extremely common, yet occurring on rocks, stones, and hills every where hereabouts, resembling

* *S. annuum. Linn. Sp. Pl.* 620, but not Ray's plant, which is *S. anglicum* of Hudson, and of *Fl. Brit.* 486.

the ordinary acrid kind in appearance, as well as in flowers, but differing in the following particulars:

1. It is annual, and rather smaller than any other species.

2. It is erect, not procumbent, nor much branched.

3. The leaves are besprinkled with unequal reddish spots.

The stamens are ten. Flower yellow, of five petals.

What do authors call this? Is it the plant of Ray's *Synopsis*, tab. 12, f. 2? The figure is good, but the flowers in mine are yellow, as in other species.

We have another *Sedum* *, exactly like the very common *S. acre*, in its whole aspect, size, colour, and structure; but the leaves are more crowded, and imbricated as it were in six rows, or longitudinal series. Is this the *S. minimum luteum non acre* of John Bauhin? I cannot find any author who properly distinguishes it.

Where does Staehelin reside? and where Emanuel Bauhin? Is the latter a botanist? What day did Professor Amman die?

We have now flowering, in the garden of the University, a plant brought by myself, last summer, from the island of *Stora Carlsoen*, which is a species of *Saponaria* †. It is *Caryophyllus saxatilis, foliis*

* *S. sexangulare. Linn. Sp. Pl.* 620.

† *Gypsophila fastigiata. Linn. Sp. Pl.* 582; see *Fl. Suec.* 145.

gramineis, umbellatis corymbis. Bauh. Pin. 211. *Symphytum petræum. Thal. Hercyn.* 113. It is also exactly *Polygonum majus erectum angustifolium, floribus candidis. Mentz. Pugill. t.* 2, *f.* 2. I wish to know how this differs from *Lychnis alpina linifolia multiflora, peramplâ radice,* which you have just mentioned only in your *It. Helvet. sect.* 80 *. Scheuchzer's *It. Alp.* 137. answers exactly, only he does not notice the rather fleshy leaves, the procumbent stems, nor the undivided petals.

Here is likewise in flower, from seeds sent by Amman, *Helleborus fumariæ foliis, Amm. Ruth.* 74. *t.* 12 †. I know not whether to distinguish this from *Aquilegia foliis thalictri, flosculis minutissimis albis, Mentz. Pugill. t.* 8, *f.* 1, which is *Thalictrum montanum præcox* of Tournefort, 271, and *Ranunculus præcox alter, thalictri folio, Clus. Hist. v.* 1. 233 ‡. But mine is much smaller than the plant of Clusius, though in form the same. My flowers are green; his white. If you have his plant, send me a morsel, or a leaflet; for I have but a single specimen of mine. Thus far is most certain — that my plant cannot be either an *Aquilegia, Helleborus,* or *Thalictrum.*

The following also are growing, but have not yet flowered: *Androsace* of Amman, *p.* 13 *(septentrio-*

* Haller, by one of those typographical errors so abundant through all his works, and especially in these letters, has *n.* 20 for *sect.* 80.

† *Isopyrum fumarioides. Linn. Sp. Pl.* 783.

‡ *I. thalictroides. ibid.*

nalis, Linn.); *Delphinium perenne, Aconiti folio,* &c. of the same author, *p.* 131 *(D. elatum)*; and *D. elatius subincanum perenne,* &c. *p.* 132 *(D. grandiflorum)*.

If you can conveniently send me any seeds, I beg earnestly that you will. I do not ask for rare exotic or Indian seeds, but indigenous ones, especially from the Alps; or of those plants you have brought from your Swiss or Hercynian expeditions. Even the natives of Germany, which do not grow in Upland, will be welcome. You may see a list of the wild plants of Upland, published by Celsius, in our Upsal Transactions.

[*Here follows a catalogue of desiderata, of no moment at present.*]

Seeds of any of these will be acceptable. I can hardly believe that you will "send me empty away," after your former favours so often repeated.

I have introduced 200 native plants of Sweden into the garden, and of these 20 are Grasses, including *Carices* and *Junci*. From this stock I shall be able to supply my correspondents with our Swedish productions.

I am teaching botany to one of the medical students, and as soon as he is competent, I intend sending him, at my own expense, to the mountains of Lapland, to collect plants for our garden. I have dispatched him this year, by way of trial, to the neighbouring country of Norway.

HALLER TO LINNÆUS.

Göttingen, Sept. 16, 1742.

After so long a silence, your last letter proved peculiarly welcome. It is more than a year since I wrote to you, with a few little dissertations. I wish you had any friend at Hamburgh who would take charge of any thing for you, as Koenig the Swedish consul, for instance. I would send any thing in my power. Seeds are not easy to procure; but I will send them, or something else, the fruit of my late excursions.

Your botanical remarks are always satisfactory. Favour me with your two *Sedums*. We have a yellow kind which is not acrid, and differs besides from *S. acre* in having more lax foliage. The *Symphytum petræum* * will be extremely acceptable. I have one of Caspar Bauhin's specimens, but it is widely different from the *Lychnis alpina linifolia et multiflora*, having a white flower, more turgid calyx, &c.

I have lately made a great collection of plants at Jena, preparatory to my edition of Ruppius, and I expect some seeds. I had 500 different sorts from Switzerland, but scarcely ten of them came up. Among those which succeeded was *Eruca cœrulea in arenosis crescens*, C. B.† I observed in the garden it had four glands. I have lately gathered the *Hyssopifolia*‡, which is very like the *Salicaria*,

* Of Thalius; see the preceding letter.

† *Sisymbrium arenosum. Linn. Sp. Pl.* 919.

‡ *Lythrum hyssopifolium.*

but has only six stamens. I found also a *Vicia*, whose seeds are here inclosed, a *Hesperis*, and other things.

Concerning the plants of which you ask me for seeds, some of them are easily to be had, but others are very difficult, though common in Switzerland. Nobody is there at the proper time for gathering them. I will do what I can, if you point out a way to transmit them. My passion is almost exclusively for European plants, allied to our own. Pray send me any of the northern species that are not to be found here. You will know what these are by the botanical work * which I lately sent you by Mr. Koenig, as a pledge of my unabated friendship. There are too many typographical errors, nor am I every where satisfied with myself. But time and repeated journeys will correct many of my mistakes.

Are you acquainted with the *Filicastrum septentrionale* of Amman †? I have it, but the flowering stems do not expand so well here as in Thuringia, where I found it. Celsius on the plants of Upland I have not; nor have I more than a single volume of your Upsal Transactions. I should be very glad to procure some Swedish publications, for which I would make a return, either in other books or in money. I want every thing of Rudbeck's, but his

* This must have been the *Enumeratio methodica stirpium Helvetiæ*, an inestimable work, by no means superseded by the second edition, or *Historia*.

† *Osmunda Struthiopteris*. *Linn. Sp. Pl.* 1522.

Lapponia illustrata; also Tillands, Frankenius, and all the rest except Bromelius. Dried plants and Mosses are likewise very desirable to me. I can more easily make a return in these than in seeds, and can send you all the plants you ask for, except only *Valeriana celtica*, *Dipsacus laciniatus*, *Passerina* * of Tragus, and the *Centunculus*, of all which my specimens are glued upon paper. I can furnish numerous rarities besides. In German plants I am very rich, such as *Coronilla montana* †, *Anemone tertia Matthioli* ‡, *Lampsana aphyllocaulon* §, *Gramen paniculatum molle* ||, &c. I have no new books on botany, but Rand's Catalogue. The new edition of your *Genera* I have not got. Dillenius on Mosses is a most learned work. Mappi has numerous varieties, but several good and new things nevertheless. What are you now employed about? I wish you would give us a Northern *Flora*. I am working at that of Germany, and propose next year to go to Halle, and to visit the salt lakes. In 1744, if I live, I will examine the Rhætian alps, not cursorily, but making some stay among them. I shall thus provide entertainment for future confinement and infirmity, which I foresee that my asthma may bring upon me. Adieu! Preserve your friendship for me, by whom you are indeed highly esteemed.

* *Stellera Passerina. Linn. Sp. Pl.* 512.

† Of Rivinus, confounded with *C. coronata* of Linnæus.

‡ *A. sylvestris. Linn. Sp. Pl.* 761.

§ *Chondrilla juncea. Ibid.* 1120?

|| *Holcus lanatus. Ibid.* 1485.

LINNÆUS TO HALLER.

Upsal, July 18, 1743.

I have delayed, longer than I ought, to write to you, my valued friend, expecting every day to receive the book you were so kind as to send me last year. Growing impatient of delay, I ordered the booksellers to get it for me at any price, but without success. At last, three days ago, came this magnificent and most valuable present, more welcome to me than any other treasure. Accept my most grateful acknowledgments for this book, which strikes me with astonishment, for I can scarcely conceive the possibility of your composing and writing so many works, from your own observations.

Whilst I see you chastise, with your correct pen, every individual botanist according to his deserts, I cannot but repeat my gratitude to you, for dealing so gently with me, and evincing so much of your usual partiality. I can easily see how often you might have reproved me, if you would. As to the work itself, it will last for ever, being entirely founded on your own observations. Would that we had more such *Floras*, finished with the same care!

Permit me to make a few enquiries and remarks.

Page 151. You say in the generic character of *Salix*, "stamens more frequently three than two." Pray point out the species in which this is the case. Certainly not in any about Upsal, besides *S. pentandra*. Perhaps this circumstance is more variable with you than with us, or possibly your species are different from ours *.

Page 343. *Rosa* and *Rubus* but slightly differ in the structure of their fructification, except the seeds of the latter being clothed with pulp, which is not the case with *Rosa*. If you open the fruit of *Rosa*, you may observe naked seeds lodged in a fleshy calyx, whose mouth is contracted, just as in *Agrimonia*, only the seeds of *Agrimonia* are fewer. If the petals of this last are removed, it becomes an *Alchemilla*. The following reasons have led me to join *Alchemilla* with *Potentilla* and its allies :

1. Because its calyx has a double set of segments. —2. The insertion of the stamens is at the margin of the mouth of the flower.—3. The style springs, not from the summit, but from the side, or base, of the germen.—4. The leaves are alternate and rough.

Page 364. You place *Oxys (Oxalis)* next to *Geranium*. You are certainly right, and I shall alter my *Fragmenta* accordingly. I have learnt this from you ; and whoever has ever looked at the African *Gerania* can have no doubts about the matter ; but mortals are blind !

Page 388. You attribute ten stamens to your ninth species of *Alsine*, but only three pistils †.

* "It happens in *triandra*, as well as in *alba* and *vitellina*." *Haller*. Such a diversity in the number of the stamens chiefly occurs in a few of the arboreous Willows, but in others it is hardly ever met with.

† *Spergula nodosa*. *Linn. Sp. Pl.* 630. The styles are always five.

Pray examine this again. Your plant must be different from ours, or you will find five pistils. Ours never varies.

Page 391. *Alsine, n.* 20 *, "has four stamens." Right. I have corrected my new edition.

Page 605. As to the *Fumaria bulbosa* of Ludwig and Moehring, I should be glad to know what reasons are given by the latter for making two species. I see that the root of one plant is hollow, and the floral leaves undivided; but there is no difference of structure in any other part, even the most minute organs of the fructification. Be so good as to examine *F. bulbosa, radice non cava, minor* of C. Bauhin †, which you have made a variety of the second species. This is found in all our woods, but its root is solid, and the floral leaves undivided, never cut. Pray examine this plant, and tell me your opinion. The question with me is, whether to establish three species or one; for it seems clear that all the bulbous Fumitories cannot be ranged under two ‡.

Page 129. You have the *Pilularia* as one and the same species with *Lens palustris quadrifolia (Marsilea).* I have long known them to be one genus, but different in species. Is Staehelin altogether right in this opinion. Why is not the latter found at Paris, where *Pilularia* is extremely common § ?

* *Sagina procumbens. Linn. Sp. Pl.* 185.

† *F. solida. Eng. Bot. t.* 1471.

‡ "They differ in the flower." *Haller.*

§ "The mistake was Staehelin's." *Haller.*

You have found out a new world among the *Fungi*. You have shown the way through this wilderness, where no one before you could find any certain path. Dillenius and Micheli are nothing in *Fungi*. We owe every thing here to Haller, who has accomplished a work of immense labour.

Saxifraga petalis latissimis luteis lineatis, tab. 8, we have also in Sweden; and I have always taken it for the plant of Breynius *.

Astragalus, tab. 13 †, I have found in our island of Oeland. All your figures are truly splendid.

Betula nana, p. 158. I wonder at your meeting with this in Switzerland. An academical dissertation of Mr. Klase, under my presidency, came out in June last, upon this same plant ‡.

What is the *Persicaria acida* of Jungermann § ?

How can I procure Staehelin's specimens of botanical anatomy?

Whilst I am writing, my colleague Dr. Rosen, physician to the king, comes and tells me of your death having taken place in April last! I hesitate, in the greatest anxiety and consternation, whether or not to send my letter. If death had deprived me of a parent, or wife, or only son, I could hardly feel more! If the news be true, I know not how I could escape seeing it in the publick papers. I hope for better things!

* "Right." *Haller. S. Hirculus. Linn. Sp. Pl.* 576.

† *A. campestris. Ibid.* 1072.

‡ "It occurs but rarely in Switzerland." *Haller.*

§ "The *amphibia.*" *Haller. Polygonum amphibium. Linn. Sp. Pl.* 517.

I wish there were any means of sending you some Swedish books, which I much want to do.

Just now I am busy removing my family and domestic establishment into the new buildings at the Academical garden.

The last part of the hot-houses is now built, so that, as all the buildings, belonging to the garden, will this year be completed, I shall be able, next season, to take a journey to West Gothland, which I have for some time projected. I have finished my account of the Swedish Animals *(Fauna Suecica)*, which, God willing, I mean to print in the spring.

I am desired, by the Royal Society of Upsal, to inform you, that the communication sent by you, about the year 1732, has been lost, owing to the negligence of a temporary secretary, in the absence of Andrew Celsius; and that the Society is very desirous to have something of yours for its Transactions, you being richer in materials than any other man of your profession.

Will you be so kind as to forward the inclosed to Mr. Moehring? I know not how it happens that I never have any answer from him.

Farewell! and believe me ever, &c.

HALLER TO LINNÆUS.

Göttingen, Aug. 25, 1743.

I am still living, my excellent Linnæus, and as much attached to you by esteem and affection as ever; nor will I permit any body, for a trifling matter, to cavil at you, who have laboured with so much assiduity and success in the study of Nature; far less would I ever undervalue you myself. I have indeed, last winter, laboured under a nearly fatal disease, the white miliary fever; but am now perfectly well.

I am happy that my book meets your approbation. I have not yet heard from Dillenius, to whom I sent a copy. My labour has, I allow, been something, especially in consulting books, and in travelling over the Alps; but any person who knows what you have done, will not think much of my performances. My defects may be pardoned in consideration of my limited time, more than half of which was devoted to anatomy. The practice of medicine also, at Berne, was very burthensome. Hence I was obliged to put forth many things in an imperfect state, rather than lose the whole of my labour, which might have been the case this winter, in the event of my death.

I live here in a bad place for collecting plants, there being scarcely any but the most common hereabouts; but I will try. I have been from home twice this summer on account of botany. In the county of Stolberg I met with *Leucojum vernum angustifolium* * of Thalius, differing, in its large flowers, from *Turritis leucoji folio* †; also a new

* *Cheiranthus erysimoides. Linn. Sp. Pl.* 923.

† *Erysimum cheiranthoides. Ibid.*

or rare species of *Galium*, and the *Digitalis lutea*, uncommon in Germany. — At Lubeck, *Tripolium maritimum & flavum* *, with other things, grow about the salt springs. At Grubenhagen, *Helleborus niger, flore viridi* †, and *Cytisus alpinus racemosus pendulus* ‡. I have introduced a few remarks into the Catalogue of my little garden, which accompanies this.

In reply to your remarks on *Salices*, I assure you they have mostly three stamens here. I find I have noted this in the *alba* frequently; in the common reddish Willow *(triandra)* constantly; in some others two; but I have not counted these organs in all. I allow *Rosa* and *Rubus* to be nearly allied. I should not willingly have referred *Alchemilla* to the Cinquefoils, on account of the want of petals, and its proximity to *Percepier (Aphanes)*. The character of *Alsine, n.* 9, was written in 1732, and I have not examined it since. There may be a mistake. The *Fumaria solida* and *bulbosa* differ even in the flowers, as you will see in my book, the scales (calyx-leaves) of the former being directed towards the spur. The observation on *Pilularia* is not my own. The figure I now send you shows some grassy leaves on the same plant with those that have four leaflets, and there could hardly be so great a resemblance in the fruit, if the plants were

* *Aster Tripolium*, and perhaps *Inula crithmoides*.

† *Helleborus viridis. Linn. Sp. Pl.* 784.

‡ *C. Laburnum. Ibid.* 1041.

different. Possibly the soil of Paris may not be favourable to the evolution of these leaflets. I never gathered the plant.

The *Saxifraga* of Breynius (*tab.* 48) has smaller flowers than ours, yet too large perhaps for another of our alpine species. The former is extremely rare in Switzerland *. *Betula nana* is indubitably a Swiss plant. I have carefully compared our specimens with your figure. *Persicaria acida* of Jungermann is founded on specimens of *amphibia* growing out of the water, with erect stems. The leaves are somewhat hairy. It never flowers.

I know of no mode of conveyance for the books you mention, and which I so much wish for, except through Mr. Koenig, who transacts the business of your government at Hamburgh. I do not doubt of his attention to this matter, as I sent my book to you through his hands. I beg you to send all the botanical books you can, except Bromelius, but especially plants. I have marked down some dissertations that I want. I shall be happy to serve you in any way in my power. I love plants even to enthusiasm; perhaps too much!

The Upsal Academy shall have some of my anatomical observations, which I have not time now to prepare, but I hope to fulfil my promise this winter. Your letter is forwarded to Mr. Moehring. He lives at Jever in Friseland. I have no news. Gagnebin the surgeon is writing a Nomenclator of

* The late Mr. Davall met with this plant, *S. Hirculus*, abundantly in a bog on mount Suchet, near Orbe.

the plants of Neufchatel, and has met here and there with some rare plants. He is very commodiously situated there, in a mountainous and boggy country.

My compliments to Professor Rosen. Wishing you all health and happiness, I am, &c.

LINNÆUS TO HALLER.

Upsal, Jan. 12, 1744.

I heard of your being alive and well by your letter, which found me at Stockholm a week ago, and which was accompanied by the *Hortus Göttingensis*. When I saw your letter and your name, my heart leaped for joy. We had all deeply regretted your loss, and I above every one. Our Royal Society had enjoined me to write a history of your life for our Transactions; from which task I am now, thank God! relieved.

I mentioned in my last the *Sonchus arborescens palustris*, which you take for a variety of the common field *Sonchus* *. I have compared the plants, and felt no small surprise. You are right; but no mortal ever made this discovery before.

I send a few seeds, such as I have. Another time I will send more. Horticulture is a new occupation to me. I hope you will receive them with indul-

* These, the *S. palustris* and *arvensis*, *Sp. Pl.* 1116, are now allowed to be distinct species. Curtis observed that the former has not a creeping root.

gence. I have noted down, on the other side, what you have that I want, in case you should be able to furnish me with seeds of any of these plants, which would be most highly acceptable. If you can, pray transmit them as soon as possible, in a cover to the Royal Society at Upsal.

Santolina foliis rosmarini. An Silphium Royeni. Bellidioides. Doria. Ranunculus (what ?). *Veronica Chia. Echinops minor*, both species. *Verbena altissima.*

Any one of these would oblige me greatly, and I will make a return when I am sufficiently rich. Formerly I had plants but no money. Now what is money good for without plants?

Upsal, May 29, 1744.

A letter from our mutual friend Van Royen mentions your not having received any letter or answer from me for a long time. I, on my part, have been grieving that I had no answer from you; but I attributed this to your numerous engagements, which devote your whole time to the publick.

As soon as I received your great work on the plants of Switzerland, I made my due acknowledgments, at least as far as I was able, by letter. But as the report of your being removed from this mortal state was universal, I wrote to an anatomical friend, once your pupil, by whom I was informed that you were alive and well. At the same time I received your *Hortus Göttingensis*; upon which I addressed a letter to you directly, begging seeds of several plants, which are among the treasures of your garden. At the same time I wrote to Mr. Moehring, under your cover, in hopes of, one time or other, hearing from him, from whom I have not had a letter since my return to Sweden. But I have no answer from either of you. Meanwhile I profit every day by your writings; and the more I turn over your Swiss *Flora*, the more fond of it I become. It is doubtless the first work of its kind, and which no botanist can do without. Your short descriptions are worth a thousand of the long botanical histories; and you are very great in synonyms.

But pray let me know how you were able, in so short a time, to introduce so many plants into your garden. I have been intent on this object during the years 1742 and 1743, as well as the present season, without obtaining half so many, though I have been assisted by Gmelin, Ludwig, Gesner, Wagner, Bielke, Van Royen, Gronovius, Sauvages, Dillenius, Jussieu, Miller, and others. All the buildings, hot-houses, &c. of the garden have been completed these two years and half.

You rightly make the (Tartarian) *Fagopyrum* * distinct from the common kind and every other.

In my last I requested of you some communication for our Royal Society, to print in the Transactions. The volume for 1740 is published, and that for 1741 will speedily be completed. We do not mean to admit hereafter any thing but valuable

* *Polygonum tataricum*. *Linn. Sp. Pl.* 521.

treatises on medicine, natural philosophy, mathematics, and such important subjects; no theological nonsense*.

The death of Andrew Celsius, our secretary, is a great loss to the Society. Jussieu, and some other botanists, have become members.

I peruse, with great avidity, your Commentaries on Boerhaave. Pray do not fail to give us the remaining parts, on diet, which subject is a favourite one with me. I have collected more materials relating to it than any body I know, and shall add to my stock from your book upon Boerhaave.

If you have any thing curious about the Fig, be so good as to let me know, for I shall publish a dissertation upon that tree in August; with another, on the Corals of the Baltic.

My Travels in Oeland are printing, but in the Swedish language. They contain numerous economical remarks on plants, insects, and birds.

I gave a dissertation on *Betula nana* last year. I know not how to send it to you, situated as you are, far from the sea.

I perceive by your *Hortus* that I have some plants you have not, and will therefore send seeds of such next autumn, if your post will carry them.

The *Flora Suecica* is finished, all but some of the minute kinds of *Byssus*, *Mucor*, *Conferva*, and a few of the *Fungi*.

Farewell! and continue your regard for me.

* Linnæus, the most religious of men, could only mean, by this indignant expression, vain cavils, or doctrines supported from interest, or party spirit, not conviction.

HALLER TO LINNÆUS.

No date.

I have long since replied to your last letter, and have sent you what seeds I had ready, but fear it will be some time before you receive them, for the new-made Doctor to whom I entrusted my packet was not going by the direct road, but by a circuitous journey, to Lubeck. If you can contrive to have your Swedish resident at Hamburgh, Koenig, take charge of any thing for you, I will send to him whatever I have to communicate. I forwarded your letter to Moehring.

I am happy that my works do not altogether displease you, who are so great a judge in the science of which they treat. I love truth, and I am mindful of order in all my pursuits, but must confess I have not had by me so great a stock of plants as my undertaking required. I was unwilling to keep it any longer upon my hands, lest, in the mean while, I might die, as was very near happening. I am ready to acknowledge my errors and defects, of which I have corrected some in the *Flora* of Ruppius, now in the press.

My little garden is somewhat richer than the catalogue, nor are any of the plants, there mentioned, lost, except what are so indicated in the appendix. I have moreover recovered some of these. The main stock of my collection consists of native species, which I have procured from all the mountainous countries around: Most of my exotics were supplied by Van Royen, Ludwig, and Gmelin. What I had from other quarters seldom succeeded.

I have answered you respecting my paper for your Society. Shall it be accompanied with figures? If you wish otherwise, I could perpetually send you descriptions of new plants, or any other communications that might appear to me worth notice.

Boerhaave's work is finished. I have added nothing on diet, &c. lest the book should swell to too great a size, and I should enter upon matters to which I am least competent. We are in great expectation of your valuable remarks in the *Lachesis Lapponica* *. I have never examined the Fig. We have not many trees; but I will try what I can do. Your dissertations are highly acceptable. They may be left for me at Lubeck, to the care of Dr. Engenhagen, as well as the Swedish Transactions since 1738, to which period I have them. By the same channel I will send you Ruppius, and any of the seeds you want, that may ripen. All you send cannot but be welcome, especially the *Flora Suecica*. I am but just returned from my Hercynian journey. On the mountain called Altstolberg I found the *Symphytum petræum* of Thalius †, in the spot where he first discovered it. This comes near *Saponaria*. It is different from *Lychnis alpina linifolia multiflora, peramplâ radice* of Tournefort ‡, if my former notes are correct, its petals

* Published in English by the editor of these letters, in two vols. 8vo. 1811.

† *Gypsophila fastigiata*; see a letter of Linnæus, p. 354.

‡ *G. repens*.

being ovate, not heart-shaped, the whole corolla smaller, and the leaves not so much turned to one side. I also saw plenty of *Rubeola flore trifido* *; also *Allium montanum, narcissi folio, minus* †; and a species of *Porrum*, not noticed by late authors, with grassy leaves, a bulbous head, and alternate stamens three-cleft ‡; as well as a species of *Trago-selinum (Pimpinella)*, or perhaps a variety only, with very long leaves, and an elegant one nearly akin. I brought all these into my garden, but several of them died.

I mean to give a few figures in the *Flora Jenensis* of the more uncommon plants. The Hungarian *Hesperis* is surely the origin of the Dame's Violet, commonly so called; for when brought into a garden, the flowers become, like that, umbellate, and the stem erect. I will give a plate of it.

I know of no botanical works in hand, except that Heister is writing about Quinces in opposition to you. I begged of him not to quarrel with you, but could not prevail. Moehring has written I know not what against my Ferns, but he does not take my opinion rightly. I do not characterize the *Thelypteris* of Ruppius by the round spots being all parallel, but by the marginal seeds. Dillenius is silent. I have the posthumous works of Zanoni, which are useful for mentioning the places of growth, and contain many little-known European

* *Valantia muralis. Linn. Sp. Pl.* 1490.

† *A. angulosum. Ibid.* 430.

‡ *A. Scorodoprasum. Ibid.* 425.

species. Many of Gmelin's plants have flowered with me, as two kinds of *Delphinium* with long spurs; the yellow *Valeriana (sibirica)*; two *Cyani* with yellow flowers; and four or five *Jacobææ (Seneciones)*. Nothing is more delightful to me than such rarities. Could I but visit the South of France in search of plants! But I mean to go to my native Alps in 1745, to examine places I have not yet explored; amongst others the favoured Valais, where the *Xeranthemum* is as common as our daisies.

Farewell!

LINNÆUS TO HALLER.

Upsal, March 8, 1745.

I am afraid this letter, on account of its bulk, will cost you much for postage, but hope you will excuse me for this once, being both partial to the writer and rich, as a young traveller the other day informed me. I beg of you to notice this little dissertation in your Nuremberg *Commercia Literaria*, and I will, next summer, collect for you seeds of the plant there described *(Peloria)*. I beg of you not to suppose it any thing else than the offspring of *(Antirrhinum) Linaria*, which plant I well know. This new plant propagates itself by its own seed, and is therefore a new species, not existing from the beginning of the world; it is a new genus, never in being till now. It is a mule species in the vegetable kingdom, propagating itself, by the transmutation of one plant into another; a totally different fructification in the same plant; a two-fold character in one and the same species*.

I beseech you to collect for me a few seeds out of your garden or your stores, and among the rest *Dracocephalum n.* 5 of *Hort. Cliff. (canariense).*

The members of our Society eagerly look to you for the essay you have promised, whether you please to accompany it with a figure or not.

Your *Flora Jenensis* is not yet arrived. We long to see it. Adieu! &c.

HALLER TO LINNÆUS.

Göttingen, June 10, 1745.

I write now by the hands of Mr. Rosen, the brother of your colleague, who will take care to deliver my letter. I am by no means rich, though my devotion to literature makes me grudge no cost that can procure me knowledge. All therefore that you send, whether seeds or any thing else, is welcome, though it may be attended with some expense.

Your dissertation containing so wonderful an account (of the *Peloria)* has been mentioned by me in

* This is no better than botanical mysticism. The *Peloria* is now known to be a mere variety of *Antirrhinum Linaria*, characterised by having a regular pentandrous flower with five spurs, curious enough indeed, but not propagated by seed, nor absolutely constant on the same root. Other species of *Antirrhinum*, and various ringent plants besides, are liable to the same variation. Nature sports with her own formalities, which we mistake for her laws.

our literary gazette, and I shall send an account of it, as soon as I am able, to Nuremberg, to the care of my good friend Trew, recently created the director of your college.

Seeds from you are always highly welcome. I have not the *Dracocephalum* you mention. If I knew what seeds you want I would put them aside. I have already sent you what I had at the time. *Phlomis tatarica* is now in flower, with a *Chamæcerasus*, and other novelties, as *Campanula echii folio* *, *Symphytum Thalii* †, &c. I will try to be not altogether useless to you. I will certainly, this summer, send what I promised to your Society; but I am overwhelmed with business, which presses upon me since my recent return from a tour to my own country. I have sent you Ruppius's *Flora* by Mr. Gentzen, and hope to have in return a few Swedish dissertations on botany or anatomy.

Nothing, I believe, has been published here in your way, except a little book by More, upon the changes caused in our globe by volcanoes, to which he attributes the great quantities of marine bodies scattered over the surface of this earth. Ehrhard has a copy to sell. I have a parcel from him directed to you. Farewell!

* *C. thyrsoides. Linn. Sp. Pl.* 235.

† *Gypsophila fastigiata.*

LINNÆUS TO HALLER.

My ever-valued Friend, Upsal, Jan. 7, 1746.

I received, in due course, your letter of the 10th of June last.

I have never received the *Flora Jenensis* you sent me, but have purchased a copy. You have given a most beautiful figure of the *Saponaria (Gypsophila fastigiata)*; far preferable to that of Mentzelius in his *Pugillus*. This plant is abundant in *Stora Carlsoen*, as you may see by my *Flora Suecica, n.* 346, and *It. Gotland.* 282. It thrives well in the Upsal garden.

Your little book on *Allium* is come safe. The collection of synonyms is a monument of immense labour. If I had known of your intention, I might have communicated some species.

I now perceive you are preparing a great work, the whole *Flora* of Germany. The Herbarium of Burser would doubtless be of great use to you, as he walked over so many of the German states, and has annexed the place of growth to every specimen.

The *Flora Zeylanica* occupies me at present; for I have in my hands the Herbarium of Hermann himself, which he collected in Ceylon. I am reducing many species to their proper genera, and I find many genera that are new.

The members of our Society present their compliments, and desire me to ask for some paper of yours to enrich their Transactions, as you have often promised.

Letters from Paris, a few weeks since, informed me that you were returned to your own country of Switzerland, and had undertaken the office of a senator; but I do not believe the report.

Wishing you health, I remain, &c.

HALLER TO LINNÆUS.

Göttingen, Feb. 3, 1746.

I have received your letter of the 7th of January, though later than I ought. I have made a list of seeds to send you. Am surprised you have not received what I sent by Lubeck. The seeds you last year promised, the publications I so much wish for, and the dried specimens of North-country plants, may all be entrusted to your colleague Rosen, who will forward them to Lubeck. His brother, as you know, resides at present under my roof.

I should be glad to have your sentiments upon the species of *Allium*. The two *Cepæ* surely cannot be separated. If I had been furnished with more species, my work might have proved more valuable; but no one has sent me any thing this long while, and I am obliged to live upon my own resources. You know better than I how much confusion existed among the species of this genus, even in the best authors. I mean now to attempt a history of the *Orchis* tribe, which is overloaded beyond measure with false species. If you could send me the *Orchis* of the sands of Zeeland *, and the Lapland *Calceolus* †, nothing could be more acceptable. I mean to subjoin some plates. You shall hear of me by my friend Rosen, who will bring you seeds of *Amaranthus siculus*, *Chamelæa*, *Boraginoides*, and *Verbesina*. It might be best to send a living plant of *Commelina flava*.

Your publications will be highly welcome to all good men. I am meditating a *Flora* of Germany, instead of a new edition of my *Stirpes Helveticæ*, of which but few copies remain on hand. Burser's Herbarium would be extremely useful, but how can I consult it?

I will without fail send something to your Society, now I have taken leave of my anatomical labours. They admit of no delay.

I do not think I have your *Anandria*, unless it be *Jacobœa cacaliæ folio*; but it has not yet flowered with me.

This summer I shall make an excursion to some of our sandy and turfy tracts, for the sake of plants. No new books in our line have fallen in my way, except a handsome new edition of Columna's *Phytobasanos*.

May you enjoy your health, and preserve your friendship for me! Write to me frequently. We may, with mutual advantage, correspond about botany, especially on the plants of Germany or Siberia

* *Ophrys Loeselii*. *Linn. Sp. Pl.* 1341.

† *Cypripedium bulbosum*. *Ibid.* 1347.

and their characters. I intend giving a description of the *Amethystina*, a new genus. Farewell! &c.

Göttingen, April 8, 1746.

I have lately had a sight of your *Flora Suecica*, and observed with pleasure the numerous plants of which you there give an account. It is not altogether so agreeable to me to find myself, in so short a work, so frequently, and exclusively, refuted; not without some rather bitter expressions, which are neither requisite nor suitable to our friendship; nor are they such as I use towards you, even when I dissent from your opinion. Believe me, my distinguished Linnæus, you gratify your enemies, who are neither few nor impotent, when you thus attack your friends, as it cannot but abate something of their affection; whilst it raises, in their own estimation, those who wish you ill. I have often pleaded your cause with more than one of these people, and have been treated with blame in return for my too great devotion to you. Yet you assault me, not without some signs of contempt, and with an evident intention to hurt me. But consider, my dear sir, how easy it would be for me to turn your criticisms against yourself, who seem to have bestowed but a very few minutes upon my book.

Fl. Suec. p. 43. You say that I have *Aparine* with *Galium*. To this I have been led by intermediate species. There is the greatest affinity between the Russian *Galium luteum (G. minutum, Linn.)* and our common European one *(G. verum)*. The smooth-seeded variety of the latter, and the seeds of *Aparine palustris (G. uliginosum)*, agree with the *G. linifolium* of *Mons Virginis (G. aristatum)*, which has a hairy seed, and which you have made a *Galium* *.

Pages 62, 63. "Haller takes this *(Campanula uniflora)* for a variety of the last *(C. rotundifolia)* in *It. Helvet.* 88, but it is a distinct plant." What occasion was there to refute me here? I have made them distinct in *Fl. Helvet. n.* 15, *p.* 495 †.

Page 105. "*Rumex, n.* 292 β, in the *Flora Helvetica* is said to have the valves of the seed toothed, but my plant has all the valves entire." I have described the *Lapathum aquaticum*, p. 171, with a smooth not toothed involucrum. You may see I reckon my *n.* 6 to 12 as varieties of the same plant, distinguished by its smooth involucrum from *n.* 13 *(R. obtusifolius, Linn.)*, which has a toothed one. You therefore charge me with a crime that I have not committed ‡.

* This statement of Haller's is not very clear. The words of Linnæus in *Fl. Suec.* are, "If any person should, like Haller, unite the old genus of *Aparine* to *Galium*, I shall not object." Here surely is no "bitterness." Linnæus did unite them ever afterwards.

† Haller's plant is not the Linnæan *uniflora*, but *pusilla* of *Jacq. Coll.* v. 2. 79.

‡ Poor Linnæus must plead guilty here, of negligence or inaccuracy at least.

Page 106. I should wish to know what "hypothetical genera" I have made. Do not the glands in the crown of *Aretia* separate that genus from *Primula*?" *

Page 111. My *Epilobium foliis integerrimis linearibus fasciculatis*, 408, has nothing in common with that very general plant the *angustifolium glabrum* of Tournefort *(E. palustre, Linn.)*; nor have I so referred it. The petals of nearly all the species are cloven, so that your specific character is not sound †.

Page 115. You say, "Let Haller examine by what rule he has removed *Chrysosplenium* (to a distance) from *Saxifraga*." Because it is an apetalous genus, which I am always very loth to mix, in my arrangement, with such as have petals; because, moreover, the calyx is four-cleft, and the fruit of one cell, both which abundantly distinguish this plant from *Saxifraga*. Besides, you know ex-

* Here Haller's agitation leads him into a great error. The words of Linnæus, under *Rumex digynus*, at p. 106, are, "This plant shows that *number* of stamens, petals, or pistils, establishes no natural genera, as the distinguished Haller rightly teaches, if assumed absolutely, without reference to *proportion*; he also teaches that figure, proportion, situation, and essential generic characters, ought to be carefully observed, lest hypothetical genera should be introduced, and in order that natural ones should be kept sacred." Where is the "bitterness" or "refutation" here?

† Linnæus says, his own variety γ is made by Haller a variety of the β, which is a correct assertion, though he himself may be wrong in referring both to *palustre*.

tremely well that all these things are mere matters of opinion *.

Page 132. The white wild *Lychnis (dioica)* is but too nearly allied to the red, and both are dioecious, as I have seen in my garden, where I cultivate them, as well as in the country; for they are extremely variable.

Page 135. I believe I have looked at *Alsine media (Stellaria media,* Fl. Brit.) a thousand times, and never found above five stamens. It has indeed three *tubæ* (styles). Your *n.* 370 *(Cerastium aquaticum, Stellaria nemorum* being confounded with it as a variety) has five styles †.

Page 168 (under *Ranunculus lapponicus)*, "*R. caule aphyllo unifloro, foliis rotundis semitrifidis. Hall. Helvet.* 326. A name (or definition) in a great measure good; but the synonyms quoted differ altogether from my plant, in the form of the flower, and its white colour." My name is in every respect good, and for my plant excellent. Your plant seems to differ from mine in having a yellow flower. Have I then misrepresented any thing in p. 326 of my book ‡?

* The expression of Linnæus may border on incivility; and he is so evidently in the right, that it could not but give offence.

† This is true of the *Cerastium* only. Linnæus does not name Haller, but merely says, "some reckon five stamens, because five fall off before the others."

‡ Haller's is now known to be *R. alpestris, Linn. Sp. Pl.* 778. He was not acquainted with the *lapponicus*.

Pages 182, 183. Linnæus says, he "sees no reason to separate the *Elephas* of Tournefort from *Rhinanthus* (the *Alectorolophus* of Haller), as their aspect is the same, and many of their characters; though some differ, as in other genera." The widely different fruit distinguishes the genus *Alectorolophus,* as is both sufficient and necessary, that the number of species may be abridged, and consequently the length of their respective definitions; for I am now possessed of as many as 30 species of *Pedicularis* *.

Page 186, at *Melampyrum sylvaticum.* "I first noticed this in 1729, taking it in *Fl. Lapp.* for a variety of the *pratense;* nor should I ever have thought otherwise, were not the authority of my friend Haller against me. I wish some other botanists would examine whether these two, as well as *nemorosum,* be not varieties of one species." But I, in 1728, in an alpine journey of above 100 German miles, with J. G. gathered this plant. Neither you nor I therefore discovered it first. I have mentioned abundance of distinctive characters. The blue one *(nemorosum)* is very different from both the others. Its spike is short and dense; the bracteas, as you call them, much more laciniated, broader, and coloured †.

* This argument does not apply; for Linnæus never meant to join the genus in question to *Pedicularis,* but to the very small genus *Elephas.*

† This decision of Haller's is generally thought indisputable, and Linnæus himself ever afterwards conformed to it.

Page 186 also, *Bartsia.* "In 1737 I gave this name to Plukenet's t. 102, f. 5, and ascertained the present plant to be exactly of the same genus. My friend Haller was afterwards pleased to establish it as a new genus, by the name of *Staehelina.* That I may therefore not prove ungrateful to that most acute botanist Mr. Staehelin, I have given his name to a new genus, which comprehends *Santolina n.* 6 of *Hort. Cliff.* and the plant of Plukenet, t. 302, f. 3." Before the year 1737 I certainly separated this plant from *Pedicularis;* and full as early as the period of your founding the whole genus of *Bartsia.* I had rather you had called some alpine plant after Staehelin, as well as after me, instead of rare African productions, with which we have nothing to do *.

Page 203. Your tetrandrous *Cardamine (hirsuta, Sp. Pl.* 915) has six stamens. No one can be mistaken who has seen the greater number, in such a case; he who has noticed fewer, may. Repeat your enquiry †.

Page 241. I cannot easily believe an alpine plant *(Artemisia rupestris, Sp. Pl.* 1186) to be a maritime one ‡.

* How difficult is it to satisfy ambition! Happily for Staehelin, some more recent species of his genus are alpine Cretan and Grecian productions.

† The character in question is now known to vary.

‡ Linnæus modestly expresses his doubts whether Haller's plant be the same as his, and subsequently judged otherwise. As to Haller's doubt, *Statice Armeria* and *Plantago maritima* answer it conclusively.

Page 243, under *Gnaphalium alpinum, n.* 673, Linnæus says, "In the *Hortus Cliffortianus* I made this a variety of the former *(dioicum)*, my specimens being all females, the males having, perhaps, not been gathered. But as that most diligent investigator of the Alps, Haller, has been pleased to reckon it a species, I leave it undecided, till somebody can revisit our mountains. The membranous margin of the calyx-scales favours the idea of its being a good species." The two *Gnaphalia*, which in your opinion ought to be united, do not only differ in sex *, but are widely different in colour of the leaves, and in habit.

Page 265. The *Orchis muscam referens (Ophrys muscifera, Fl. Brit.)* is not very rare here about Göttingen. It has six petals, though two of them are thread-shaped and narrow. Pray examine whether all the Orchideous family ought not to be reduced to two genera, *Orchis* and *Helleborine* †. The want of a spur cannot be of any more consequence here than in *Antirrhinum;* and besides, there are gradual distinctions between the two extremes.

Page 303. I do not perceive what separates *Rhodiola* from *Anacampseros*, or the latter from *Sedum*. That the dioecious flowers are not sufficient, we learn from the Dock and Campion tribes, as well as many others.

* Haller means that *G. alpinum* is not dioecious. These species are now allowed to be distinct.

† About sixty genera are now made of this natural order!

Page 324. My *Bryum alpinum viridissimum*, &c. *t.* 3, *f.* 7, is not a mere variety (of *B. pomiforme*, now *Bartramia pomiformis)*. Its longer leaves and more vigorous habit, even on the Alps, distinguish this moss *.

Page 391. Your *Embolus* is most evidently my first species, in which I have said that there is a brownish powder; and not my second, which is very different.

Page 392. The Narbonne *Lathyrus (latifolius)* has much more ovate † leaves than *Urtica*, some species of *Chamœnerion (Epilobium)*, or numerous other plants which you have so defined.

I could easily add more, and could prove that, if I were so inclined, it would not be difficult for me to be troublesome to you ‡. Far from me, however, be that sad necessity! I mention these things, as becomes a good man, to yourself in secret, in order that, if your disposition be friendly, as I hope, you

* Muscologists of the present day have not distinguished this from *B. pomiformis*; but they may perhaps not know Haller's plant perfectly.

† Linnæus says "lanceolate." We should rather term them elliptical.

‡ This captious displeasure of Haller might partly be excited by Dillenius, Rosen, Heister, Ludwig, and Siegesbeck; for their letters to him justify what he tells Linnæus, that the latter had many enemies. But there is one indignant letter of Gmelin, *Epist. ad Hall. v.* 2, 213, which justly says, "such critics are unworthy of the notice of an honest and truly learned man;" and one cannot but feel that Haller ought to have thought so, instead of holding them up *in terrorem*.

may take any opportunity you please of showing it. If, on the contrary, you are determined, without any cause, to depreciate me, I shall then take refuge in my habitual disposition, and avoid, as long as possible, all occasions of enmity.

With regard to the seeds for which you enquire, I now send all I could procure. The *Solanum spinosum* flowered, without forming fruit, and the yellow *Commelina (africana)* spreads so much by its creeping root, that it produces no seed. This will be conveyed to you by Mr. Rosen, who has lived near a year in my family. Farewell! may all prosperity attend you!

LINNÆUS TO HALLER.

Upsal, Aug. 23, 1746.

On my return from a journey of 250 Swedish miles, this summer, to the hills and sea-coast of West Gothland, I found your highly welcome letter, for which and all your friendship I return my best thanks. The *Flora Zeylanica* will be completed and printed by March next, and I shall there publickly declare my high respect and esteem for you*.

* In the preface to his *Flora Zeylanica*, Linnæus thus speaks of Haller:

"In Germany, among the Swiss botanists, Haller is distinguished as a second Boerhaave. He is engaged in the illustration of the Alps, and the whole of Germany."

To this passage is subjoined the following note, written apparently after the receipt of the foregoing querulous letter,

Aparine with a seed like Coriander * is an extremely different species from the common kind †, and accords with *Valantia*, or *Cruciata*, in having some male flowers, distinct from the perfect ones.

Can you see the stamens and pistils in your specimen of *Juncaria (Ortegia)?* Pray inform me concerning them.

I have forty new genera of plants, principally from the Indies.

Two dissertations only of mine appeared last summer; one entitled *Sponsalia Plantarum*, the other *Museum Adolpho-Fridericianum*. I will send you these, and what I have besides of the same kind.

to which, as well as to any present or future suggestions of their common enemies and rivals, it is intended as an answer.

"The distinguished Haller is known to every body as the most indefatigable of men, being at present without a rival in the theory of medicine, and in anatomy. With regard to botany he has, certainly, examined and described a greater number of plants than any other person. It is therefore very unjustly that I have been charged with writing uncandidly on the subject of this excellent man. I wish the world to know that I have ever held his publications and remarks in high estimation, and am continually engaged in studying them; more valuable productions not having hitherto appeared in any part of Germany."

With what unruffled composure does Linnæus, in the letter before us, continue his accustomed correspondence with his friend, disdaining to poison it with altercation!

* *Valantia Aparine. Linn. Sp. Pl.* 1491. *Galium verrucosum. Sm. Prodr. Fl. Græc. Sibth. v.* 1. 93.

† *Galium Aparine. Linn. Sp. Pl.* 157.

In the course of my late tour I collected fishes, and new species of insects, in plenty; and made out the complete history of a certain species of *Cantharis*, which is the cause of much destruction to the oak timber destined for ship-building *.

I have seen the flowers and fruit of the *Alga vitriariorum*, which forms a new genus *(Zostera)*, agreeing in natural order with *Arum*.

Two youths have met with the *Peloria* in two different places in Sweden, and each preserved it in his herbarium.

My *Anandria* has this summer borne one flower of a remarkable kind, having a radius, of deeply three-cleft petals (or florets), whilst all the rest were such as described in the dissertation. This seems to me very curious.

I will send you the *Fauna Suecica*, if you have it not already.

All the plants sent me from New York have fallen into the hands of the Spaniards, along with those that Dr. Mitchell has for many years been collecting in Virginia. He himself is returned safe, though in a destitute condition, to England.

This spring I have observed that *(Convallaria) Polygonatum* with an angular stem is a distinct species from the common kind, whose stem is round *(C. multiflora)*.

* *Cantharis navalis*, by tracing the economy of which, and directing the timber to be laid in water during the short period in which the insect lays its eggs, Linnæus saved the Swedish government some thousands a year.

I am pleased with your remarks on the *Orchis* tribe. The species in question are rare with us, and found in some distant provinces only.

You have doubtless, long ago, seen Seguier's *Flora Veronensis.* He has sent the book to me, but it is not yet come to hand.

Limnia of the Russian botanists is a true species of *Claytonia,* with ovate leaves, in which it widely differs from that of the *Flora Virginica,* 25, whose leaves are linear. I have described the former in our Stockholm Transactions for June.

I must, in the name of our Upsal Society, once more request you to send something or other for the Transactions.

Various shells have fallen in my way in my journey, one of which, of marine origin, but found under ground, is to be traced out by two perpendicular openings in the sand *.

Alsine polygonoides, Pluk. Phyt. t. 75, *f.* 3, proves a plant of a new genus in *Diandria Digynia* †.

Wishing you may long enjoy your health, and retain your friendship for me, I am, &c.

* *Mya arenaria.* See *Fauna Suecica, ed.* 2, 516.

† *Buffonia tenuifolia. Linn. Sp. Pl.* 179. It was afterwards found to have four stamens.

HALLER TO LINNÆUS.

June * 27, 1746.

I rejoice to hear of your return home in good health, knowing how indefatigably you are always employed in the promotion of natural history. On this account I have ever loved you, commended you, publickly justified you against your enemies, and praised you even when I was not writing in my own person. Such being my disposition, I feel the more severely that in a book where you attack nobody else, you criticise me alone, of all men, on many occasions, and those, as you yourself must be conscious, not urgent ones. You do not seem to be aware, my distinguished friend, what mischief your censure may do to my fame; nor how gratifying it will be to my rivals and enviers. Even those who know nothing of me will be liable to form an unfavourable opinion. Few people examine into the merits of a quarrel, or are capable of seeing to the bottom of its causes. Almost all who may observe me to be found fault with by a man of eminence, will depreciate me themselves. For my part, I have always admonished you privately, not publickly, not in anger, but with anxiety to preserve your friendship, lest any thing should arise, in the course of our dispute, that might lead to an irreconcileable breach between us. I hope therefore that, as the effect of your censures are now made known to you,

* This date is a manifest error. It should perhaps be September. The letter is in answer to that of Linnæus, dated August 23.

they will be more sparingly thrown out in future, as well as in less offensive terms. Such expressions as "let him examine" indicate a degree of contempt inconsistent with the continuance of friendship, and, still less, of respect. You privately express to me the demonstrations of your good opinion; but what are these to your publick remarks? Will the conviction of your Upsal calumniator restore me to what I shall lose in the opinion of a thousand readers of your Flora? As to Rosen indeed, I have no desire to quarrel with him, he having no concern whatever in the question between us.

I have no inclination to dwell on every particular cause of displeasure; but what grieves me most is, that you should so frequently indicate a difference of opinion from *me alone.* Such a difference may indeed be expressed, consistently with friendship, otherwise I myself have offended, against you as well as others. I shall therefore only make the following observations by way of answer.

Rhodiola has five petals, and five seed-vessels, with ten stamens, as often as four, and therefore differs from *Anacampseros (Sedum)* merely in being dioecious; nor does the latter, as far as I see, differ in any respect from the smaller kinds of *Sedum* of authors. *Alsine media* here has but five stamens, even before the calyx expands. The species of *Galium,* one of which has a hairy seed, and the yellow Russian four-leaved kind, which is smooth, are perhaps varieties of the common *Aparine,* which is in flower in my garden, with a very smooth seed.

Now I entreat you not to be angry, nor to charge me with enmity, which is foreign to my nature, for I have always felt and written concerning you in a widely different manner. It remains therefore for you openly to testify your esteem for me, as I have ever, on all occasions, shown mine for you; and here let there be an end of these altercations!

I have the *Juncaria* of Clusius *(Ortegia hispanica)*, but only dry, and glued down, in my usual manner, upon paper. If a flower of it would be of any use to you, let me know, and I will pick one off.

Some Thistles from Switzerland are flowering in my garden, drawings of which are making under my inspection, for the Catalogue of the garden, now in hand. The *Allium (arenarium)*, of which I have published a figure in Ruppius *(tab. 2)*, and which you gathered in Oeland, is now likewise flowering, having acquired in the garden a great increase of luxuriance and beauty. *Phalangium parvo flore, non ramosum* * I have gathered wild here. It differs in leaves and flowers from the branched kind †. So also the *Polygonatum* ‡ with simple flower-stalks differs from the common one, in having an angular and firmer stem, not half so tall, much fewer leaves, and a broader, differently shaped, flower. It is common in this neighbourhood. My *Hesperis* §, delineated in the edition of Ruppius

* *Anthericum Liliago. Linn. Sp. Pl.* 445.

† *A. ramosum. Ibid.*

‡ *Convallaria Polygonatum. Ibid.* 451.

§ *H. inodora. Ibid.* 927.

(tab. 1*)*, whose calyx is hairy, the flowers few, and of a pale red, when removed from its native rocks into the garden, greatly changes its appearance, and very nearly approaches the cultivated *Hesperis (matronalis)*. I have again investigated the flowers of *Helleborine, Ophrys,* and *Nidus Avis*; nor do I think they can be made separate genera, as they accord in their firm stalked stamen, having a divided summit under a scale, and differing besides in very trifling particulars only. The *Limodorum* certainly belongs to this genus, and not to *Orchis,* which latter has two distinct stamens, inclosed in membranous hoods. I am not acquainted with the structure of the *Corallorrhiza.* My *Orchis (N*° 9*) rotunda* * of Dalechamp is widely different from that of Dillenius, *purpurea spicâ congestâ* †, in having a dense globose head, with awned petals. The *Orchis muscam referens* ‡ grows here, and is, in my opinion, an indubitable *Orchis,* on account of the structure of its stamens; nor do I remove from this genus such species as want a spur, any more than I would separate *Antirrhinum* and *Linaria,* there being gradations between the two extremes.

But I have elsewhere treated of these matters more at large. I sincerely wish you well; and if you are still disposed to cultivate my friendship, send me some of your smaller works, such as those

* *Orchis globosa. Linn. Sp. Pl.* 1332.

† *O. pyramidalis. Ibid.*

‡ *Ophrys muscifera. Fl. Brit.* 937.

academical dissertations which are not to be purchased, along with a few dried plants and mosses, peculiar to your northern climes. You may entrust a parcel every year to the bookseller. I promise you, in my turn, whatever rarities I may pick up in my various solitary excursions.

Professor Schmidel of Erlangen is said to have met with the same deformity in *(Antirrhinum) Elatine* * that you have described by the name of *Peloria*. Adieu!

LINNÆUS TO HALLER.

Upsal, Sept. 24, 1746.

By order of our Royal Society, I return the best thanks of that body to you, Sir, for the two treatises with which you have been so good as to favour the Society by Mr. E. Rosen. Every body was delighted with your most curious account of the tunic of the eye in new-born children. The *Amethystina* † requires to be described somewhere. If you do not wish to add a figure, I will furnish one which I have prepared. If your admirable artist has made a drawing, it will probably be better, for our draughtsman is not of the first rank. I have understood the corolla otherwise in my specimens; but I may be mistaken, though there were plenty of flowers. I refer all the *Verbenæ* to my second class, *Diandria*, being unwilling to divide a natural genus. Those

* Similar instances are not very uncommon in England.

† *Amethystea cærulea. Linn. Sp. Pl.* 30.

which make the *Sherardia* of Vaillant have all two stamens, as well as three other species, of which I am possessed of dried specimens. Your *Amethystina* is a distinct genus, as I found on seeing the living plant. From a dried specimen I took it to be a *Lycopus*.

I am now extremely busy every day in getting together the species of each genus, with their synonyms. You have cleared the way for me greatly amongst the German and other European plants. Unless all the synonyms are collected, no certain knowledge of species can be expected. I have finished a third part of the genera, with respect to their species; but their number increases upon me without bounds. My object is attentively to ascertain their respective differences, and my joy will be great if I can ever accomplish my undertaking.

Mr. Everard Rosen, who was here lately, tells me you have given him your engraved portrait, in a book containing the portraits of various learned men. Such a work is not to be had here, and I earnestly beseech you, by our long established friendship, to send me this likeness of you by the first post, as I shall look out for it with eagerness. I have already the portraits of several distinguished persons, among which I cannot suffer yours to be wanting. I would purchase it, if I could, at any price, but that is not at present possible.

I have entrusted to Mr. Everard's care two dissertations for you, one on the King's Museum, the other entitled *Sponsalia Plantarum*, which he has promised to forward the first opportunity. This Museum will be less interesting to you, as I think you scarcely attend to the study of animals. I shall be glad of your opinion upon the other dissertation, as well as upon the fecundation of plants.

I have got Seguier's *Flora Veronensis.* I wish he had described his plants in your manner, as it would have rendered his work much more valuable.

Dr. Mitchell is returned from Virginia, where he has been closely occupied for six years in collecting plants; but he was plundered in his voyage home by Spanish pirates, to the great misfortune of Botany.

I have lost, in the same ship, numerous specimens and descriptions, sent by Governor Colden from New York.

In return for your portrait, I shall send you a medal of myself, but as it weighs about an ounce, I doubt whether or not the post will take it.

May you enjoy long life and prosperity! Farewell!

HALLER TO LINNÆUS.

Göttingen, Oct. 17, 1746.

I am happy that my little essays proved acceptable. I could furnish you with more remarks of the same sort, if I had but time. I have a most exquisite drawing of the *Amethystina*, but I fear you can hardly have in your country an engraver capable of doing it justice. Bergquist has an elegant manner enough; but he perhaps is too rapid, and does not well express the shadows. I would have had the plate engraved here, at the expense of your Society, but the younger Rosen said this was not allowable. I therefore kept the drawing.

The flower is precisely as I have described it. I have a very perfect microscopical delineation of its parts. The plant comes next to *Verbena,* but differs in the bell-shaped calyx. I could wish to arrange the whole genus of *Verbena* according to the European species, and those that resemble them; by which means they would remain with their associates, the *Ringentes* of Van Royen. Possibly even *Salvia* might be placed with them, for many of its species bear a larger or smaller polliniferous cell *.

The study of synonyms is, indeed, a difficult task, unless you are furnished with dried plants from every different country, to enable you to form a judgment of what every writer has described. I have been but too sensible of my deficiency in this particular, whence I have been led to range as varieties many things which I have subsequently discovered to be good species. I would advise you, therefore, to mention all the synonyms. Descriptions indeed, made after nature, cannot well mislead, any more than the plants themselves. In mine I have been anxious to guard against introducing any thing doubtful, without mentioning my doubts. Among your plants, pardon me for saying so, several genuine

* We presume Haller means, on the transverse, usually abortive, filament, which supports the perfect one.

species are expunged; which I wish you would alter in another edition of your Swedish *Flora*. I cannot but think that we, by doing away with real species, render a great disservice to our science, and tend to reduce it to that poverty from which learned men have, with great efforts, raised it. I am well aware, nevertheless, that much depends upon opinion in these matters; so that we cannot, in all cases, say what is a species and what a variety; at least not without culture and repeated observations. I have four totally distinct species of *Phalangium (Anthericum)* with white flowers, three of which, but for culture, might be taken for one; and so of other plants.

I have not done much in botany of late, except my journies, and the examination of characters in my garden. Next year I mean to return to my books. I have collected no animals of any kind. I should be overwhelmed with too great a variety of studies. Anatomy is my chief occupation, combined with physiology. To these I am obliged to devote the greater part of the year. Your happier fate spares you such interruption of your pursuits. But if ever I can return again to my own country, I shall seek no other pleasure than botany. To that pursuit, in the investigation of what Switzerland produces, I hope to dedicate the remainder of my life.

I inclose my portrait, in which there is nothing worthy of the solicitude you express.

I shall be very glad to receive your medal; and, if you please, I will take care to have it noticed in a book which my colleague Koehler is preparing, under the title of *Munzbelustigungen*. An account is subjoined of certain persons who have had medals struck in their honour.

But I cannot refrain from begging of you dried plants from the north, in case you find it difficult to supply me with seeds. There can be nothing to hinder this. I would send you the productions of the Alps, and am now quite ready to communicate the more uncommon plants of Germany, gathered in the course of my various journeys. Among them are some mentioned by Ruppius, from their original places of growth, which I presume you are not likely to have. If there be no other way, the elder Rosen, or your bookseller, Kieservetter, will forward them.

Seguier's work has not fallen in my way. Italian books are long in reaching us, as well as all botanical publications. I expect Blackstone soon from England.

Not to leave the rest of my paper blank, I shall subjoin a few botanical matters.

You blame me for having subjoined the *Granadilla (Passiflora)* to the *Cucurbitaceæ*; but where have I done this? If I had, indeed, I should not have done wrong. They agree in their calyx, united to the five petals, their soft fruit, and their habit. There is a resemblance even in their styles.

But it is not requisite, in a natural class, that every thing should be the same. *Melothria*, for instance, has both organs in one flower.

The name *Trientalis* is an adjective. *Herba trientalis*, &c. (an herb three inches high) says Cordus. I have united two perfectly distinct species of *Gentiana*, by the name of *pratensis flore lanuginoso*. The orifice of the flower in each is furnished with a filamentous ornament. But one of them has a bell-shaped calyx; the other an irregular (unequal) calyx, with two larger and two smaller segments. Both of them grow here, and in the neighbouring country *.

In your Stockholm Transactions, as well as in your Travels, you confound the *Rubeola flore trifido* † with the *R. quadrifolia lævis* ‡ of Tournefort. They are distinct. The former is a cubit high, lax in habit, with dark flowers, universally three-cleft, and loosely scattered over the branches. The latter is of very humble stature, hard, and rather shrubby, with more umbellate flowers, which are always four-cleft. The first occurs in a few woods of Germany ; the other grows almost every where hereabouts, by the road sides. The former is a plant proper to your country; the latter belongs rather to a milder climate, being found not only at Göttingen, but likewise at Jena, and in Switzerland.

* The former is *G. Amarella*, the latter *G. campestris*.

† *Asperula tinctoria*. *Linn. Sp. Pl.* 150.

‡ *A. cynanchica*. *Ibid.* 151.

As to what you write on the subject of *Valantia* and *Aparine*, I should not easily be disposed to separate plants merely on account of sexual differences. The cause of sterility in some of the flowers of the *Aparine* with seeds like Coriander comfits * may possibly be the great size of those seeds. I have remarked the same thing in the umbelliferous family, where those whose seeds are large always bear a considerable number of male flowers interspersed.

The *Myrrhis* figured in Ruppius † proves, on examination, a perfectly distinct plant. I will send you some of its seeds.

The common wild white *Lychnis* is as truly dioecious as the red one. *Geranium moschatum* and *G. myrrhidis folio (cicutarium)* possibly are varieties of each other. But *G. creticum acu longissimo* and *G. coriandri folio acu longissimo* ‡, are very different from them, though both likewise pentandrous. The whole structure of the seeds, which are also very much larger, shows a wide difference; as well as the flowers, which are very unlike the first-mentioned, and not so irregular.

I long very much for some seeds of your little *Thalictrum fœtidum (Sp. Pl. 768)*. I hope this year to collect seeds of *Commelina lutea (africana)*.

* *Galium verrucosum. Prodr. Fl. Græc. v.* 1, 93.

† *Chærophyllum aureum. Linn. Sp. Pl.* 370.

‡ These are both one species, *G. ciconium. Linn. Sp. Pl.* 952, *n.* 15. *Hall. Helv.* 370.

The genus *Aloe* bears various kinds of flowers, but there is a general agreement in habit.

Gramen typhoides, spicâ instar limæ dentatâ, is a mere variety *. (Of what?)

The common *Rheum (Rhaponticum)* and the Siberian one *(undulatum)* are perfectly distinct.

Acetosa montana, lato ari folio is a different plant from our common one *(Rumex Acetosa)*, and I have too hastily combined them.

May you enjoy your health, and retain your affection for me! which I will neglect no opportunity of deserving. See, in the *Biblioth. raisonnée*, the account of your *Iter Oelandicum*, written after I had perused your *Flora;* and also what follows relative to the Stockholm Transactions.

I have this very day made a preparation of the membrane of the pupil, in the eyes of a new-born infant.

LINNÆUS TO HALLER.

Upsal, Oct. 21, 1746.

I have this very hour received, with the greatest delight, your most welcome portrait. A painting shows at once more than the most perfect description, and hands down the living forms of men, through successive ages, to their latest posterity.

* If this means *Phalaris aspera, Willd. Sp. Pl.* v. 1, 328, *Phleum paniculatum* of Hudson, and *Fl. Brit.* 70, it is a very distinct plant.

In collecting together the species of plants, I search out all their synonyms as far as I possibly can. Those of Dillenius in his Mosses and *Hortus Elthamensis*, as well as your own in the work on Swiss plants, I take in full confidence, without examination, by which my labour is greatly diminished *.

I had likewise combined your two Gentians. I now see they are perfectly distinct, but I doubt whether they have hitherto ever been distinguished.

If varieties be not reduced to their respective species, the number of species will be more and more multiplied, new ones arising every day. Some mark or other may be found to characterize every variety, well or ill.

The *Rubeola*, to my surprise, has always white flowers with us, never purple ones. I will enquire further.

I beseech you not to throw blame upon me in every letter. I lament much more than you could do, even if it were so, that I should offend you. I will explain myself publickly in such terms as that you yourself shall be perfectly satisfied. Where have I found fault with you for joining the *Granadillæ* with the *Cucurbitaceæ?* I really do not recollect. It could not be in a letter, for I have always avoided controversial points; nor have I proposed

* This confidence may account for many errors in the works of Linnæus; for Dillenius, at least, was often incorrect in his synonyms.

any. Be but gentle and amiable, as you are accustomed, and I will be devoted to you in any manner that you may order me.

A few days since I obtained a specimen of the *Sterna, Faun. Suec. n.* 129 (*Larus parasiticus, ed.* 2, *n.* 156), and have had a drawing of it made. I never had this bird in my hands before. The two middle feathers only of the tail are extremely long. I mean soon to give a history of the bird.

I have also received, from the north, the *Muscus Norwegiæ, umbraculo latissimo purpureo,* of Petiver, mentioned by Dillenius in the appendix to his Mosses. This was found 20 Swedish miles above Tornea. The orbicular part, like a hat, is a receptacle, as in *Polytrichum,* supporting the head (or capsule), and forms a very broad, flattish, membranous expansion, yellow in my specimen. I never saw this plant before. It must, I suppose, be referred to *Sphagnum* *.

Is your *Alsine tenuifolia, Helvet. t.* 7, the same plant as that so called in Seguier, *t.* 6, *f.* 2, and the *Alsine polygonoides* of Ray's *Synopsis, ed.* 3, 346? Has your plant but two stamens and as many pistils? The aspect of my dried specimen from Magnol seems different; but I have only one. Seguier represents a greater number of stamens. The Montpellier plant looks more like a *Polygonum.* I ask

* The specimen in question was *Splachnum luteum. Linn. Sp. Pl.* 1572. The plant of Petiver is the *rubrum.*

these questions, not in the way of objection, but for my own instruction *.

I send the medal, for you to do what you please with. Charles Tessin is a senator of the King of Sweden. Mr. Horleman is the principal surveyor of the royal palaces and gardens, very eminent in his profession. Claude Ekeblad is one of our richest noblemen. Andr. Hoepken is a young man of extraordinary talents, one of the founders of the Royal Academy of Sciences at Stockholm.

I am now interrupted by publick business, but would not omit writing by this post, to express how highly I value your portrait. May you long enjoy health and happiness, and never acquire any worse enemy than myself!

We take, at the publick expense, for our Academy, the *Commercia Norimbergica*, and *Bibliothéque raisonnée*. Perhaps you can assist us in procuring the numbers of each for the ensuing year.

I neither have, nor ever had, the *Thalictrum fœtidum*.

The two Rhubarbs are very different in appearance, but it is difficult to find any specific character to distinguish them, except their comparative size.

* The plant of Haller is *Arenaria fastigiata*. *Engl. Bot. t.* 1744; that of Seguier *Arenaria tenuifolia*. *Linn. Sp. Pl.* 607; the "Montpellier plant" *Buffonia tenuifolia, ibid.* 179. What the English plant of the *Synopsis* may be, remains doubtful.

Has the *Acetosa montana lato Ari rotundo folio* of Tournefort stamens and pistils in the same flower?

Pray mention in your next whether this letter comes safe to hand. If it miscarries, I will write again.

The seeds now sent should be sown this autumn, in a situation neither too hot nor too cold. The Rice is quite fresh, but ought not to be sown before the spring, and requires a peculiar treatment. There are two *Thapsiæ*, the *Cyperus odoratus (longus?)*, and three others.

HALLER TO LINNÆUS.

Göttingen, Nov. 17, 1746.

Your very beautiful medal arrived safe, for which I return you my due thanks. I shall value it as a pledge of your regard.

Varieties of colour, often very trifling and variable, I do not admit; but such as are not changed by culture, and depend on the proportion of different parts, I would not have overlooked. Too much attention to varieties is injurious and degrading to botany.

You will find your *Rubeola* with a three-cleft flower has a taller, more lax and pliant, stem, and is altogether different from the *R. quadrifolia lævis*, (see the preceding letter).

You take me to task about the *Granadillæ*, in your dissertation on *Passiflora*, unjustly; for I have not admitted them into the order of *Cucurbitaceæ*, though I believe I could justify myself if I had.

I thank you for your account of the *Muscus Norwegicus*, a plant I never saw.

My *Alsine tenuifolia* has always five stamens, or more. The *tenuifolia muscosa* of Caspar and John Bauhin is widely different, and has eight stamens*. The latter is a tender, delicate plant, with four petals; whilst mine is so rigid as to be almost spinous, and has closely-crowded leaves, five petals, and three styles. It grows in rough or barren places.

I have seen in the publick papers an interesting account of persons who have been commemorated by medals.

Where the *Bibliothéque raisonnée* is to be bought I know not. You may doubtless get it from the Dutch booksellers, by means of Mr. Wetstein. I have lately given an account of your Travels, and afterwards of the *Stockholm Transactions.* You will find nothing there unbecoming an honest man and your friend, though I had just been reading your *Flora Suecica.*

The *Acetosa* you mention is dioecious, and allied to the *pratensis (Rumex Acetosa)*, though differing in its broad, spreading, not undulated leaves, from that variety of the latter which grows in Friesland.

Many thanks for the seeds.

I am now more drawn aside from botany, and become a mere anatomist; but I have educated

* *Moehringia muscosa. Linn. Sp. Pl.* 515.

several young men, in whom the love of plants is very ardent. They will by degrees supply me with what is rare and beautiful in my old age, to which I begin to look forward; for I have grown wonderfully lazy in body within these two years; nor can I take long walks as I used to do.

Once more farewell! Forgive the emptiness of my letter, my chief object being to acknowledge the receipt of your medal.

LINNÆUS TO HALLER.

Upsal, April 10, 1747.

I have lately published my Travels to West Gothland, in which you will see a character of the *Alga (Zostera)*, and the repeated encomiums which you deserve.

The *Flora Zeylanica* is printing daily. In the preface to that work I have spoken of you as I have done of no one else. Dr. Rosen has promised to forward the book to you.

In my dissertation upon some new genera I have, likewise, commended you as I ought. This is printing, and in the course of a month will be distributed.

By these three publications my devotion to you, in preference to any other person, will be made manifest. I have endeavoured, by every possible means, to conciliate your favour.

I am, at this time, deeply engaged in the history of diseases, on which I give publick lectures, as also private ones on mineralogy.

Is it true that Dr. Doppelmeyer, professor of the mathematics at Nuremberg, has brought a paralytic complaint upon himself by electricity?

All the world, as well as myself, are anxious to know your mode of applying electricity to medical use.

Do you think I meant, in my dissertation, to blame you for referring *Passiflora* to the *Cucurbitaceæ*? By no means. I have done the same in my fragments of a natural system; and I continue in that opinion, without any alteration *. I am sorry that you are angry with me; otherwise you could not have so understood me.

Wishing you long life and happiness, I am, &c.

* Linnæus, in his dissertation *de Passiflora*, pp. 4, 5, indicates the place of that genus in the systems of different botanists, and refers it to the order, or section, *Cucurbitacea*, of Haller's *Isostemones*. This, so far from being controverted, is confirmed in the next line, by a reference to a similar order of his own, then called *Cucumerinæ*. But in p. 6 he says, in allusion to Hermann, the only author who had referred a species of this genus to *Cucumis*, "*Passiflora* cannot be a *Cucumis*, as the germen is not below the receptacle of the flower, nor are the stamens combined." Haller must have been strangely blinded by passion to take this to himself; nor could Linnæus suspect him to have done so. He is no where alluded to but in the place first mentioned.

HALLER TO LINNÆUS.

Göttingen, May 25, 1747.

It is not without pain that I have read your last letter, of the 10th of April. While you declare your love and praise of me, you say I am angry, because I see in your dissertation that you censure me for adding *Passiflora* to the *Cucurbitaceæ;* and you write me a letter full of sharp expressions*. If this be your manner of showing your friendship, how do you express your anger? Do I wrong you by reading your works as they are? I beg of you to be convinced, my distinguished Linnæus, that I neither wish to rival nor to envy you. Pray do not treat me as if I were desirous of detracting from your reputation. As a proof of the contrary, I now send you an engraving of your medal, with an explanation by my friend Koehler, to whom I communicated the original, along with your publications. I love every man who is a lover of nature, but I find myself neglected by you; our correspondence and our interchange of books are interrupted; the communication of any of your northern plants and mosses is evaded; I am blamed in very many parts of your *Flora Suecica,* and the names of all the plants which I first determined, except one, are changed. Now see how I revenge myself, and you must allow me to have shown myself, not only totally free from enmity, but most studious of your reputation, avoid-

* Can the reader find any such in our faithful translation of the preceding letter?

ing all publick mention, even of my concern on the occasion. Dillenius is no more; but I have letters in my hands which manifestly prove how solicitously I laboured to clear you in the opinion of that greatest of botanists. Such being my character, see whether I do not deserve your affection, rather than sharp words, or the overthrow of every thing I have done in botany, that is, the determination of above 400 species of alpine plants *. Would it not have been better, and more pleasing, that our labours, altogether insufficient for the illustration of nature, should have been united, instead of being allowed to degenerate into trifling contentions, vain emulation and controversy, destructive of our health and our happiness? I can very easily bear a difference of sentiment; but when that difference arises between friends, it may be mollified by a suitable mode of expression; and care should be taken to give our opinion in such a manner that it may not seem aimed at our friend.

I have not done much in the service of Botany lately, but shall now resume that study. This summer I propose to take a journey to Eisleben, and the neighbouring warm parts of Saxony, for the enriching of the *Prodromus* of my German *Flora*. I hope to ascertain the *Hippomarathrum* of Rivinus, and several other plants. Some of the productions of the Alps have grown well from seed in my garden. I shall give plates of some of them, not yet

* Where has Linnæus even aimed at this overthrow?

engraved. Lindern has published a *Flora Alsatica* *, with some original figures, and distinguishing notes of species. I am every day in expectation of Ludwig's *Definitiones.* I foresaw that he would make many changes, as I now find is the case. He has adopted many things from you, but would not follow your classes, lest he should differ too widely from Rivinus, by not taking his characters from the petals. He is entirely a compiler. I have heard of nothing new in botany besides. A pupil of mine has found the *Napellus cæruleus* †, and some other things, wild in Germany. Farewell! and retain your regard for me so long as I deserve it.

LINNÆUS TO HALLER.

Upsal, May 12, 1747.

When I was at Stockholm, about a week since, the Royal Academy of Sciences ‡ there was engaged in the appointment of foreign members, as Gesner, Jussieu, yourself, Gmelin, Sauvages, Clayton, Collinson, and Van Swieten. I am charged by the Academy to communicate this intelligence to you, and respectfully to invite you to become one of its body. I hope this testimony of my own regard for you will, though slight, not be disdained by you, who

* *Hortus Alsaticus, a Franc. Balthazare von Lindern, M. D. Argent.* 8vo, 1747.

† *Aconitum Napellus. Linn. Sp. Pl.* 751.

‡ Of which Linnæus was the institutor and first president.

were born for the reformation and restoration of science. If you are pleased to accept of this compliment, be so good as to signify your acquiescence by a letter to the Academy, acknowledging the receipt of my communication. I am, &c.

HALLER TO LINNÆUS.

Göttingen, June 8, 1747.

I am greatly indebted to you, my dearest Linnæus, for the election you have honoured me with. Let us thus continue to love and to be loved. We can neither of us do any thing more wise or more useful. Farewell! and continue your friendship for me.

I write in the greatest haste, on account of the pressure of academical business.

LINNÆUS TO HALLER.

Upsal, Oct. 23, 1747.

I yesterday received, from my colleague Dr. Rosen, your very kind letter of the 25th of May, with the account of my medal, in which I perceive, and gratefully acknowledge, a fresh instance of your friendship *.

* With what forbearance and good sense does Linnæus abstain from all allusion to the angry parts of his friend's last letter!

I hope Mr. Rothman will, before long, convey to you my *Flora Zeylanica*, in which I have referred a considerable number of species to their proper genera. I wish the work may meet with your approbation.

I wish you would put down the numbers, from the *Flora Suecica*, of all the plants of which you want dried specimens, and I would send them without delay. The Lapland ones I can scarcely furnish you with. Lapland is further from me than your residence at Göttingen; nor has any botanist been into that country, to gather plants, since I was there. What I collected have long been distributed among my friends, except a single specimen of each. I may be able to procure nearly all the rest of the Swedish species, if I know what you want. I am not furnished with seeds, except those of Siberia; and I observe most of those, if not all, in your *Hortus*. I will, however, if you wish, look out a few for you.

Kalm is lately gone to England, from whence he will proceed to Canada, in order to collect seeds for the Upsal garden, as the plants of that country bear our winters perfectly well. He is a pupil of mine, and has just been appointed œconomical professor at Abo, in consequence of my recommendation.

Mr. Ternström, who at my instigation went, about two years since, to China, is not yet returned, but is waiting for the proper season in the East Indies. He is an excellent collector of plants, but nothing more.

Yesterday another friend of mine set sail with the East India fleet, as a chaplain. He is likewise an able collector, but otherwise not remarkable for information.

The *Hortus Upsaliensis* is finished, and you shall have it next summer; but the collection, as yet, is a poor one.

I am now collecting all the species of plants I can meet with, and referring them to their proper genera, intending to publish them as a part of my *Systema Naturæ*.

Ludwig's *Definitiones* I have not yet seen. Lindern's book has been sent me, but it is of little value. De Gorter's *Flora* has not yet come into my hands, nor the Siberian *Flora* of Gmelin. Barrere's Ornithology is an elegant work, but wants descriptions.

I have nearly brought to a conclusion my work on the *Materia Medica*, but I am not acquainted with the *Agallochum, Lignum Rhodium, Bdellium, Caranna, Elemi, Sagapenum, Ammoniacum, Myrrh, Thus, Balsamum Peruvianum, Anisum stellatum, Myrobalanus chebula, Bellerica,* or *citrina;* I mean the plants which produce them. If you can give me any information, I shall be much obliged to you.

The Belvedere * flowered this season with me, but was destroyed last night by frost. I was unable to discover the stamens. Pray let me know if any of the plants have separate organs.

* *Chenopodium Scoparia. Linn. Sp. Pl.* 321.

Enclosed are seeds of the *Arbutus, n.* 340, *Fl. Suec.* * But perhaps you have it from your own Alps. You are aware that these seeds must be sown in poor sandy earth. They are quite fresh, having been gathered from the living plant this very day. I would advise you to sow them immediately, in the open ground.

We have nothing new here. I have given a description of the *Ursus caudâ elongatâ,* in the Stockholm Transactions †. The Upsal Transactions are printing, in which your very admirable, and perfectly new, account of the membrane in the eyes of the fœtus will appear.

Adieu !

HALLER TO LINNÆUS.

Göttingen, Oct. 27, 1747.

I have just received your very kind letter, with seeds of the *Arbutus (alpina),* which however I almost despair of raising. I have also received your very admirable *Flora Zeylanica,* and have to thank you for the additional honour you have done me. I subjoin a list of my *desiderata* from the *Flora Suecica.* You shall have an interchange of German plants, when anatomy will allow me time to breathe. We may promise ourselves many fine things from Kalm, and from the other botanists

* *A. alpina. Linn. Sp. Pl.* 566.

† *U. Luscus. Linn. Syst. Nat. v.* I. 71; the Quickhatch, or Wolverene, from Hudson's Bay.

educated in your school, whom you have inspired with so much ardour. I was long before I could excite any love of botany in the young men here; but within four years past I have succeeded better. Here are now several, from whom, when they become dispersed over Germany, I am promised specimens of the productions of various districts.

Ludwig gives you credit for a great deal, though he durst not, at Leipsic, entirely desert Rivinus. The *Flora* of Gmelin will lay open to you many interesting things, and is a most excellent performance. Its estimable author lately passed this way, and showed me the elegant figures designed for his following volumes. There are many new species, and even new genera. Gmelin is remarkable for his modesty, and rarely makes any innovations.

You must expect little from me on the subject of exotics, which, in the tumult of so many various pursuits, I have never dared to venture upon. Our Kæmpfer has something relative to the *Agallochum*, as well as the Skimmi, or *Anisum stellatum*. I presumed there was no doubt respecting the *Elemi*, after the observations of Catesby, whose work I suppose you have; but if not, I will draw out for you a description of the plant which yields the *Elemi*. The old writers have something on the subject of *Lignum Rhodium*, which appears to be a *Cytisus*, or something akin. Clusius, and other writers upon Cretan plants, mention it. *Bdellium* is a kind of palm, and, if I mistake not, is also described by Kæmpfer. As to *Sagapenum*, *Thus*,

Myrrh, *Opobalsamum*, *Balsamum Peruvianum*, or *Caranna*, I know nothing of any consequence about them.

I find nothing sufficiently certain in my notes relative to the Belvedere. The cold has now destroyed it. I have a Persian plant very like this plant, but with a smooth glaucous leaf, forming a similar bush. I have also a new Persian *Astragalus*, with incurved spines; and some other things, which I mean to introduce into the description of my garden. This book will comprehend, besides, all the indigenous plants of Germany. The preparation of drawings and plates now chiefly requires my attention. Winter is devoted to dissections. I have lately written on physiology, and am at present engaged in completing a work on the practice of physick, or *Methodus medendi*. Farewell, my indefatigable friend! Do not cease to love me.

LINNÆUS TO HALLER.

Upsal, May 28, 1748.

I have been reading your account of the *Flora Zeylanica*, in the Literary Gazette, and am sensible of your kind partiality, which has awakened my feelings towards you in no small degree. I shall take care, in whatever I may publish hereafter, that you shall experience the same treatment from me.

I sent you a few days since, by a student going to Göttingen, my *Hortus Upsaliensis*, consisting of a short catalogue of the garden plants, for the use of students, without any remarks*.

I regret that I have not yet got a copy of the Upsal Transactions, as I could now have sent it to you. This volume contains your valuable papers on the *Amethystina,* and on the new-discovered tunic of the eye, both of them long since printed. You shall have the book by the first opportunity.

Before the summer vacation I propose to give out for discussion three academical dissertations:

1. *De Surinamensibus Grillianis,* which treats of Serpents and other animals.

2. On the economical uses of Swedish plants.

3. On Man, as an intelligent admirer of his Creator's works.

I am about beginning to print my *Materia Medica,* which will make but a small book. Botanical demonstrations and excursions occupy all my time. A building is preparing in the garden, to receive our academical museum. If Mr. Ternström, who went to China two years ago, is living, we may expect him back this summer.

We have heard nothing of Kalm since his arrival in London; but I have no doubt of his having, long ago, pursued his voyage to Canada.

Just now I am occupied with the travels of Mr. Hasselquist in the Holy Land. I wish some botanist

* This is scarcely an adequate character of the book in question, which is rich in synonyms, and abounds with excellent observations, sufficient to establish the reputation of any botanist.

would furnish us with a *Flora* of Palestine, that philologists and divines might, with more ease and certainty, understand the plants of the Bible.

Next summer, if I live, I shall travel over the province of Scania, by command of the king.

I sent likewise, by the student above-mentioned, a dissertation on the Formation of Crystals.

We have no news in this part of the world. Every thing comes to us from your countrymen.

The seeds you sent have come up very well. If you wish for any of my plants, except those marked with a cross, let me know, and I will take care you shall have them.

My father, Nicholas Linnæus, born July 1, 1674, died on the 12th of the present month. He was a great admirer of curious plants, and always kept a choice garden of the more uncommon kinds.

Adieu! Believe me, &c.

HALLER TO LINNÆUS.

Göttingen, Aug. 12, 1748.

This will be delivered to you by your friend Missa, a young man extremely attached to you, and to every branch of botany, but especially desirous of studying generic distinctions. He has been here a few days.

I was never hostile to you, nor am I unjust; but I have been obliged by Hamberger, who gives your method of arrangement a degree of weight which you yourself would not sanction, to explain to him that I am obliged to differ from you on some points respecting Grasses, and a few other plants; that I might set an ignorant man right. I hope this will not interrupt our harmony, which you will see is unimpaired on my part, by my review of your *Hortus Upsaliensis.* This book, I know not by what means, is just come into my hands, and I have eagerly perused it. I find therein many alterations with respect to particular species. Your dissertations will be highly acceptable, and may be sent by Mr. Bergstraten, or any other conveyance. I would send you something likewise, but have nothing worth your notice. There is, indeed, come out a dissertation on a natural method of arrangement, which however is not mine, but Mr. Trentdenburg's. I will reserve it for you.

I have lately found plenty of *Mariscus marinus* *, and have re-examined its character, which is peculiar, differing both from *Schoenus* and from Micheli's account. In *Hieracium fruticosum angustifolium* † I have found the *placenta* (receptacle) furnished with little scales between the florets. In *Arctotis* the ligulate florets differ from your description, being abortive, while the androgynous (perfect) florets of the disk alone produce seeds.

You deserve highly of botanical science, for you are always furnishing something towards its advancement. That is denied me. The visit of the

* *Schoenus Mariscus. Linn. Sp. Pl.* 62.

† *H. umbellatum. Ibid.* 1131.

king to Hanover has involved me in much business, injurious to botany and its objects.

I will speak of the seeds hereafter. I have but a few, and I fear they daily become fewer. We have nothing new in Germany. You have Guettard's work. Farewell!

LINNÆUS TO HALLER.

Upsal, Sept. 13, 1748.

I received by Mr. Missa your most delightful letter, replete with candour and the most exalted friendship. This is what I wish to cultivate, and what I hold in the highest estimation. Only say what you would have, for your will shall be a law to me. Nothing shall be out of my reach that can secure your friendship. I would be at enmity with no man; least of all with a leader in my own science, and my old friend.

If you will but listen to me as a friend, I would advise you to write no answers to Hamberger and such people. He is not on a level with you; and the more he is your inferior, the more consequence you give a man who would otherwise remain in obscurity, known only to those immediately about him. Our great example, Boerhaave, answered nobody whatever. I recollect his saying to me one day, "You should never reply to any controversial writers. Promise me that you will not." I promised him accordingly, and have benefited very much by it. Your time is too valuable to the publick. You could do more than a hundred others for the advancement of the sciences. The majority of mankind judge without knowing any thing of the matter. Only observe how kings make war upon each other, and reap victories with the ruin of thousands of their subjects! So a man of letters comes off conqueror, with the loss of much of the credit that he really deserves. Our late friend Dillenius was liked by very few people, because of his dispute with Rivinus *; and yet he came off conqueror. He subsequently much regretted this quarrel, as he once told me. If what we do be correct or otherwise, so it will remain, though we fight for it with our lives. After we are dead and gone, boys, who now praise us, will be our judges. What Hamberger has done will stand for nothing, if it be wrong, though he may, for the present, come off conqueror †. Consider the botanical disputes of the early writers. Does it not turn your stomach to read them? Many read controversial writings with delight, who never love the acrimonious writer. They laugh and are disgusted. You will, of course, act in your own concerns as you please. I merely counsel you as a friend. A great general ought not to carry on the war too long. He often urges the enemy to what he might not else have thought of, and excites fresh hostility against himself. So it may easily happen that Hamberger will gain some

* See p. 84 of the present volume.

† Haller had at this time a vehement dispute with Professor Hamberger, about the theory of respiration.

ally, who will begin to call in question every thing you have done. Possibly he may lay hold of something, which he now knows nothing about, where you have made a slight mistake. I have said thus much in the confidence of friendship; and may I perish if I do not mean it in all sincerity. You will be very wrong to take it in any other light.

I have examined the *Mariscus*, and I perceive clearly the difference you have pointed out between it and the *Schoenus* of *Fl. Lapponica*, which difference I have mentioned in the *Iter Gotlandicum*. But pray observe the other species, and then see whether it be proper to found a new genus, considering the habit, mode of flowering, &c.

We are looking out for your reformed system, which you promised Missa, that I may introduce it into a new edition of my *Classes Plantarum*. When I meet with a fit opportunity, I will send you my dissertation *de Curiositate naturali* *. I request of you, as a particular favour, and for certain reasons, that you would write a review of this treatise.

That this letter may not come to hand empty, I will give you a very rare Moss, which perhaps you have not already. It is the *Muscus norvegicus umbraculo ruberrimo* of Petiver, mentioned in the Appendix of Dillenius †. I have met with the same, bearing a deep yellow *umbraculum* (or umbrella). What you relate concerning the *Arctotis*

* On the Use of Curiosity, translated by Stillingfleet in his Tracts, of which there are several editions.

† See a preceding letter, p. 407.

is very extraordinary. That the florets of the radius are abortive, and those of the disk fertile, I could not believe even with my own eyes or from my own pen. I therefore have, this very hour, examined them over again. But I find the florets of the radius in mine have large villous seeds, with a crown of obtuse chaffy scales. Those of the disk, only, bear very narrow dry seeds, or mere rudiments. Is not yours, therefore, some other genus, or perhaps species, which I have not seen?

I have been examining the *Tænia*, and have met with fourteen entire living specimens. I sought for the head, which all physicians have endeavoured to find in the *Lumbricus latus* (Common Tape-worm, or *Tænia)*, but in vain. What Tulpius has represented, in his remarks, as the head, is a gross imposition; for a real head is in each joint, and a mouth also, in one species underneath, in another at the side. No mortal could possibly understand this worm, unless he understood the nature of polypes, of which late writers have said so much. The *Tænia* partakes of this nature, and is propagated by the joints separating from each other, each joint continuing alive, and growing to a complete body. I have given a paper on this subject in the Upsal Transactions now going to the press.

At length I have ascertained the manner in which pearls originate and grow in shells*; and I am able

* "For this discovery the illustrious author was splendidly rewarded by the States of the kingdom." *Haller*. Specimens of pearls so produced by art, in the *Mya margaritifera*, are in

to produce, in any mother-of-pearl shell, that can be held in the hand, in the course of five or six years, a pearl as large as the seed of a common vetch.

You have, doubtless, seen the VIIIth volume of the *Naturæ Curiosorum*, and Mitchell's new genera there described.

His *Cynorrhymbium* is certainly *Mimulus*.
Diconangia - - - - - *Itea*.
Memecylon - - - - - *Andromeda*.
Pentstemon - - - - - *Chelone*.
Melilobus - - - - - - - *Gleditsia*.
Spondylococcus - - - - *Callicarpa*.
Trilopus - - - - - - - *Hamamelis*.
Trixis - - - - - - - - *Proserpinaca*.
Aphyllon - - - - - - - *Orobanche*.
Chamædaphne - - - - *Lonicera*.

Perhaps his *Aronia* may prove an *Acorus*.
Corion - - - - *Arenaria*.
Elymus - - - - *Zizania*.
Erebinthus - - - *Cracca*.
Hedyosmus - - - *Ziziphora*.

True and genuine genera are *Acnide*, *Angioplaris*, *Viticella*, and *Malacodendrum*. Imperfect ones are *Garosmus*, *Leptostachya*, *Myrrha*, and *Helix*.

All these things were very difficult for me to make out.

I did not misinform you, my dearest friend, when I gave you a list of the (new) members of the Royal Society of Stockholm. All whose names I men-

the Linnæan cabinet. The shell appears to have been pierced by flexible wires, the ends of which perhaps remain therein.

tioned were elected, and I received orders to inform them of it by letter. The Society indeed finally admits those only who give an answer, signifying their acceptance of this honour.

I have wished to amend many of the specific names, or definitions, in the *Hortus Upsaliensis*, in consequence of having become acquainted with more species of each genus. A person who does not take into consideration all the characters, errs as often as he writes such a definition, because he leaves some of them without their due weight.

Wishing you health and long life, I remain, &c.

HALLER TO LINNÆUS.

Göttingen, Oct. 17, 1748.

Why, my distinguished friend, do you preach peace to me, who is as desirous as any man to preserve it? who, you must be conscious, has ever shown that disposition towards you, and who is capable of being excited to anger by the injustice of but very few men. Many things have I borne without the slightest word of reply, which, in many cases, would have been as easy as possible.

Why do you request me to review any thing of yours with indulgence, when I contemplate with admiration all the productions of your most fertile genius, and ever shall? I eagerly expect your work, but, as yet, have seen nothing of it.

I have dissected and described the *Anemonospermos foliis & facie taraxaci incanis* *.

I am very glad to hear it was our friends' own fault that they were not received into the Society. If you value me, pray contrive to have John Gesner admitted, who is, as you know, greatly attached to you. He is president of the Physical Society of Zurich. It was my fault that this most amiable and modest of men called himself a member of your Academy. I am anxious that my negligence should not cause him any discredit.

Pray, my dearest Linnæus, let me not have to regret the advice that I here give you. Do not strike out species, and reduce them to varieties, so frequently as you are accustomed. Do not irrevocably determine the characters of plants till you have examined every particular. *Ophioscorodon* is not the *Allium* figured in Ruppius. It differs in having abortive flowers, much larger bulbs, and a convoluted stem; all which circumstances are different in the *Allium amphicarpum.* Its flowers are manifestly of a violet colour, the bulbs ten times smaller; nor is there any curvature of the stem. There are also some others of this genus, which you have erroneously combined. Thus you oblige me to differ from you, and to express our disagreement, as I cannot, without concern, see good and genuine plants perish, as it were, and become lost to botanists, under the title of varieties.

* *Arctotis tristis. Linn. Sp. Pl.* 1306.

My *Prodromus* of the *Flora* of Germany will be completed, I hope, in the course of a year; sooner it cannot.

Farewell! and grant me your accustomed regard.

LINNÆUS TO HALLER.

Upsal, Sept. 26, 1749.

I congratulate you, my illustrious friend, on the recent honours and distinguished rank which have been conferred upon you, who have so long acquired, by your learned writings, a reputation that never can decay. It eminently tends to the advancement of science, when princes bestow these splendid distinctions, as rewards of pre-eminent merit.

A packet from Professor de Sauvages, of Montpellier, directed for you, is in my hands. It appears to contain some printed pamphlets. You shall have it by the first opportunity, of which Dr. Rosen has promised to give me notice.

My printer has called for a new edition of the *Flora Suecica*, which is now preparing. I have corrected or expunged every thing of which you formerly took notice in your letters. If you happen to recollect any thing else, pray let me know time enough. I will never have any dispute with you.

Wishoff has asked me for a new edition of the *Classes Plantarum*, which I shall furnish him with in the spring. I shall insert your system. If you will be so good as to write it out for me, I will print it word for word. If you rather wish me to extract it from your book, I will do so, altering whatever you please to point out, nothing else.

In Scania I have met with the flowers, male as well as female, of both species of *Calamistrum*, figured by Dillenius, in his *Historia Muscorum*, by the name of *Calamaria* *.

Wishing you long life and happiness, I remain, &c. †

HALLER TO LINNÆUS.

Göttingen, Oct. 23, 1749.

You are extremely kind, my dear Linnæus, to pay so much attention to all my wishes, and more especially in your intention of striking out of your *Flora Suecica* whatever might have any appearance of inconsistency with our friendship. As to the *Classes Plantarum*, I will transmit, if I can, a view of my system, which has necessarily many gaps in it, as embracing but few genera; but I have made some slight improvements. I have no time to write it out at present, being, as you know, busy with my anatomical occupations.

* *Isoetes lacustris. Linn. Sp. Pl.* 1563. These supposed species are now considered as varieties, with longer or shorter leaves.

† This is the last letter of Linnæus, published by Haller.

I beg of you to give to our friend Rosen what you have received for me from Sauvages, and he will forward it along with his usual communications to me. I have sent you several things, some from Barere, and plates of the veins of leaves, published at Nuremberg; but you do not appear to have received any of them. I am under great apprehensions respecting the fate of this whole parcel.

I congratulate you on having made out the character of the *Calamistrum (Isoetes)*. I have been confined all this summer, first with a dangerous attack, and subsequently a chronic disease. My stomach still continues very much disordered, and my digestion is greatly impaired. I have neither found, nor indeed been able to search for, any thing new. I have however completed the characters of such German plants as do not belong to Switzerland, and these you will see in the *Prodromus Floræ Germanicæ*, if God preserves my life.

Several of your dissertations, a most acceptable present, have been forwarded to me by Rosen.

We are looking for news by the publick papers, but I have scarcely any thing in botany to mention. Ludwig has at length finished his work, where your writings are discussed; and the Figures of Rare Plants, edited by Trew, are perhaps ready. The publication of Hannæus, with a microscopical detail of the characters of *Tussilago*, has not appeared. In another volume (or number) he has given the *Galanthus*, but with a double flower.

I long to know what acquisitions we shall have from Kalm, who is likely to make many valuable discoveries. You cannot but derive a great influx of treasures from your travelling pupils.

May you be no less happy in the enjoyment of health! and may you continue your accustomed regard for Your HALLER!

MY NOBLE FRIEND, Berne, April 10, 1766.

I have received, within these few hours, a little book, accompanied with other pledges of your regard. Among them is your *Diapensia*, and a kind of *Gentiana* *, both which are different from any thing we have in Switzerland. I will take care to make manifest some proof of my gratitude.

I am now about preparing a second edition of my *Enumeratio Stirpium Helveticarum*, and shall devote myself to this work as long as I live. My publick labours indeed leave me but a small portion of my time for this purpose.

Nothing can be more acceptable to me than dried specimens of your rarer plants, that I may compare them with mine. I wish, with this assistance, again to go over all my alpine plants, of which I have got together a very considerable number. They may be sent either to Professor Murray at Göttingen, or to Ludwig at Leipsic.

* Probably *G. aurea*. *Linn. Sp. Pl.* 331.

This short letter of mine will be conveyed to you by your countryman who brought me yours.

Farewell! and believe me, &c.

We have no letters from Haller to Linnæus between that of Oct. 23, 1749, and this last. But there are a few from his son, Theophilus Emanuel de Haller, between Nov. 4, 1750, and March 20, 1751, apologizing to Linnæus for the publication of a dissertation of his, under the title of *Dubia contra Linnæum.* To the first of these letters it appears that Linnæus returned a kind and indulgent answer. He commissioned his pupil Loefling to remove these "Doubts" of his young correspondent, who has printed, at the end of a second dissertation of the same tenour, Loefling's letter and his own reply. On the 27th of June, 1753, young Haller wrote again to Linnæus, lamenting that he had ever been led, by an "itch for scribbling," to publish these "cursed works." He attributes his error to youth, and says his opinion is changed, by many causes, but principally by his veneration for the name of Linnæus. The doubts themselves are, generally, of the most trifling and inaccurate nature, and are conclusively answered, with candour and politeness, by Loefling, to whom young Haller returned as civil a reply. Here the matter ended.

It is to be lamented that Haller published so many confidential letters, unjustly reflecting, here and there, on Linnæus; and that he betrays, in his prefaces and notes, so much petulance towards this old and distinguished friend. He pretends, indeed, to have excluded, from all the letters he published, every thing personal or confidential. But there are few more disgraceful chronicles of ill humour than this collection of letters of various persons to Haller. He leaves chasms truly, in many places, which, like Madame Dacier's asterisks, is "hanging out lights;" for they serve to aggravate the force of what remains. Above all, he is censurable for printing letters from this very son of his, after his death, reflecting severely on persons who had, as the young man says, shown him the greatest favour at Paris; and abusing the Academy of Sciences, which had just elected him into its body as a corresponding member.

M. CELSIUS TO LINNÆUS. [Latin.]

Paris, Feb. 22, 1742, N. S.

MY BELOVED FRIEND AND BROTHER,

Your letter written last year was long in coming to my hands, through the neglect of my merchant; but the longer I had looked for it, the more welcome it proved. I wish, as sincerely as a friend can do, that you may long enjoy life and health, to fulfil the duties of the academical office, which your talents and your ardent love of Flora have merited, and which the publick opinion has long since allotted you. If what Mr. van Royen told me at Leyden be true, you have enriched your family with a new sort of plant, which I hope will conduce no less to the honour and increase of botanical science than of the said family *. The two brothers, Messieurs de Jussieu, your sworn friends as well as mine, desire their best remembrances to you. I have seen here, in the hands of Bernard de Jussieu, as also lately in Holland, the compilation of Seguier †, to which is subjoined the *Bibliotheca* of Bumaldus, the latter being a rare work, likely to assist the sale of its companion. The bulk of the volume is much increased by the enumeration of writers on horticulture, husbandry, and cookery. But Jussieu assures me the whole is but a very crude performance. The author, who accompanied Maffei in his travels through several parts of Europe, had opportunities of collecting the titles of books in various libraries, especially in the Royal Library at Paris. He had indeed begun this labour before you published your *Bibliotheca;* and not being able to bear the mortification of yielding to you the honour of originality, he attacks your little volume, in his preface, in very scandalous terms. But while hunting out errors, chiefly of the press, in your work, he has committed serious and disgraceful mistakes himself, as M. de Jussieu confesses. These it will be worth while for

* Alluding to the birth of Linnæus's first child, Charles, afterwards his successor in the Upsal professorship.

† *Bibliotheca Botanica, sive Catalogus Auctorum & Librorum' &c. à Joanne-Francisco Seguierio, Nemausense, digestus. Hagæ-Comitum,* 4*to*, 1740.

you and your friends to expose to the learned world, in the new edition of your *Bibliotheca**, of which I have already heard mention. Seguier's work was published at the Hague in 1740, in quarto.

Jussieu is anxious to receive the books which you have entrusted to the care of Count Ekeblad. He has given me the seeds you asked for. As I conceive it to be important that they should reach you in due time for sowing, I send them by post directly to Mr. König at Hamburgh, whom I have by letter requested to forward these seeds immediately to your Academy of Sciences, as being intended for the Upsal Botanic garden. I know of nothing published here in Botany. I presume you have already seen the fifth volume of Reaumur's Memoirs upon Insects, which appeared in 1740. M. Bolduc, a celebrated Parisian apothecary, member of the Academy of Sciences, died in January last. Mr. Olreich is lately arrived here from Germany, entirely devoted to medicine, or rather to chemistry. Mr. Berck has received five or six hundred pounds of silver, by draughts from Sweden, for the purchase of anatomical preparations illustrative of the muscular and glandular systems, for the use of your School. But this is probably no news to you. I lament the conduct of Mr. S——, whom I have two or three times seen here. To live upon other people's property does not become a man of honour. He is anxious to get away as soon as possible, and

* Linnæus generally declined such controversial or critical occupation, and certainly employed himself better.

is in daily expectation of money, which, if you can induce his relations to send, I trust you will readily afford him that assistance. He talks of returning early this spring.

Monsieur Obriet* is very well, but Isnard is confined to his bed. I have presented your compliments to both. Farewell, my good friend; make my best respects to your amiable wife; and believe me, no less attached by friendship than by other ties, ever faithfully yours, M. CELSIUS.

I had entirely forgot the *Parkinsonia,* which is here inclosed.

MR. MARK CATESBY† TO LINNÆUS.

SIR, London, 26 March, 1745.

On board the Assurance, Capt. Fisher, is a case of American plants in earth. They are a present to you from my good friend Dr. Lawson.

Knowing this his intention, by his consulting me what plants I thought would be acceptable, I selected these, as being hardy and naturalized to our climate, and consequently somewhat adapted to endure your colder air. Yet I wish they may not require as much protection from the severity of your winters as plants from between the tropicks do with us. Possibly you have already some of them; yet if but a few of them be acceptable, I shall be much

* The famous botanical draughtsman, usually called Aubriet.

† Author of the Natural History of Carolina, and Fellow of the Royal Society. He died in 1749, aged 70. *Pulteney.*

pleased. And whatever other American plants in this inclosed catalogue will be acceptable, you may freely command any that I am possessed of. In regard of that esteem your merit claims, I am, Sir, your most obedient, humble servant, M. CATESBY.

P. S. This case of plants were intended to be sent last May, and they were sent to the ship with the consent of the skipper, yet they were refused to be taken on board.

Cupressus americana (1).
Arbor tulipifera (2).
Cornus americana.
Populus nigra carolin. (3).
Periclymenum (4).
Barba Jovis arborescens (5).
Phaseoloides (6).
Arbor virgin. citri folio (7).
Euonymus americanus.
Aster americ. frutescens (8).
Lychnidea flore purpureo.
Styrax aceris folio (9).
Bignonia Catalpa.
Angelica spinosa (10).
Pseudo-acacia (11).
Anapodophyllum canadense (12).
Rubus americanus (13).

(1) *C. disticha.* (2) *Liriodendrum.*
(3) *P. balsamifera.* (4) *Bignonia sempervirens.*
(5) *Amorpha fruticosa.* (6) *Glycine frutescens.*
(7) *Laurus Benzoin.* (8) *A. linarifolius.*
(9) *Liquidambar.* (10) *Aralia spinosa.* (11) *Robinia.*
(12) *Podophyllum peltatum.* (13) *R. occidentalis.*

JOHN MITCHELL, M. D. TO LINNÆUS. [Latin.]

LEARNED SIR, London, April 16, 1747.

Your last letter to Dillenius is just come to hand, and I am commissioned to answer it *. I the more readily undertake this, that I may have an opportunity of expressing my esteem for your excellent writings, and their eminently learned author; as well as of returning you my thanks for the kindness and compassion you have been pleased to testify for the misfortunes of myself and those who belong to me.

In the first place I am sorry to inform you of the death of your friend Dillenius, which happened before the arrival of your letter! I understand his disease was a severe stroke of apoplexy, which carried him off in the end of March last. Such is the inevitable consequence of intense application, which ought to serve as a warning to others!

It is doubtful how far this most learned botanist has accomplished his object. The Sherardian *Pinax* is said to be still incomplete, though he has, for several years past, devoted himself so assiduously to this work, that he appeared to consult the ease of other botanists far beyond his own. His loss to science is unspeakable, and perhaps, at this time, irreparable, unless you could be induced to lend a hand to the great work, which is so much wanted!

You ask, what known genus is allied to *Chionanthus*? I answer, *Olea*. They each bear racemose

* See a letter of Mr. Collinson's, vol. I. p. 18, which accompanied this.

flowers from the bosoms of the leaves. The calyx differs merely in duration, and the corolla in length. The stamens, pistil, succulent *drupa*, and rugged nut, are the same in both. The stamens are naturally two; by great accident I have found three, but it was only in luxuriant flowers of *Olea sativa*. Hence I define *Chionanthus* to be an *Olea*, with ovate deciduous leaves, and very long flowers.

What a wonderful plant is that which you call *Anandria!* We know nothing of it, at least by that name. Does the name express its nature? Is it a plant without anthers?

Catesby has shown me, in the Swedish Transactions, the character of *Stuartia*, which I wish you would re-examine. I found it answer to the *Monadelphia Polyandria*, and *Pentagynia*. The fruit when ripe is a capsule of five cells and five valves, as in other Malvaceous plants. But I have neither your description nor my own by me at present.

What our learned people, about whom you enquire, are after, I know not; but I believe they are doing nothing. Perhaps you have exhausted the riches of Nature; or the Muses are silent in the din of arms. Martyn has only completed the Abridgment of the Philosophical Transactions. Catesby has finished the Appendix to his work. Mr. Hill, an apothecary, who published Theophrastus on Gems and Stones, with notes, is expected soon to print a History of Fossils. I am inclined to give the publick something on the natural and medical history of North America, if not a history itself, at which I have long laboured. I wish I could profit by your assistance.

The periodical publications here afford nothing commendable at present. A short catalogue of what books are printed is given in the Magazines. There has lately appeared a monthly work on literary subjects, under the title of "The Museum."

Are you acquainted with the manner of preparing what are commonly sold by the name of Russian Ashes? By what art, or peculiar process, do the Russians convert ashes into a salt?

I have not yet received the American seeds I expected, or I would have sent you some. I shall be extremely happy to be favoured with your correspondence; and if I can do any thing to deserve such a favour, pray command your devoted servant,

JOHN MITCHELL.

SIR, London, Aug. 10, 1748.

I was favoured with your letter, by the hands of Mr. Kalm, a very short time before he left this country; since which I have been almost constantly out of town, and have therefore not been able to write sooner.

I had sent you a parcel of seeds by Mr. Burgmester, before I received your letter. I hope they arrived safe, as well as several living plants from his

Grace the Duke of Argyl. They consist of various kinds of Larch, Fir, Walnut, Birch, *Rhus*, *Myrica*, &c. from America. The Duke promises to send you more hereafter, as well as the kind of Bear called the Raccoon, when he can find an opportunity.

I am extremely obliged to you for the dissertations you sent by Mr. Kalm. They all interest me very much, but particularly that "On the Virtues of Plants," which is a part of botany to which I have paid particular attention. I am also glad to obtain that upon Ashes, "*de Cineribus clavellatis.*" If you have any thing more on these subjects, I beseech you to inform me.

Have you any treatise on the mode of making Pitch, by the Danes, Norwegians, or Swedes? It is a curious art, and Tar-water is, at present, highly commended by many people. We have numerous publications relating to it.

Your undertakings, of a *Species Plantarum* and *Systema Naturæ*, cannot but prove highly acceptable to the learned world, and they are anxiously expected. I have scarcely time at present to transcribe any remarks I have occasionally made about such matters; I have indeed referred a considerable number of American plants to their places in your system; but I have nothing worthy the attention of your readers.

I should have been happy to send you a few plants, if they had not been so much damaged by pirates, as well as injured by their sea voyage; so that, among more than a thousand specimens, I have scarcely a perfect flower. They came from Virginia to France, and thence to Hamburgh, Rotterdam, and London. I hope, however, to get a fresh supply, and to resume a study which I was by degrees obliged more and more to neglect.

I can now only send you a few small seeds, just arrived from America, and some dissertations of Mr. Tennant upon the *Polygala*, two of which only have come out, among his latest publications. His former ones, of inferior merit, are not now to be had. I send also the *Polygala* root, recently gathered. If you want any thing more, I shall be very glad to procure it.

The descriptions I drew up, of new genera of plants, have been sent by Mr. Collinson to persons in various parts of Europe, so that I scarcely know at present where to procure a copy. I will try shortly whether I can find any others worth submitting to your inspection.

You have heard from Mr. Kalm himself that he had left England, since which time we have no news of him. I hope his journey will be prosperous. I have rendered him every service in my power, as I shall always be happy to do to you, or any one connected with you.

Your obedient servant, J. MITCHELL.

SIR, London, Sept. 20, 1748.

I have just received your letter of the 12th of August, written since my last, which was accompanied by various seeds and plants, and forwarded to you, with several other things, by Mr. Collinson. I now proceed to reply to yours.

I am happy that we agree about the genus *Napæa* and the *Althæa* (rather *Malva* of Hermann) *ricini folio.* I see that you have observed both; which at first escaped me.

We are anxious to see a fuller exposition of your Botanical Clock, *Horologium Floræ*, which cannot but be very curious and useful, if it can be brought to perfection; as also the experiments you have made on the vegetable food of domestic animals.

I long above all things to peruse the last edition of your *Systema Naturæ*, as well as many other things. Where was it published? I wish your bookseller would send some copies to London. They would, doubtless, sell readily.

You say Steller has, in his journey, penetrated into Canada from Russia. This, if true, has not come to our knowledge. Be so good, therefore, as more fully to explain by what route, and into what parts of America, he is said to have travelled, and what is your authority.

The report which has prevailed in Sweden, of the death of Mr. Miller, arose probably from Joseph Miller, superintendant of the garden, having died in the early part of this summer. Philip Miller, the gardener, is alive and well, and, as he informs me, is about publishing some Figures of Plants.

You desire us here to examine the *Linaria Scoparia* * with a microscope. I know not whether this plant is to be found in the English gardens, nor have I ever seen it, otherwise I should be glad to make this examination.

Mr. Catesby desires his compliments, and wishes to know whether the plants he sent you by our friend Lawson are alive and flourishing.

Mr. Trew, to whom Mr. Collinson sent a few small papers of mine, informs us that they have appeared in the last volume of the Nuremberg Transactions †. They consist of a dissertation on a new botanical principle, derived from the sexual theory, which I think accords with your ideas, and, if I mistake not, our systems support each other; also characters of several new genera of plants, sent, seven years ago, from Virginia. I long to know your opinion upon them, which I hold in high estimation. Some of these genera have, I believe, appeared in your last publications and those of Gronovius.

May you long enjoy life and health! Do not forget your sincere friend and servant,

JOHN MITCHELL.

* *Chenopodium Scoparia, Linn. Sp. Pl.* 321, a common annual, in old rustic gardens, by the name of Belvedere.

† *Ephemerides Academiæ Naturæ curiosorum.*

SIR, London, May 1, 1750.

Since your last letter, accompanied by several highly acceptable presents, reached me, I have taken a long and laborious journey, with the Duke of Argyl, to the uttermost parts of Scotland, over the mountains and wilds of that country, as well as through various counties of England. On this account I have been unable to reply to your letter, or to communicate many things that I have had in my mind. Since my return, I have not enjoyed very good health; so I must rely on your indulgence for my apparent negligence.

We collected in our tour many Lapland and Swedish plants, mentioned in your Floras, but met with nothing new.

Meadows in Scotland abound with the *Anthericum*, or *Pseudo-Asphodelus*; not only the common kind *, but also the *Phalangium Scoticum minimum* of Ray †, apparently a variety of the other. The cattle fed in these meadows are very subject to that fatal disease described by Simon Paulli as a dissolution of the bones, caused by this herb ‡. Have you observed any thing of this kind in Sweden, as proceeding from the cause alluded to?

Rubia minima § of Lobel and Gerarde grows plentifully all over the sandy islands on the west of Scotland. The inhabitants make great use of the

* *Narthecium ossifragum. Fl. Brit.* 368.

† *Tofieldia palustris. Fl. Brit.* 397.

‡ *Paull. Quadripartitum Botanicum,* 529.

§ *R. peregrina. Linn. Sp. Pl.* 158. *Fl. Brit.* 181.

wild plant for dyeing wool red. The root is large and creeping, of the colour of coral, which is the part used.

I must beg you to allow me a little time to collect my other observations, as well as my replies to your queries. Time presses just now, but I would not let slip the opportunity of sending you something. While the navigation of the Baltic was stopped, I could convey nothing to you. Nothing can be sent from hence by the post, not even a single letter, without paying half the postage, which would be more than the value of this whole packet, which will still cost you too much.

I entrust to the mate of the ship two small parcels, one containing various American seeds, just come to hand, the other two publications on pitch and tar-water, written by Dr. Hales and other able men.

The learned here differ about the manner of preparing tar, on which, it is to be hoped, you can give us information. The chief question is, whether it is obtained from old wood, or from the fresh sap, or the growing tree, and whether from the roots, or the whole trunk? Is the wood in the furnace allowed to burn with a gentle not violent flame, or is it only surrounded by fire? and, if the latter, how is it contrived? We learn, from the periodical publications, that a work has lately appeared in Sweden upon this very subject of making tar, which I beg of you to send me by the first opportunity, in whatever language it is written. Our people are daily swallowing this article, as a medicine, especially its acid spirit described in the publications I send you, and yet we are ignorant of its true preparation.

We are daily receiving abundance of new plants from America, many of which, I hope, may, this autumn, afford seeds, and of these you may expect to partake.

Mr. Catesby is dead, after having completed his book, which is already fallen in price, being now sold at seventeen guineas and a half.

Pray continue your kindness to me, and honour me with your letters. If I can be of any service to you, I beseech you to inform your devoted and most humble servant, JOHN MITCHELL.

Direct to me F. R. S. at the apartments of the Royal Society.

HIS EXCELLENCY CADWALLADER COLDEN, GOVERNOR OF NEW YORK, TO LINNÆUS.

Coldengham, in the Province of New York,
SIR, Feb. 9, O. S. 1748-9.

You have done me so much honour by your two letters, one of Aug. 6, 1747, the other without dàte by Mr. Kalm, that I cannot otherwise account for it than by your willingness to encourage every attempt to promote knowledge; for I am so sensible of my want of skill in the botanical science, that I can no way deserve the praises you are pleased to bestow on the little performances I have made.

When I came into this part of the world, near forty years since, I understood only the rudiments of Botany; and I found so much difficulty in applying them to the many unknown plants I met with every where, that I was quite discouraged, and laid aside all attempts in that way near 30 years, till I casually met with your books, which gave me such new lights that I resolved again to try what could be done, with your assistance. If then I have been able to do any thing worth your notice, it is entirely owing to the excellency of your method. However, I still find myself at a loss in a fundamental point in Botany; what it is that certainly distinguishes one genus from another of the same class, so as we may not be in danger of confounding plants of different genera by reducing them into one, or make different genera of such as are really one. This difficulty I find puzzles sometimes the masters of the art, by the disagreements in judgment on that head, which appear among them. I mentioned this to Gronovius, but I have not heard from him since. I am persuaded that if this difficulty can be removed by any method, it may by yours.

It is observed that animals of different genera seldom copulate, unless they be of the next adjoining genera in the chain of nature; and that though they produce young by their copulation, the breed of these are never capable of continuing their species. So it is observed of the mule proceeding from an ass and horse. We observe the same of our wild geese and household geese. They

26.

Viro Clarissimo, Experientissimoque D. D. Carolo Linnæo

Sal. pl. dis. J. C. Mutis.

Quærenti mihi jam diu, quid esset, quod ad me præter expectationem scribere intermisisses V. H. neque inconstantiæ, neque negligentiæ causam ullam occurrere videor.

27.

Viro Celeberrimo et Sapientissimo Naturæ Scrutatori Dno Mutis s. pl. d. Carolus à Linné

Acerrima mihi orta est dies 10 januari hujus anni; dum jacturam Carissimi Parentis ferre debui; per duos annos vixit post tres insultus paraplecticos B. m. parens debilitato corpore, at tandem illa die a Dysuria abeunte in gangrænam terminante mortem.

28.

Adieu, Monsieur, continuez d'ouvrir et d'interpréter aux hommes le livre de la nature; pour moi, content d'en déchiffrer quelques mots ~~à~~ votre suite dans le feuillet du regne végétal; je vous lis, je vous étudie, je vous médite, je vous honore et vous aime de tout mon coeur.

J J Rousseau

J. Martin, Lithogr. Rochester St Westm.

by copulation produce young, but these never continue their species any further. And yet we have several species * of the household geese, which to all appearance differ as much, or more, from each other, than some of them do from the wild geese. It is likewise certain that throughout the whole genera of both animals and vegetables, the distinction of the male and female is every where observed, and that this distinction is necessary for the propagation of the kind. But what it is in the one and in the other that makes it necessary they should be distinct, so that the species cannot be produced by one alone, I know not, nor do I find that any philosopher has attempted to explain it. What adapts the female of one species to the male of the other, is not merely the size, shape, or number of their respective organs, as is plain in the case of the wild and tame geese, and the horse and ass, but something else likewise which I know not. What I would infer from this is, that a perfect similitude in the figure, number, &c. of the organs of generation, is not a certain characteristic of their being of the same genus; but that something else must frequently be added to distinguish the genera. You have on your principles made the apple and pear of the same genus, and yet I am persuaded they are of different ones, because there is something so different in the nature of the pear and of the apple, that a stock of the one is not proper for

* Rather varieties.

the scyon of the other in grafting or inoculation *, whereas, stocks of the same genus, though of different species, equally serve for the scyons of any other species, as those of the same species do. We observe the same in different species of animals. The males of any species of one genus, equally serve the females of the other.

I did not intend to say that the *Zea* is not a native of America. If you have seen all I have written to Gronovius on this head, you will evidently perceive my intention was otherwise, however I may have expressed myself to occasion that mistake. On the contrary we have many different species (varieties) of it, which, as far as I can learn, are not in Europe or Asia. From what I have observed of this plant, I think it necessary to take in the magnitude of the plant, and the time of its producing the seed in full ripeness, as a distinction of the different species of this genus; for after the most careful examination, I can discover nothing else to distinguish them, and yet they are certainly distinct species. Give me leave on this occasion to make one observation, though it be a very obvious one. There are some plants, and those the most necessary for human life, which grow no where but when sown by men's hands, and in cultivated lands, such as our

* This is a mistake. Apples and pears may certainly be grafted upon the same stock. On the other hand, the *Chionanthus virginica* is very successfully grafted upon the common ash, a tree indeed of the same natural order with itself, but not of the same genus.

Indian Corn, or *Zea*, Wheat, Barley, Rye, &c. that never were the spontaneous produce of the earth, without the art and labour of man, otherwise they must be found somewhere growing spontaneously. So the household animals, dogs, cats, dunghill fowls, &c. seem to have been concomitants with man from the beginning, and unable to live without him. For the species at least, and I believe I may venture to say the genera of household animals, are no where wild, but have from the beginning been dependants on man. Man therefore has a natural right over them; whereas he seems to be in a state of war with other animals.

As I had but little skill in Botany when I began first to examine the plants of this country according to your method, it is no wonder I fell into many mistakes. Most part of them I afterwards discovered by myself, and sent proper corrections to Gronovius, the most material of which I shall copy for your use. I likewise sent several dried plants to him, but they had the misfortune to be taken by the French. I had directed the packet, in case of capture, to be delivered to *Messieurs du Jardin Royal à Paris*, so that perhaps you may still hear of them.

I have been obliged for near three years past to lay aside all botanical amusements, the publick affairs of this government having obliged me to be, during the summer season, on the frontiers, where we could not go out of the fortified places, during the cruel and barbarous war with the French Indians, without danger of being surprised by the skulking enemy.

I hope, now we have peace, it may be in my power to make some return to the obligations you have laid on me, by sending some dried plants as you direct, with such descriptions as I can make of them. I cannot hope for any great reputation from what I do of this kind; but if you think that my observations or descriptions can be of any use to the publick, you have my leave to make use of them in whatever manner you shall think proper.

I received the dissertation you sent by Mr. Garden, and likewise your *Fauna Suecica* and *Flora Zeylanica* from Mr. Kalm. They are very acceptable, as I could not otherwise have procured them in this country. They shall remain with me as tokens of your esteem for me, which I hope my children will value after I am gone.

Mr. Kalm arrived so late last fall in Pensylvania, that the season of the year did not permit him to proceed in his intended voyage. He tells me he designs to go to Canada next spring. I hope to see him at my house in his way thither, and to have the pleasure of his conversation for some days. It will give me pleasure if I can be of use to him in making his voyage more convenient or safe. If you will please to continue your favours of writing to me, Mr. Collinson of London will take care of your letters, to transmit them to me; or if they be sent to Philadelphia, to the care of Mr. Benjamin Franklin, postmaster there, they will come to my hands. God preserve you in health for the benefit of mankind. But before I conclude I must inform you that the title of *Summus Præfectus* no way belongs to me. I know not what has led you into this mistake. I am, with great regard, dear Sir, your most obedient humble servant, CADWALLADER COLDEN.

SIR, Coldengham, Feb. 1, 1750-1, English style.

Mr. Kalm being now on his return home, I cannot suffer him to go without acknowledging the favours which I have received from you, a person so highly esteemed and distinguished in the republic of letters, and to whom the world owes so much, for the vast acquisition it has gained in Botany and all parts of Natural History, by your wonderful skill and indefatigable labours. My last to you was entrusted to Mr. Kalm's care. In it I gave you several corrections of the botanical observations which Gronovius sent to you from me, and which Mr. Kalm told me are published in the *Acta* of your Academy. I hope you received them before those observations were published, and that the proper corrections were made. I am very desirous to see them, as they are printed, and I must beg the favour of you to send them to Mr. Collinson in London for me, if you have no opportunity of sending them to Mr. Benjamin Franklin in Philadelphia, because I know not otherwise how to obtain them.

Mr. Kalm has so much more knowledge in Botany and Natural History than any in this country can pretend to; he has been so industrious, and has undergone such great difficulties in travelling through a great part of this vast forest, and risked such dangers in his person from its savage inhabitants, that as, on the one hand, his zeal in the pursuit of knowledge cannot be sufficiently applauded, so, on the other hand, I have no hopes left me that I can be of farther use to you. However, Sir, if there remain any thing in which you think I can give you information, you will give me the greatest pleasure in receiving your commands. And though it be too probable, that you may have no such inducement to write to me, yet I hope you will so far favour a person who has the greatest esteem for your merit, as to let me sometimes know that you live and continue an ornament to your country, by giving some account of the fruits that you daily produce. I am, with the greatest respect, Sir, your most obedient humble servant,

CADWALLADER COLDEN.

P. S. Mr. Kalm was so industrious, I could not persuade him to stay above one night at my house in the country, though the fatigues he underwent required his taking some ease and refreshment. And my happening to be deeply concerned in publick affairs deprived me, in a great measure, of the pleasure of his conversation while he was in the city of New York.

20

And though it be too probable, that you may have no such inducement to write to me, yet I hope you will so far favour a person who has the greatest esteem of your merit as to let me sometimes know that you live & continue an ornament of your Country, by giving some account of the fruits which you daily produce

I am with the greatest respect

Sr

Your most obedient
humble Servant
Cadwallader Colden

21.

Noailles Duc d'Ayen

22.

Salutem tibi plurimam dicit. cc. B. jussieu. Vale; negotia Botanica prosequere, rei herbariae lumen adjice; ego quoad nova eladhuc incerta, officia mea praestandi occasionem laeto animo arripiam.

Parisiis 28.a Junii 1754.

Tibi Deditissimus et
Devotissimus cultor.
Adanson

R. Martin's Lithog.y Rochester St Westminster.

The material originally positioned here is too large for reproduction in this reissue. A PDF can be downloaded from the web address given on page iv of this book, by clicking on 'Resources Available'.

LINNÆUS TO THE MARQUIS OF GRIMALDI, AMBASSADOR FROM THE KING OF SPAIN TO THE COURT OF STOCKHOLM. [Latin.]

MY LORD, No date — written probably in 1750.

His Majesty the King of Spain having been pleased, as I learn from Dr. Raibaud, to express a wish that some young man, well versed in natural history in general, should be sent to investigate the productions of the various Spanish provinces; in order to further, as much as possible, a design so beneficial to the inhabitants of those countries, as well as to all lovers of natural science, I beg leave to submit to your Lordship the following remarks.

Spain is well known to be more favoured by nature than any other part of Europe, vieing with the Indies in the warmth and mildness of its climate, so as to be well calculated for bringing to perfection every production of nature, that can serve to the use, or even the luxury, of civilized society.

It is now generally understood that the powerful sovereign of this fortunate country gives all possible encouragement to every useful art and pursuit — to agriculture and planting, as well as to manufactures and commerce; all which his Majesty is anxious to have brought to the highest perfection that human industry can attain.

Whatever the bounty of nature affords for the benefit of mortals, except the elements, comes under the denomination of Stones, Plants, and Animals. Nothing else is to be met with in this terraqueous globe; and in order that man should render these things subservient to his own use, it is necessary that he should study their distinctions and properties.

The first step of science is to know one thing from another. This knowledge consists in their specific distinctions; but in order that it may be fixed and permanent, distinct names must be given to different things, and those names must be recorded and remembered.

Without such a knowledge of the individual productions of nature, no solid principles for the improved cultivation of fields, meadows, or woods, no certain information relative to plants, whether for the purposes of medicine, food, dyeing, or other useful arts, can be taught or acquired.

In order for the promotion of this kind of natural science, upon the best foundation, I can recommend a young man named Peter Loefling as one of my favourite disciples, and indeed the most able among them at present; for whose thorough knowledge of plants, fossils, and animals, as not inferior to that of Kalm, or of Hasselquist, who have been sent from Sweden on similar expeditions, I will be responsible.

The object of this young man will be, in travelling through Spain, to collect every species of plants and trees, not neglecting even the smallest mosses; as well as to observe all animals, birds, fishes, insects, and the minutest worms. He will also investigate all kinds of stones, minerals, fossils, and earths, which the surface of the ground may afford.

He will describe, or define, these several productions, for which he is well qualified, noting down the names given them by authors as well as by the natives of the country where they are found. He will record where every specimen was gathered; its use in medicine, dyeing, or any other respect; not omitting the advantages or mischiefs which may arise from the superabundance of any particular species.

A dried specimen of each plant, preserved, as much as possible, according to its natural appearance, and glued upon paper, with its scientific as well as vulgar name, principal places of growth in your country, and its useful or noxious properties, will be transmitted by him to the king.

The result of this undertaking cannot but be highly beneficial. Some useful plants, natives of Spain, or capable of being cultivated there, though hitherto perhaps purchased from abroad, may be discovered; while, on the other hand, means may be pointed out for eradicating such as are injurious. Soils and situations fitted for the cultivation of particular species with advantage may be pointed out. Popular errors in husbandry may be detected, and improvements suggested. Lands, now barren, may be found capable, by some means or other, of cultivation. The traveller will every where notice such medicinal plants as grow wild, and which may perhaps be exported to other countries with advantage; and he will extend these and similar observations to animals and minerals, not overlooking the meanest insect.

The example of Sweden, if I mistake not, proves the utility of studying the natural economy of one's own country. Agriculture has scarcely been attended to at all amongst us for more than 20 years. During the first half of this period we endeavoured to improve our farming by travelling to England, or by translating English books. But the climates of Sweden and England are different. We therefore have found more advantage in attending to the good things which our own country produces; and we have succeeded so well as not to want foreign aid. So your countrymen, if they will but make themselves once acquainted with their own natural productions, may not find it necessary to recur to England or elsewhere for arts or merchandize.

If your truly beneficent Sovereign would be pleased to establish a garden for the purpose, I would direct my pupil above-mentioned to furnish it with roots or seeds of all your native plants, in order that those who are studious of such matters may there learn what their country produces. Or if his Majesty would appoint a certain number of young men to accompany Loefling, he might point out to them every important production in its native soil, and teach them, at the same time, the principles of botanical discrimination.

I have only to request that my pupil may be allowed to send seeds of the more uncommon plants for the Upsal garden, and specimens for my herbarium. I rely on your Lordship's favour, to provide for the safety and due accommodation of this really deserving man, on as economical a plan as possible, not omitting any thing that may be essentially useful to his undertaking. I shall not neglect to suggest or to communicate whatever may occur to me for the promotion of the important objects in view.

"*Felices agricolæ sua si bona norint.*"

DR. RAIBAUD TO LINNÆUS. [French.]

SIR, Stockholm, Jan. 7, 1751.

The Marquis de Grimaldi has received an answer from his Court, on the subject of what I communicated to him from you, relative to the sending one of your pupils to botanize in the provinces of the kingdom of Spain, under the conditions which you have proposed. The King of Spain has granted the request of the Marquis de Grimaldi, and will allow 100 ducats annually to the person you shall send, with every assistance that he can require. The Marquis has even told me that he will cause this salary to be augmented every year, provided the person sent should prove worthy of it by his learning and merit. It therefore only remains with you, Sir, whenever you judge proper, to prepare the person you shall recommend for his departure. The Marquis de Grimaldi will furnish him with the necessary letters of recommendation.

I have the honour to wish you and Madam de Linnæus a good and happy new year.

I am, with much attachment, Sir, your most humble and most obedient servant, RAIBAUD.

THE DUKE D'AYEN * TO LINNÆUS, THE GREAT AND LEARNED RESTORER OF BOTANY. [Latin.]

SIR, Paris, July 1, 1752.

I am favoured with your two dissertations, entitled *Nova Plantarum Genera* and *de Plantis hybridis*, and have perused them with no less pleasure than the rest of your works. How greatly Botany, my favourite study at present, is indebted to you, more especially for the immutable establishment of genera, and the certain discrimination of species, I am so ready to agree with all the learned world in acknowledging, that I am ardently desirous of entering into correspondence with you upon these matters. I shall do my utmost to furnish you with the rarest kinds of plants from my garden, more particularly, as far as possible, with such as you have never seen. If, however, there is any thing of

* Afterwards the Marechal Duke de Noailles, who died, at the age of 80, in February 1793, about a month after the death of Louis XVI., to whom he was strongly attached by affection as well as duty. He was one of the first four honorary members of the Linnæan Society.

European growth which would be acceptable, I beg you to point it out, and I will procure it as soon as I can. I return you thanks for what you were so kind as to send me, along with the above dissertations, by M. d'Havrincourt.

Farewell, Sir, and continue your learned writings, for the benefit of all good men! I remain, &c.

NOAILLES DUKE D'AYEN.

ADANSON TO LINNÆUS. [Latin.]

SIR, Paris, June 28, 1754.

Being just arrived from the west coast of Africa, I beg leave to acquaint you with my return to France. I have been devoted, from a child, to natural history and philosophy, and the persuasion of my friends induced me to devote five years to the investigation of a tropical country at Senegal. M. Bernard de Jussieu, by various letters, has informed me of my being not unknown to you, in consequence of numerous communications relative to the above science, and especially to botany, which I have sent every year to the Academy at Paris. I am therefore unwilling that you should any longer remain ignorant of my return, or of my zeal and affection for the highly important study of botany. I am anxious for your acquaintance and correspondence, more especially that we may thus discuss the hidden secrets of Nature, and that, as far as my present occupations will permit, I may lay before you what I have remarked in foreign countries. I have there discovered many new genera, as well as many that authors have imperfectly described; but still more numerous are the instances of plants referred to wrong genera; as you will see by a work which I mean next year to publish. More of this another time. It will be more agreeable to you at present to receive some seeds of Senegal plants, hitherto unknown to any botanist. I send you at present nine new genera only, as my letter cannot hold any more, in order that you and other botanists may judge how much still remains to be done in botany, to perfect the principles of an artificial method. The genera in question are thus numbered in my list. N° 15, *Bdellifera arbor*, the tree from which flows the gum-resin called *Bdellium*. N° 33, N° 42, *Kahower*, which the Europeans call a Cherry. N°s 53, 76, 98, 101, 109, and 213.

Please to inform me if any thing is known concerning them, especially the *Bdellium*. The Europeans know it by the name of *Thus* (Frankincense), instead of which this substance is sent to Europe for the use of churches. But though not very different, it is easily proved by examination to have all the properties of *Bdellium*. Time will not allow me to say more. What I now send will serve as a pledge of my strong attachment, till I am enabled to communicate more.

The celebrated Bernard de Jussieu desires his best compliments. Farewell! May you continue your learned labours, to illustrate the science of Botany! I shall be happy to seize every opportunity of communicating any thing in my power that is new, or hitherto not well understood.

Your most attached and devoted admirer,

ADANSON.

Paris, Oct. 2, 1758.

I have received your long-wished-for letter, without a date, but not the less welcome, and was preparing an answer, when innumerable impediments at once beset me. In the first place we are lamenting the death of my dear friend Anthony de Jussieu, a man of rare virtue and vast erudition, the Professor of Botany at Paris, who was taken off suddenly, with scarcely a day's illness. At the same time Bernard de Jussieu laboured under an obstinate inflammation in his right eye, the weakest of the two, which greatly endangered his sight. This happened when the botanical demonstrations in the Paris garden pressed upon him, so that he was obliged to devote his whole attention to European plants. In order to afford some assistance to so good a friend, I undertook the herborizations. Indeed I have been entirely occupied, from June to the present month of October, in collecting, drying, and describing plants, though but too frequently interrupted by domestic affairs. These are the causes of my silence; to which I may add the having been obliged to attend the Academy ten, or perhaps 100, times, in the place of Anthony de Jussieu.

Among numerous new observations in natural history which I have formerly communicated to the *Academie des Sciences*, is a complete description of the *Bahobab*, which Bernard de Jussieu has named *Adansonia*, and of which I had long ago given a description before your letter reached me. B. de Jussieu had refrained from sending you this description during my absence, that he might not deprive me of the opportunity of giving you pleasure. I therefore now send the essential parts of the character which you ask for, taken from the Memoirs of the Academy intended for publication, or rather from my own Latin manuscripts, according to the plan of your *Genera Plantarum*, as I mean to give them to the publick.

ADANSONIA.

Calyx. Perianth simple, of one leaf, cup-shaped, divided half way down into five revolute segments, deciduous.

Cor. Petals five, nearly orbicular, ribbed, revolute, united by their claws to the stamens and to each other.

Stam. Filaments numerous (about 700), united in their lower part into a conical tube, which they crown at the top, spreading horizontally. Anthers kidney-shaped, incumbent.

Pist. Ovary nearly ovate. Style very long, tubular, variously twisted. Stigmas from 10 to 14, prismatic, shaggy, spreading from the centre.

Peric. Capsule oval, very large, woody, not bursting, internally separated into from 10 to 14 cells, filled with dry pulp and with seeds; the partitions membranous and longitudinal.

Seeds numerous, almost bony, kidney-shaped, lodged in friable pulp.

Hence you may perceive how much this genus differs from the rest of the Mallow tribe. First, by the calyx falling off immediately after flowering;—second, by the number and situation of the filaments at the top of a monadelphous tube; — third, by the number and form of the stigmas; — fourth, by the woody and close capsule, with its pulp and cells;—fifth, the compound fingered leaves; — and sixth, by the tree itself, which of all hitherto discovered is the most prodigious in the size of its trunk and branches, being as it were the stupendous vegetable monster of Africa. This tree is found in the country of Senegal only, from whence its fruit, with that of the *Agihalid*, is sent every year, as an article of commerce, to Egypt. Some of its seeds having been planted there, in a garden, one or two trees were raised, which appear, from Prosper Alpinus, to have attained no remarkable size, nor perhaps to have flowered, if we may judge by the figure in that author, which in every particular, except the fruit, is erroneous. In the West Indian island of Martinico, a single tree of this kind, already full grown, bearing flowers and fruit, is carefully preserved. It was formerly sown there by the negroes. These and similar remarks are detailed, with my authorities, in the communications to the Parisian Academy.

A second volume of my History of Senegal is going to the press immediately. As soon as it comes out, I will beg your acceptance of it, if I can meet with a safe conveyance *.

I am not able to determine whether the Birds of which I have spoken, as armed with spurs on their shoulders, and screaming at the approach of men, are the same as what you have called *Charadrius spinosus*, because the latter is not to be found in the sixth edition of your *Systema Naturæ*, dated 1748, the last that has reached this country; nor are we better acquainted with the Travels to Palestine. With respect to the other bird, which you say is *Fulica spinosa*, it is certainly not Edwards's tab. 48; for that has a solitary tubercle in front, and a very long posterior claw; whereas mine has a pair of wattles on its face before the eyes, a very short hind claw like that of a *Tringa* or *Gallinula*, and differs widely in colour. Nothing can be more distinct in species, if not in genus, nor can I find my bird any where described.

It will be difficult, if not impossible, to furnish all the shells you want, as many of mine are solitary specimens, or chosen varieties. Some, however, I will send readily, when there is an opportunity, as *le Pietin*, tab. 1, and fig. 24 and 32 of tab. 9.

Agihalid of P. Alpinus is not, as you suppose, a species of *Ximenia*. The latter, originally noticed by Plumier in America, is frequent by the sea-side

* This volume never appeared.

at Senegal, and has eight stamens. The *Agihalid* of Alpinus, which is also a native of Senegal, but cultivated in Egypt, has ten or twelve, being moreover an entirely new genus, scarcely reducible to any natural order of European plants. That these plants therefore belong to two very distinct genera, I can, of my own knowledge, affirm. I must further remark, that similar errors are to be pointed out, relative to almost all the exotic genera hitherto established by travellers, scarcely any one of them being well and clearly described, or founded on distinct solid generic characters. At least this is the case with above 100 American or Indian genera, which I have met with in my travels in Senegal. The same country has furnished me with 100 more, altogether new, which from their novelty cannot but prove acceptable to the botanical world.

B. de Jussieu desires his compliments. I return you my thanks for your offer of procuring, as far as lies in your power, my election into the Academy of Upsal *. I am extremely anxious to enter into a correspondence with that celebrated body, respecting the observations I have made on the natural history of Europe. As to what concerns Senegal, you know, from my Prospectus, that my hands are tied, by a prior engagement to the publick. I must conclude with recommending myself to your favour and friendship. MICHAEL ADANSON.

* This election did not take place.

DR. JAMES LIND TO LINNÆUS.

SIR, Edinburgh, 23d Sept. 1754.

Although I have not the honour of being personally known to you, I have the pleasure of being well acquainted with your excellent writings.

From a high esteem of your talents, and an opinion of your being one of the greatest promoters of useful knowledge, I have ventured to send you a small performance of mine. Perhaps it may serve to entertain you at leisure hours; and if, in reading it over, any thing remarkable occurs to you, I shall take it extremely kind to have your observations communicated to me in whatever language you think proper. I have indeed chosen to address you in the language in which my book is written, as, without a tolerable acquaintance with the English language, it will not be easy for me to obtain the pleasure I expect from your friendly and critical remarks upon it.

I am not a little surprised that, after the most diligent search made for all books whatever that have been written on the Scurvy, I never have been able to meet with any in the Swedish or Danish languages; although in those cold countries, as also in Norway, I am persuaded the Scurvy is a very frequent distemper.

You will perhaps observe, that, however small the merit be of my poor performance, that yet it must have cost a good deal of labour; *viz.* the third part of my book, where a *Synopsis* is given of all that has been published on the Scurvy.

May I presume to ask it in the name of the learned world, and for the benefit of mankind, that, in order to bring so useful a subject to still greater perfection, you will please to favour me with an information of any observations or writings upon the Scurvy which you find, by the Index of my book, have not come to my knowledge? Perhaps, in your noble library at Upsal, such writings may be found as I mention p. 446 of my book, and many may be published in your language which have not come to my knowledge.

You promise, in your *Flora Lapponica*, to account why the Laplanders are never afflicted with the Scurvy, meanwhile their neighbours are greatly distressed with it. When may we expect to have your sentiments made publick on so curious a subject?

I shall be happy in hearing that this has come safe to your hands, and finds you in perfect good health; and have only further to entreat that, after pardoning this trouble, for which many apologies are due (and which nothing but the interest of learning could have prompted me to have given you), you will do me the justice to believe that I am, with great esteem, Sir, your most obedient and very humble servant, JAMES LIND.

P. S. Direct to Doctor James Lind, Physician in Edinburgh, Scotland.

I have sent you two books published here by Dr. Alston, Professor of Medicine and Botany in this University.

LINNÆUS TO DR. JAMES LIND. [Latin.]

SIR, No date.

I lately received from you three books, accompanied by a letter. Your work evinces your very extensive learning, and your letter displays the greatest benevolence and candour. I beg of you to accept my grateful acknowledgments for both; wishing it were in my power to make any adequate return.

I have perused your publication in a cursory manner, not being well acquainted with the English language, but I intend to have the whole translated to me by a friend. Even the little I have already seen, shows the infinite labour and continued attention you have bestowed on the subject.

You ask my opinion on the Scurvy. This disease is extremely frequent in Sweden, especially among the common people. Those who dwell near the sea are particularly liable to it, especially about the gulph of Bothnia; and it prevails among country people, smiths, and miners. This description of persons live principally upon salt provisions during our long winters, partly salted meat, but more generally pickled herrings. Their humours become, by this means, impregnated with sea salt, which not being expelled, produces the Scurvy. This corrodes the teeth and gums, injures the texture of the lymphatic system, and brings on a general languor. As the lymphatics become still more affected, œdematous swellings and dropsy supervene, or asthma from dropsy of the chest, which usually closes the scene.

Those who labour under this disorder lose, as it may be said, the light of their countenance; the face becoming literally less bright, and assuming an aspect of sadness. They walk slowly and laboriously, especially up-hill, like asthmatic people. They lose their appetite; and if they take any food, require rest after it. They awake from their sleep in a morning more tired than when they went to bed the evening before. Blueish spots appear on the legs and thighs; or the gums bleed when touched. If the patient drinks spirits, he falls a more speedy victim to his disease, as also if he gives up his usual labour. The inhabitants of low, fœtid, boggy situations, or the neighbourhood of the sea, are sooner attacked than those who live on high ground, in a pure air.

We daily see workmen in the iron forges, bathed in sweat, all of whom live, in this country, upon salt provision, and whose perspiration is very strongly saturated with sea salt. So long as this perspiration is kept up by their laborious occupation, they enjoy tolerable health; but if they receive any external injury, so as to be confined in bed, the wound indeed is easily cured, but the patient becomes so affected with scurvy, as scarcely ever to recover. The same thing happens every day to those who work in the mines.

The Laplanders, who eat no salted food, are entirely free from Scurvy. All people of superior rank among them are, indeed, accustomed to lick up a portion of salt from some substance or other, and such persons, taking little exercise, are consequently not subject to much perspiration. Particles of salt, taken into the blood, are expelled either by perspiration or by urine. The cure of the disease in question is best promoted by labour, to produce sweat, and by the subsequent use of diuretics. The plants which chiefly serve this purpose amongst us, are Scurvy-grass, Water-cresses, and Horse-radish; above all a syrup of the last mentioned, prepared cold, which moreover sweetens the blood. Diluting medicines are likewise useful, such as a decoction of young shoots of the Fir, with Stone-crop *(Sedum acre)*; to which may be added acidulous liquors, whey, or new milk.

Many scorbutic patients in this country, after having long had œdematous swellings, are afflicted with itching sores on the feet and legs, daily discharging salt. If these ulcers are healed, the patient is attacked with asthma. Nor has any cure been found for this state of the disease, except recently in the root of the Water dock, called *Herba Britannica (Rumex aquaticus)*, which I have introduced on the recommendation of your countryman Colden, who was taught its use by the country people of New York.

I return you thanks for Dr. Alston's book, nor am I at all offended by any difference of opinion between us. He doubts the sexes of plants. The female of the *Rhodiola* grew in my garden for thirty years without producing fruit. At length I received from the mountains a small male plant, which flowered in a pot, in a considerably distant part of the garden from the former; and yet the female bore fruit that very year. Gleditsch had at Berlin a female Palm, which blossomed every year without ripening seed. He obtained, by the post from Leipsic, some male flowers, which were suspended over the females, and ripe fruit was the consequence. The same thing happened with a female *Terebinthus (Pistacia)*, which had flowered for 20 years without bearing any fruit, but which became fertile in consequence of the accession of male blossoms. But it is useless to talk to the deaf. This writer has very ingeniously distinguished plants from animals, by the former deriving their nourishment through external vessels, the latter the contrary; but is it so with the Misseltoe *(Viscum)* ?

M. DE REAUMUR TO LINNÆUS.

à Paris, ce 12 Nov. 1754.

C'est un présent, Monsieur, bien à mon goust, et dont le prix est beaucoup augmenté pour moy de ce que vous le dois, que les œufs du papillon que vous avez nommé *Alpicola**. Vous auriez contribué à l'ornement de nos campagnes, si je parvenois à le

* *Papilio Apollo*, said by Linnæus to be common upon *Sedum Telephium* in Sweden, where it is vulgarly named *Alpicola*.

naturaliser aux environs de Paris, et ce seroit un ornement que sa chenille ne leur feroit pas acheter trop cher, puisqu' elle se contente de peu de plantes. Je n' aurois pas tardé si longtemps à vous en faire mes remerciments, si j' eusse eté à Paris, lorsque votre obligeante lettre y est arrivée. J'étois alors en Poitou, ou je passe regulièrement les vacances, et dont je ne suis de retour que depuis peu de jours.

Mais, Monsieur, ces œufs, pour lesquels je vous fais des remerciments, et dont je viens de vous parler comme si je les eusse recus; je ne les ai point encore. Auriez vous chargé de me les remettre quelq'un qui ne s' est pas acquitté fidellement de sa commission? N' auriez vous point eu intention de les renfermer dans votre lettre, et quelque sujet de distraction ne vous l' auroit-il pas fait oublier? Me voila inquiet pour leur sort. S' ils étoient perdus par quelqu' accident, ce seroit pour moy un avanture desagréable. En cas qu' ils soient heureusement restés chez vous, vous pouvez me les envoyer dans une lettre, dont ils n' encheriroient pas beaucoup le port, et qui m' en seroit rendue franche, si vous voulez bien la mettre sous le couvert de Mr. de Gerseuil, intendant general des postes et relais de France.

La remarque que vous ont donné lieu de faire les insectes apportés depuis peu de la Palestine et d' Egypte, ne me paroit pas devoir s' étendre aux autres contrées du monde. Parmi ceux qu'on m'a envoyés de l'Amerique, d'Afrique, des Indes orien-

23.

J'ai a vous demander pardon de vous ecrire en francois, mais une lettre en latin seroit pour moy un ouvrage, n'aiant pas ecrit depuis plus de quarente ans deux lignes en cet Langue. Le francois ne sçauroit mesme me fournir des termes pour exprimer combien j'admire l'etendue de vos connoissances, et que vous puissiez suffire a tant de travaux si utiles au progres de l'histoire naturelle, ni pour exprimer la tres grande estime et le parfait et respectueux attachement avec lequel j'ai l'honneur d'etre

Monsieur

votre tres humble et tres obeisant serviteur
de Reaumur

24.

Viro Illustrissimo Excellentissimo aeque ac
Doctissimo Domino
C. Linnaeo Equiti &c &c.
S. P. D.
Emanuel Mendes da Costa

25.

Descriptiones novorum generum Plantarum quas exara... ad varios per Europam a Collinsono transmissae fuerunt, ade... ut denuo comparare nequeo. videbo tamen brevi, si no... & alia mihi sunt, tuo examini subjicienda.

De discessu Dni Kalm ab ipso audivisti, ex quo nihil d... ex audivimus. Spero faustum iter, & quicquid officii potui ipsi praestiti, quod semper Tibi vel tuis praestabit

Tibi devinctissimus
Joannes Mitchell

London. august. 10. 1748.

R. Martin, Lithographer, Rochester St, Westm.r

The material originally positioned here is too large for reproduction in this reissue. A PDF can be downloaded from the web address given on page iv of this book, by clicking on 'Resources Available'.

tales, de la Chine, &c. j' en trouve tres peu de ceux de France.

J'ai à vous demander pardon de vous écrire en Francois; mais une lettre en Latin seroit pour moy un ouvrage, n'ayant pas écrit depuis plus de quarante ans, deux lignes en cette langue. Le Francois ne scauroit meme me fournir des termes pour exprimer combien j' admire l' étendue de vos connoissances, et que vous puissiez suffire à tant de travaux si utiles au progrès de l'histoire naturelle, ni pour exprimer la très grande estime, et le parfait et respecteux attachement avec lequel j'ai l' honneur d' être, Monsieur, votre très humble et très obéissant serviteur, DE REAUMUR.

DR. PETER ASCANIUS * TO LINNÆUS. [Latin.]

London, April 7, 1755.

I received your welcome and long-expected letter through the hands of Mr. Collinson, which informed me of your having received my last from Holland, accompanied by a box of plants.

A few remarks on the present state of Natural History in England may not be unworthy of your notice, considering the celebrity of this country for its rich collections of every kind.

Our Pontopidan's Nat. Hist. of Norway is published in English. That author, in the second part

* A celebrated Danish zoologist and mineralogist, afterwards superintendant of mines in the northern part of Norway.

of his work, gives an account of a marine monster, *Siöe ormen*, or *Microcosm*, as he calls it*, supported by evidence that might almost satisfy a historian. Nevertheless I suspend my opinion. The book will please the English, though the translation is bad enough.

In October last I visited Oxford, that famous seat of the Muses; nor is it wonderful that the sacred nine should choose such a residence. No seat of learning in the world contains more splendid palaces, nor richer libraries, galleries, museums, &c.

A natural history of Jamaica is expected to appear in about 12 months, by a Dr. Browne, who, after residing nine years in that island, is just returned to England. He is well skilled in natural science, and his work will be much superior to that of Sloane. His attention has been particularly directed to plants, and I believe he has near 150 new genera, examined in their native situations. This able man follows the sexual system, and his book will be enriched with figures by the celebrated Ehret, who still retains his love of plants, and is truly a botanist. He desires his best respects to you. He had, some years since, the care of the Oxford garden, but having more ardour than the Professor, he was obliged to quit his station. It is not impossible that he may become the draughtsman of our intended *Hortus*. Mr. Miller gave me a packet of seeds for you in February last, but I had no opportunity of sending it till now.

* The famous *Kraken*.

I saw nothing of Prof. Sibthorp at Oxford, he being absent from thence; nor of the manuscripts of Dillenius or Sherard, of which, I am sorry to hear, he takes little care. When he has been spoken to on the subject of their publication, he replied, that such an undertaking would require much time, and would not suit the taste of the booksellers.

Mr. Watson *, an apothecary, and Fellow of the Royal Society, in an English periodical publication for December, has given a review of your *Species Plantarum*, in which he has controverted many points, without saying any thing to the purpose. The English chiefly find fault with your exclusion of Catesby's generic name of *Meadia;* nor do I find myself able to give them a sufficient reason. Dr. Mead is celebrated by every body, and especially by Ehret, for his great attention to Natural History. He left 200 drawings of rare plants, for the doing of which he paid Ehret 400 guineas.

The British Museum, consisting of the immense collections of Sloane and the Royal Society, will soon begin to be placed in Montague house, but the whole undertaking can hardly be accomplished in the space of ten years. When complete, this museum will alone well repay the trouble of a visit to England. Both these collections however are at present in the greatest confusion, and many articles have been lost, either through neglect, or from being placed in a bad situation; but they receive acquisitions daily from every part of the globe.

* Afterwards sir William.

Mr. Ellis, F. R. S. has just published a treatise on Corallines, *Sertulariæ*, which, by means of an excellent microscope, he has discovered to be entirely the work and crusts of Polypes, by which they are inhabited throughout their whole length. He possesses many specimens in which *tentacula* (feelers) are protruded from the divarications and summits of the branches, in the same manner as Trembley relates. To this tribe also belong the productions called the Dead man's hand, Sea Fig, &c. Mr. Ellis asserts the same thing of the *Lithophyta*, or true Corals, and especially of the Sea Fan, *Flabellum Veneris*; but to this I hardly dare, as yet, assent. The original author of these experiments is Dr. Buttner*, who has lately left Paris for Berlin. I mean to repeat his curious observations at the sea side the first opportunity. The opinion of Bernard de Jussieu, relative to these matters, has not yet prevailed here; but rather Baker's doctrine of crystallisation. This last is a very worthy man, whose microscopical enquiries have great merit, though very simple. He has confirmed your opinion of the formation of crystals.

Da Costa is a jew, who has long laboured at a history of fossils, in English. He certainly possesses an excellent collection of minerals; or rather, I should say, he did possess it; for he is at present in prison for debt. But his collection is in the hands of a friend, who allows him a partial use of it. Da Costa is certainly well versed in this

* See vol. I. pp. 170, 177.

study, and will make us acquainted with more species than any other writer has done.

Dr. Hill, the too famous naturalist of England, is in the lowest possible condition. I do not think any mortal has ever written with more impudence or more ignorance. His only excuse is that he must write in order to exist.

I have a letter, dated March 24, from Dr. Gronovius, who is just recovering from a very severe illness. His preface to the *Flora* of Rauwolf *, with the life of that celebrated traveller, are ready for the press.

I am about to bid farewell to the English, whose kindness I have reason to acknowledge, but I must be in Paris early in June. I beg of you to remember me to my countrymen, especially Dr. Holm.

Charlottenburg, Copenhagen, April 24, 1761.

I neither have forgotten, nor can I ever forget you, my distinguished friend. Nevertheless, as I have been, for many years, hindered, by various accidents, from writing to you, I received, with peculiar pleasure, your last letter, confirming that good will towards me, which you have formerly, in so many different ways, testified.

I am happy to learn that the plants I formerly transmitted you from Holland, proved acceptable. At that period I studied botany, as well as other

* *Flora Orientalis, Lugd. Bat.* 1755, 8vo.

branches of natural history. At present Mineralogy exclusively occupies me; nor is it given to man to excel in every thing alike. I have begged of my colleague Oeder the enclosed seeds, hoping they may not prove unacceptable, as they are not all to be found in the *Species Plantarum*. Please to inform me if they succeed, and we shall be glad of any thing in return, for our botanic garden. The advertisement of our *Flora Dano-Norvegica* is just printed, and shall soon be sent to you.

The numerous notes which I have made in the course of my seven years' tour, have never been reduced to order, on account of various more urgent occupations; I hope however, before long, to publish what relates to Italy. It is remarkable enough that mineralogical descriptions have hitherto scarcely been extended beyond the productions of the northern parts of Europe. The volcanic class is altogether wanting, though it includes very numerous species of earths, stones, and minerals. Italy especially abounds in these; having been formerly so much under the dominion of fire, though the dates of the changes it has undergone are necessarily out of the reach of our chronology. The mountains in that country chiefly consist, from top to bottom, of pure, white, impalpable limestone; the lesser hills of lava, of the structure and hardness of *Saxum*, micaceous and containing iron (from the abundance of *Mica* about volcanoes, this also is probably to be considered as a volcanic production); the plains, particularly that of Rome, 30 Italian miles broad, of various species of Pumice. This last substance, which, according to Pliny, works wonders*, is now called *Giura* in its native station near Puzzuoli. I passed many days upon Vesuvius, which I acknowledge as my great teacher on these subjects.

You perceive, therefore, that we are less acquainted with the fossil than the vegetable productions of the south of Europe; which is owing to the greater facility of studying one than the other.

Charlottenburg, Dec. 4, 1767.

For several years, in which I have, by the Royal command, been engaged in travelling through various parts of Denmark, chiefly with a view to mineralogical and economical observations, I have scarcely ever had the benefit of a draughtsman. But in the vicinity of the metropolis such assistance is more easily to be obtained. I have here, therefore, collected together a number of figures, chiefly of fishes, some of which I am induced to publish in a small work, now offered to your acceptance†.

The easiest and most likely plan for the advancement of natural history towards perfection, appears to me for every man to devote himself exclusively to his own department. By such means the world

* See *Plin. Hist. Nat.* book 35, chap. 13.

† Ascanius, Figures enluminés d histoire naturelle. Copenh. 1767, fol. oblong.

has been put in possession of your *Flora* *, and of Oeder's *Flora Danica.*

Among these figures you will find two new species of fish, a *Labrus* and a *Gadus,* as well as two new birds. I perceive that Dr. Kœnig has, without my knowledge, sent you a description of the *Tringa ferruginea.* It might indeed have been called *islandica,* but I find, from the observations of Mr. Teilmann, that it abounds in the north part of Jutland. I am informed also, by the same authority, that the grey variety is merely the plumage of the first year, as happens in many other birds. Thus the opinion of the Icelanders is confirmed.

Our Arabian expedition is at length concluded by the return of Niebuhr. The literary party employed herein was less suitably composed latterly, whence disputes and misfortunes arose. In Egypt various observations and collections were made, at a great expense, but owing to these misfortunes, a very small portion only of what was sent home reached its destination in safety. After the travellers reached Yemen in Arabia, the place of their final destination, though all came there in good health, a kind of malignant consumption attacked the whole company, who had been sent out at greater cost, and with a more complete apparatus, than had ever been known, except in the case of the French mathematicians. We still however expect much from Niebuhr in the mathematical department. In collecting natural produc-

* This application, as far as Linnæus is concerned, seems not very correct.

tions, especially marine ones, besides what has been done by Forskall, it appears to me, from a box now under examination, that Kramer has been very useful. I procured his appointment for this purpose, Professor Michaelis having been chiefly solicitous to send out an able observer, without thinking of a draughtsman or a collector of specimens. I know not as yet what may be contained, relative to natural science, among the manuscript journals and remarks of the deceased; but these will speedily be examined by a committee.

The articles of natural history which have been sent home, chiefly relate to Conchology, and contain many new and interesting subjects. The *Strombi* vary prodigiously in their finger-like lip, according to the age of the animals. Sometimes these prominences are numerous, sometimes entirely wanting. In this tribe it séems that the established genera altogether want confirmation from observations upon the living animals; but how few naturalists can enjoy such an opportunity!

A very beautiful brownish species of *Cardium*, formerly sent from Egypt, a duplicate of which I have met with in one of the parcels last examined, appears to have been communicated to you by Mr. Sprengler, being your *retusum;* but the description does not exactly answer to my specimen.

Our museum already contains a large number of rare, and partly new, shells, chiefly from the Red Sea and Persian Gulph, of which I obtain permission to send a part to you, in the hope of your favouring us with some rare or exotic specimens in return.

Dr. Koenig, formerly sent as a missionary and naturalist to Iceland, is lately gone to India, with the appointment of physician to the Moravian brethren at Tanjore.

Mr. Oeder, with his compliments to you, requests that you will appoint some one belonging to the Swedish embassy to receive, on your account, the first fasciculus of the *Flora Danica,* as well as any that may follow, sending him a receipt under your own hand.

May you long be preserved for the benefit of science, and to honour me with your favour!

P. Ascanius.

LINNÆUS TO MR. EMANUEL MENDES DA COSTA.

[Latin.]

Sir, Upsal, Nov. 9, 1757.

Your letter of the 5th of August has been read at a full meeting of the Royal Academy of Sciences on the 5th of November. Your unparalleled knowledge and rare learning have excited so much esteem and respect in all those who were present, that I am commissioned by them to testify to you how highly they value your communications. We had long ago heard, by publick report, of the publication of your important work, and were the more desirous of seeing it, that we might profit by your information. Hitherto we have looked for it in vain, the copy which you were so good as to send us, through Mr. Brander's hands, not having yet reached Upsal. I therefore request, in the name of the whole Society, that you would enquire how Mr. Brander has forwarded the book, as they are all extremely anxious to possess what is likely to be so useful to science.

I, who am occupied in preparing the 10th edition of the *Systema Naturæ*, with numerous additions, cannot dispense with your work, as I intend to quote it, with due commendation, throughout the fossil kingdom. I therefore earnestly intreat you to forward it to us as soon as possible. I will take care that your present shall be received with due respect, and gratefully acknowledged.

May God preserve you to complete successfully the succeeding volumes!

DA COSTA TO LINNÆUS. [Latin.]

SIR, London, Feb. 10, 1758.

I duly received your most welcome letter of the 9th of November last, which I should certainly have answered, as I ought, before now, but a tedious illness has prevented me. As soon as I was restored to health, I called on Mr. Brander, who assured me that my book, destined for your acceptance, had been forwarded long ago by a safe conveyance. He suspects, however, that the early setting in of the frost might have delayed its arrival. Meanwhile, I learn with the greatest pleasure, that your illustrious Society is disposed to receive, with indulgence, the little book I have sent, and I rely on the favour of that learned body to enhance this kindness, by any admonitions or corrections where I may have erred. Nor am I less flattered by your kind intention of referring to my work, and honouring me with your approbation, in the tenth edition of your most learned *Systema Naturæ*, which performance may God favour with his assistance! You may be assured that all the powers of my mind are devoted to truth and the good of mankind, as well as to a most ardent desire of reducing this noble study to some certain methodical rules. This is the chief end I have had in view, and the foundation of my whole work. Allow me therefore to cultivate an epistolary correspondence with you, especially on the subject of Natural History; and at the same time I must request you to present my best respects to the excellent Mr. Wallerius, with whom I have already communicated. You may both of you depend on my ever conducting myself with due respect towards correspondents of such distinction, and on my readiness to render you any service in my power.

I beg leave to recommend myself to your illustrious Society in general, as well as to each of its members in particular. Nothing could be more agreeable to me than to have my performances so far sanctioned by their approbation, as that I might be thought worthy to become a member of their illustrious body. I trust you will sanction the wishes expressed in my preface, and favour me with such learned remarks and corrections, that the sequel of my work, at which I am continually labouring, may be brought to the greater perfection, nor shall I fail to acknowledge, with the sincerest gratitude, the assistance I hope to receive. I can never be ignorant how much the study of fossils is indebted to the learned Swedish nation.

If, nevertheless, contrary to my hopes, the book in question should not yet have reached you, I beg to be speedily informed of a safer mode of conveyance, that I may supply its loss by sending another copy.

Hoping to be reckoned among the chief admirers of your learning, virtue, and reputation, I remain, &c.

LINNÆUS TO DA COSTA.

Upsal, Feb. 27, 1759.

Your work upon fossils, so long expected, has come to hand within these six days only, having been put aside in the warehouse at Stockholm, where it was at last found more by accident than intention.

I presented the book at the last meeting of our Royal Academy, where it was generally admired for the abundance of its materials; the dexterous selection of, often very intricate, synonyms; the highly finished descriptions and excellent remarks; and, finally, for the method of arrangement, which is altogether new and singular. I was ordered by the Academy to transmit to you, by the first post, the thanks of all the members for your present of so handsome and learned a work. We all concur in one wish, that your life may be preserved to complete, with equal success, the history of the whole fossil kingdom;- by which the knowledge of this department of nature cannot but be brought to great perfection, and conduce to the glory of the Creator, as well as the benefit of mankind.

DA COSTA TO LINNÆUS.

SIR, London, Oct. 5, 1759.

I am truly happy to learn, by yours of the 27th of February, that your highly-distinguished Academy has received my book, such as it is, with so much favour. Indeed I consider it as a peculiar honour, that the thanks of this learned body, which you have kindly conveyed, have, however unworthily, been bestowed upon me; and the more so, as I well know how very far the Swedes excel all other people in the knowledge of fossils. I shall continue my work with all possible assiduity, and if my second volume, or any other treatise of mine, shall ever see the light, I will not omit to offer it to your learned Academy, as a testimony of my respect and devotion. I cannot but own, that the high honour of becoming a member of such a Society is the object of my most earnest wishes, as I have mentioned in a former letter. But as you pass over this subject in absolute silence in your last letter, I cannot but conjecture that I have been judged unworthy of the honour in question, which does not diminish a particle of my respect for your Society, though I repeat my request in hopes of an explanation from you. I have moreover aspired with earnestness to the high gratification of your correspondence, on the subject of natural history, with an interchange of specimens of fossils, especially as I am in want of certain Swedish fossils, for which I am ready to communicate English ones in return, and shall be glad to learn your wishes and inclinations on this subject.

Our mutual friend, the worthy and learned Mr. Edwards, knowing that I was about writing to you, has requested me to acquaint you with his having received your letter of September last. He says there is a full and ample account of the *Balanus (Lepas)* figured in his tab. 286, in the last volume of the Philosophical Transactions, which is vol. 50th, part 2d, p. 845. He promises moreover to publish accurate and complete descriptions of the subjects in the rest of his plates. He further mentions, in reply to your enquiries, that the protuberances on the beaks of the *Mituporanga* and *Pauxi (Crax, Linn.)* are hard, and almost bony, to the touch, being solid and covered with skin; and that the Tortoise, tab. 287, is the size of life, and comes from Pensylvania.

He sends you as a present, by Dr. Biörkin, the skin of a Chimpanzee, in excellent preservation, which is the same species with that he has figured, though different in sex. That of Tyson appears to differ in no respect from this, except in colour, the Doctor describing his as having black hair. By the same conveyance Mr. Edwards also presents you with Dr. Tyson's plates; hoping you will receive the whole in safety. You will also receive proof impressions of 25 plates of the second volume of his Gleanings, which, with what you have already, make up 100, or one entire volume, which, God willing, he intends to publish next March. The work of Dr. Tyson treats, moreover, of the Rattle-snake, and the *Tajacu*, or Musky Boar of Mexico, as well as of Pigmies. He gives a long and tedious dissertation on the Pigmies, Satyrs, &c. of the antients, supposing them all to have been Monkeys. The original edition of this book appeared in 1699, in folio; the second, in quarto, in 1751. Mr. Edwards returns his best acknowledgments to you for pointing out to him those books, treating of natural history, which are likely to be serviceable to him, and which he will do his utmost to procure. He joins with me in hoping that Dr. Biörkin will reach you in safety, and we unite in charging him with our best wishes and compliments.

So much for our worthy friend Edwards! I have no further literary intelligence, as my Muse, if not altogether asleep, is at present taking breath. I shall be the more glad to receive your answer to this, abounding with literary news. Please to direct to me at the Bank Coffee-house, London.

It only remains for me now to offer my prayers to the Great Ruler of all things, that he would grant you, my illustrious friend, all publick and private happiness, and preserve the vigour of your admirable mind, as well as the health of your bodily frame, unimpaired to a very late period; and that under his supreme direction you may never cease to esteem your friend and servant, &c.

We have found no answer to this letter among the papers of Linnæus, nor in the correspondence of Da Costa. The latter appears to have taken great offence at not being chosen a member of the Upsal Academy, and conceived an antipathy to Linnæus, which the writer of this has often heard him express, but could never before account for. That Academy was always very select in its choice of foreign members, and subsequent events too amply justified its conduct in the present instance.

MR. GEORGE EDWARDS*, LIBRARIAN OF THE ROYAL COLLEGE OF PHYSICIANS, TO LINNÆUS.

SIR, No date.

I received your compliments by our good friend Mr. Ellis, and am very glad to hear you still pursue the studies of nature, in which you are already so far advanced as to draw on you the eyes of all the naturalists in Europe. I wish you heartily to go on and prosper. As it will be some months before I can publish a work I have now in hand, and knowing your thirst for the earliest insight of what is going forward in natural history, I have herewith sent you 75 prints, in full confidence that you will make no use of them to my disadvantage. I believe, before the next winter is over, the letter-press will be ready to be delivered, with the sets of prints I am now colouring; yet I thought this small present of black prints, might be acceptable to you in the mean time. I am, Sir, with the greatest respect, your most humble and most obedient servant,

GEORGE EDWARDS.

LINNÆUS TO EDWARDS. [Latin.]

SIR, Upsal, March 20, 1758.

I received three days since, your valuable present of 75 elegant plates of Birds and other rare animals,

* The celebrated author of a work on birds, and some other subjects of natural history, in quarto, with coloured plates, published between the years 1743 and 1764. Mr. Edwards died in 1773, aged 79.

17.

Viro claro, & Nobili,

Carolo Linnaeo, medico,

H Boerhaave.

Gratias ago pro literis, et libro. ex prioribus perspexi, quo sis in me, prorsus non meritum, animo. amico certe, benevolo, & humano. utinam esset in me aliquid, quod hisce dignum, utinam esset, unde provocatus beneficiis vicem possem reddere!

18.

And whatever other American plants in this inclosed Catalogue will be acceptable, you may freely command any that I am possessed of — In regard of that Esteem your merit claimes

I am Sr

Your most Obedient Humble Servant

M Catesby

London 26. March 1745

19.

I beleive before the next winter is Over the letter press part will be ready to be deiverd with the Setts of Prints I am now colouring, yet I thought this ~~small~~ smal present of black prints might be aceplable to you in the mean time

I am Sr with the Greatest respect your Most Humbe and Most Obedient Servant

George Edwards

R. Martin, Litho.

The material originally positioned here is too large for reproduction in this reissue. A PDF can be downloaded from the web address given on page iv of this book, by clicking on 'Resources Available'.

which I have contemplated with great admiration. Nothing can more conduce to the advancement of solid natural knowledge, than such beautiful and excellent figures, accompanied by such exact descriptions. How much I have always valued your publications, even before I enjoyed your correspondence and friendship, you may easily perceive by the first volume of the *Systema Naturæ*, published not long since. I shall subjoin, in an appendix to the second volume, whatever I can collect, for the advancement of science, from these last plates. I am particularly pleased with your admirable figure of the *Satyr*, which, if I mistake not, will prove a different animal from the *Ouran Outang* of Bontius's Java, p. 34, though it may be the same as the Indian Satyr of Tulpius, book 3, chap. 56. I wish you would inform me whether there be any space between the canine teeth and the rest; *item si modo tua sit fœmina, num nymphis et clitoride instruatur, nec ne.*

Your portrait in my study brings you every day before me, and reminds me of your indefatigable assiduity in collecting, delineating, and describing.

Your *Jerbua* is the *Mus Jaculus*, described by Hasselquist in the Upsal Transactions for 1750, p. 17, the Stockholm Transactions for 1752, p. 123, t. 14, f. 1, as well as in his Travels 198, (Engl. ed. 186), *Aldrovand. Quadrup.* 395. The Spur-winged Plover, Rus. Alep. t. 11, and your 280, is *Charadrius spinosus, Hasselqu.* (Engl. ed.) 200. Your Long-tailed Duck, 280, is frequent with us all winter long. The lower figure represents *Chætodon ciliaris, Syst. Nat. ed.* 10, 276. *Mus. Reg. Adolph. Frideric,* 62, *t.* 33, *f.* 1. *Osbeck,* 273.

Farewell, may you long enjoy life, not unmindful of your friend, &c.

EDWARDS TO LINNÆUS.

SIR, London, June 2, 1758.

I am glad the little present I sent you is safely received. Your commendations of the figures are greatly beyond their desert. As to the Satyr, Ourangoutang, or Chimpanzee, as they are called by different authors, I believe them to be different species of animals. The subject from which my figure was drawn is the Pigmy of Edward Tyson, M. D. F. R. S. who wrote a treatise, with a great number of copper-plates, explaining all its particular parts in an anatomical manner. His subject was a male. He makes the teeth agree very nearly with those of men, as I have observed them to do. I can give you no information as to the parts you enquire about; for my subject being dead, and the parts beginning to be rigid before I drew them, I could not inform myself. Tyson has collected in his work the figures and descriptions of all the authors on the Pigmy, &c. who had written before him, wherein they are, undoubtedly, different species of animals. There is a very small space between the canine teeth and the four flat teeth in the front of the mouth in Tyson's figures; but no space between the canine teeth and the grinders. My figure is from a female. The first 50 of the plates I sent you are published, with their descriptions in French and English. The other 25 will remain unpublished till I can make them up 50. I give you my thanks for the notes with which you have obliged me, which will serve as improvements to my descriptions. I shall always think myself honoured by your correspondence.

LINNÆUS TO EDWARDS.

Upsal, Sept. 29, 1758.

Your letter, written in June last, is come to hand, as well as three plates, engraved, after your usual manner, from nature, all welcome pledges of your friendship, which I shall be happy to deserve.

The *Conchœ anatiferœ* (Common Barnacles), upon the Fleshy Barnacle, strike me as altogether extraordinary, nor can I satisfy my eyes with contemplating them. The first figure in the same plate is a long-armed hairy Crab, which I cannot correctly define, without knowing whether there be any teeth, and how many, concealed by the hairs on the thorax. I suspect it to be tab. 5, f. 1, of Plancus. Next to this Crab, on the right, is an *Oscabrion* of Petiver, which, in the *Systema Naturœ, (ed.* 10,*)* 667, *n.* 1, I have called *Chiton hispidus,* nor have I found it noticed by any other writer. I wish to know what number you will affix to this plate, that I may quote you in my Appendix. (The No. is 286.)

With regard to the *Mituporanga*, or *Crax (tab.* 295), I am very anxious to know what is the nature of the ball above the nostrils. Is it solid or hollow, membranous or fleshy?

The Tortoise with an awl-shaped tail (*tab.* 287), I have seen in a dried state, but do not recollect where. (It is now in the Linnæan museum.)

I am never tired of contemplating those figures of the Chimpanzee; and much regret that I can no where in this country meet with Tyson's treatise on the Pigmy, which I presume is printed.

Mr. Clerck has just begun to publish his collection of *Phalænæ* that are not elsewhere delineated. His figures are in 4to, coloured after nature, one plate often containing 15 different species. I know nothing else published here of late, elucidating the history of animals.

I have recently noticed, upon plants from the Cape of Good Hope, two remarkable species of insects. One is a *Coccus*, with an aperture in its back; the other a *Chrysomela*.

My pupil Martin, who went with the ships engaged in the whale fishery this summer to Spitzbergen, brought back with him several stuffed birds. Amongst others, *Alca Arctica* (the Puffin), Albin's Birds, v. 2. t. 78, 79; *Alca Alle* (Little Awk), Edwards, t. 91; but especially *Colymbus Troile* (Foolish Guillemot), the Lumme of Marten's Spitsbergen,

7. t. M. fig. A. This bird resembles your tab. 97, but is only half as large, and has no coloured mark under the throat. The palmated feet have, moreover, only three toes. He has also brought the whitish *Procellaria* with a grey back (*P. glacialis*, or Fulmar), Malle mucke, Marten's Spitsb. 68. t. N. f. C. This agrees in size with your tab. 90. f. 2, but is an entirely distinct species. It is said to seat itself on the whales as they are cutting up, and to pick out and devour vast quantities of their fat.

Your little bird, tab. 30, is also found in Spitsbergen, which surprizes me.

Mr. Martin procured the *Phoca*, called by Anson the Sea Lion, and noticed its power of inflating the crest on its head like a bladder. As every kind of Seal is easily killed by the slightest blow above the base of its snout, this animal, when in danger, takes care to inflate the above-mentioned bladder, on which it receives any blow, as on a cushion, without injury.

I am now more convinced that the Crab above alluded to is the *Cancer hirsutus personatus* of Plancus, *de Conchis minùs notis liber, Venet.* 1739, 4to. pag. 36. t. 5. f. 1, having compared your figure with my Italian specimen, as your representation is more correct than that of Plancus.

I have this summer received about 68 species of stuffed birds, well preserved, a few of which I had never seen before, and others but rarely. What chiefly pleased me was the *Anas Fuligula forte*

Gesneri, Ray's Birds, 141, (*A. Marila*, or Scaup Duck.)

Kramer has just published a *Flora Austriæ inferioris*, in 8vo. a book of a small bulk. In an appendix he has noticed the animals, especially birds, with which he seems to me to be very conversant. I wish you could see the work. It was published at Vienna in 1756. He has one new genus, figured in his only plate, by the name of *Pratincola*. This comes very near your Little Grous from Aleppo, t.249 (*Tetrao Alchata*), though not the same. What is most curious in his bird, is the nakedness of the thighs above the knees.

EDWARDS TO LINNÆUS.

SIR, London, Aug. 12, 1760.

I herewith deliver to the obliging Mr. Solander, to be conveyed to you, a complete copy of the letter-press of my Gleanings of Natural History, together with 25 black prints. These, added to 75 prints formerly sent you, make 100 prints, which answer to the 100 chapters contained in the Gleanings. I have also sent such part of the letter-press of the second part of the Gleanings, as will complete what was formerly delivered, and paid for by Mr. Biörkin.

I have received the very great favour of your most kind and obliging epistle, by the hands of the highly accomplished Mr. Solander, whom I shall do all in my power to oblige. I wish I were a little younger, that I might be the more able to answer and return my obligations to my friends in many parts. But they must excuse the slowness of old age, because it is an infirmity of nature. I return you my thanks for all former favours, and hope for a continuation of your agreeable correspondence.

P. S. You will receive herewith a few unpublished prints of new subjects. You will know which they are, because they are not numbered.

LINNÆUS TO EDWARDS.

DEAR FRIEND, Upsal, April 13, 1764.

I received, a month since, your new plates, t. 235 and the following ones, all excellently delineated and coloured according to your usual manner. You may depend on my sending the amount of their value, as soon as I learn from Baron Alstroemer's brother the rate of exchange.

I wonder you have not put a number to that fine short-tailed *Simia*, sitting upon another*. Is it different from the Baboon or *Sphinx*.

But what chiefly induces me now to write, is your tab. 336, in which you represent some " Ve-

* Linnæus seems to allude to the Sloth, tab. 310, mistaking the bank of earth for another animal, in the uncoloured plate.

getating Wasps," and which appear to be *Vespœ*, provided they have four wings, a circumstance I wish to be informed of. My thoughts are so taken up with these productions, that I cannot sleep without dreaming of them. I conjure you, by the Author of Nature, to write to me the first day you can spare, to explain this phæmomenon. What is the shrub, or branch of a shrub, which grows out of the back or breast of the insect? Is it a small branch of a Rose? What connection is there between them? Is the branch of the plant grafted upon the insect, so as to grow out of it? You are very cruel if you do not speedily relieve me from this puzzle. I never saw any thing in nature like this production. If indeed the Wasp merely cuts off the tip of a branch to build its nest with, the mystery is solved; otherwise it is altogether wonderful. Whence did you procure this Wasp? Is there no deceit in the specimen?

Is the Monkey, tab. 312, furnished with an elevated cap, as it were, of hair only? or is there a protuberance of the head itself.

With regard to the Butterfly tribe, I beg of you always to represent both sides of the wings, one side being not always sufficient to distinguish the species.

The fly in t. 355, cannot, for this reason, be ascertained.

t. 342, is *Papilio Helenus.*

t. 343, I think is *P. Machaon*, provided it be an European specimen.

t. 299, I do not remember to have seen. Perhaps the other side might help to determine it.

t. 333, is *Libellula depressa.*

t. 346, *Papilio Deiphobus.*

t. 297, *P. Leda.*

I congratulate you on the acquisition of such beautiful and innumerable rare birds, beyond what any other person has seen, or is likely to meet with; still less is any other hand likely to equal your representations, in which nothing is wanting to the birds but their song. Yet even these will sing your praise, as long as birds or men endure. Your performances are an ornament to the age in which we live. May you long reckon me among your sincere admirers! Please to remember me to all my friends in England who are devoted to the study of Nature! Farewell!

Biographical Memoir

OF

DON JOSEPH CELESTINE MUTIS,

WITH

His Letters

TO

LINNÆUS, BOTH FATHER AND SON.

Don Joseph Celestine Mutis, to whom Linnæus was indebted for the knowledge of the finest botanical productions of South America, was a learned and highly-enlightened Spanish ecclesiastic and physician, settled, as professor of philosophy, mathematics, and natural history, at the university of Santa Fé de Bogotá, in New Granada. His instructions and the labours of his pupils at length produced the magnificent *Flora Peruviana*; a work, indeed, more rich in materials than classical in execution, but, nevertheless, a valuable accession to botanical literature. Its imperfections, whatever they may be, are not attributable to Mutis; and, considering the general state of science in Spain and its dependencies, such a work, from any of them, is a phænomenon. This able man was many degrees above the standard of his natural or his adopted country; and on his first teaching mathematics and astronomy, in the old university of Santa Fé, was treated as a conjuror, or ally of the devil, because he could measure the distance of the sun and moon, and foretell eclipses. The Inquisition confirmed these censures. "It was natural," as we have elsewhere remarked *, "that the hardy adventurer, who disturbed the dreams, and scared away the slumbers, of the monks, should, like those who first roused the bats and owls in the caves of Elephanta, be bespattered with their filth." Many good people withdrew their children from the school of an alleged necromancer; and it required all the influence of an enlightened and powerful Viceroy, as well as of an Archbishop of similar character, finally to overcome the absurdity. This could hardly have been accomplished, had not the manners and virtues of Mutis himself co-operated, to remove prejudices which were apparently honest, and not, as in many other instances, assumed to cover more base and interested motives. When he became better known, he enjoyed the esteem and even veneration of all around him, for the benefits he conferred by the exercise of his medical, as well as his religious, profession. His physick and his philosophy were alike guided by wise and simple principles. On these

* See the article MUTIS in Rees's *Cyclopædia*.

subjects his letters speak for themselves; and they display his character better than any thing we could here relate.

From one of these letters, it appears that Linnæus had been puzzled to ascertain the precise place of residence of this highly-valued correspondent. By the earliest mention, in his works, of the communications of Mutis, which occurs in the second *Mantissa*, p. 200, we perceive that he considered those communications as from Mexico *. Accordingly, the discoveries of Mutis have, by all subsequent writers, without examination, been attributed to that country. We have ourselves fallen into this error. Mutis and his pupils were known to have travelled extensively, and every body took it for granted that Linnæus was correct. A careful perusal of these letters has cleared up this point, though the cause of the mistake could hardly have been detected, but by the writer of the present remarks. He finds in the old Dutch Atlas of Frederick de Wit, which Linnæus always used, the name of New Granada applied to New Mexico, the principal town of which is Santa Fé, in latitude 39 north. This Linnæus undoubtedly took for the residence of Mutis; the real New Granada, with its capital Santa Fé de Bogotá, lat. 2 north, being so much less conspicuous in the same map as to have escaped observation.

* See also our vol I. p. 275.

With regard to the following series of letters, it is imperfect, the earliest letter in our possession alluding to a former correspondence. Nor have we any copies of the answers of Linnæus. Nevertheless, the communications and expressions of Mutis are so interesting, that, however incomplete, they cannot but be acceptable; nor was the peculiarly laborious task of decyphering and translating them ever found irksome.

Correspondence.

JOSEPH CELESTINE MUTIS TO LINNÆUS. [Latin.]

Santa Fé di Bogotá, Oct. 6, 1763.

I have long been considering what can have, so unexpectedly, deprived me of your correspondence, as I cannot attribute it to inconstancy or negligence on my part. I am perfectly certain that I have done nothing to break it off. I therefore venture to trouble you with another short letter, to express my anxious hopes that my former ones have reached you, and my fears lest you should not be aware how much I value your good opinion. That you may have reason to doubt this, may have arisen from the shipwreck of the Thetis, and the disastrous fate of many of our ships during the late war. I have lately requested Mr. Bellman, the Swedish consul at Cadiz, to mention me to you, that you may not think I have forgotten you. Now therefore, that I may preserve your friendship, I intend very shortly to forward at once copies of four letters that I have at different times written to you. I am so anxious that nothing should interrupt the kindness which your earliest letters to me expressed, and which I have more than once endeavoured to return, that I will in future endeavour to send you duplicates of every letter. Farewell!

Santa Fé di Bogotá, Sept. 24, 1764.

I firmly expected to hear of you, my honoured friend, by this opportunity; for though you have left unanswered several letters which I have written since July 1761, I hoped what I wrote in October last might have obtained a reply. But, disappointed in all my expectations, I should be quite in despair had I not kept a copy of every letter that I have, at different times, sent you; by transcribing which, and sending them again, I am able to remove all suspicion of the negligence which you may perhaps have laid to my charge, notwithstanding my repeated assurances of the high value I set on your correspondence. I have very fully replied to that letter of yours, so full of kindness, which I first received, and which came to my hands when I was thinking of opening a correspondence with you myself. I recollect writing to you a second time, in March 1762, with some remarks of mine upon the American Ants. I mentioned in that letter my anxiety to procure the second edition of your *Fauna Suecica*, of which I had for some time been in great want. In the course of July, in the same year, I wrote to you again, by way of the Caraccas; understanding that, owing to the interruption of our commerce by the attack of the English upon the Havannah, there was scarcely any hope of my letters reaching you by the latter conveyance. I subjoined some descriptions of plants, with an account of my recent journey to Carthagena, in company with the Viceroy. At length, when the war was over, and our commerce restored, I wrote a fourth time, in May 1763, from Carthagena; for I began, with great concern, to think my labours hitherto were likely to be all in vain. I thought it right also to accompany this last letter with a sketch of our intended return to the Viceroy's original station, by a new route, and a prospectus of an American journey which I had long had in contemplation. This last was sanctioned by a most honourable testimony in my favour from the Viceroy to the King of Spain.

In October of the same year, after a long journey of 50 days to my usual abode, I wrote to you in great haste, chiefly to enquire why our correspondence had been thus put a stop to, almost as soon as it was begun, and to ask for an explanation of what had justly given me so much concern. In January 1764, while just recovering from a severe illness, and labouring under the consequences of a nearly fatal relapse, I wrote a sixth letter, in which I recollect giving a description of a species of *Crax*, which I was induced to consider as entirely new, on account of the beautiful crown, resembling a fig, with which this bird is distinguished. Once more, for the seventh time, I have determined to write to you; and I rely on your great indulgence for me, that you, my much esteemed friend, will not think me too troublesome.

Having, at the departure of our friends Alstroemer and Logie, as I faithfully promised in my first letter, caused all my papers to be fully and accurately copied, so I will take care to do in future, as I have already signified, sending duplicates of my letters, that they may be the more likely to reach you. I can indeed send copies of all I have sent already; but if this should seem useless or tiresome to you, be so good as to let me know, that my labour also may be spared. Indeed I am so limited for time at present, in consequence of the troublesome practice of physic into which I am plunged, and my new undertaking of giving lectures on natural philosophy, that I have not yet been able to finish my inaugural discourse, in defence of the Newtonian philosophy against the peripatetics, to be delivered in the schools before the Viceroy and a very learned audience. You will therefore readily perceive that I could not possibly send you, by this opportunity, as I had intended, a copy of a paper to be laid before your Academy, which is already begun, *de glebis aureis memorabilibus*. But that my present letter may not seem entirely unprofitable, I send you a figure, with some of the flowers, of the Peruvian Bark *. I am not certain whether the celebrated Monsieur de la Condamine has given any figure along with his description, nor whether you have had an opportunity of examining a dried specimen, as I find no mark indicative of this, in the generic description of *Cinchona*, in your Stockholm edition of 1754.

I have little more to add. A letter of a recent date, from my very learned friend Dr. Salvarezza of

* Both preserved in the Linnæan herbarium.

Cadiz, informs me, that he had in January last entrusted one of my letters for you to the care of Mr. Bellmann. Which of the above-mentioned this might be, I am unable to guess, so many of my letters having miscarried. But you can easily inform me which of them have come to your hands. I long indeed impatiently for a letter from you, as well as for some of your printed works. Farewell! forget not your sincerely devoted though far distant friend and admirer, J. C. Mutis.

Near Cacota of Surat, May 19, 1767.

I am extremely desirous, as I have always been, that all my observations should, in due time, be communicated to you, my much esteemed friend, as I remember to have long ago promised; but especially what relates to things either entirely overlooked, or imperfectly determined, by travellers. My intentions, however, are often frustrated, either by the few opportunities of communication, or my frequent excursions, as well as by the great distance between us, which is liable to so many casual impediments and accidents. I think I have mentioned, in a former letter, that, having finished my travels through the whole country of Bogotá, I had come to this province of Pampilona, chiefly for the purpose of investigating the silver mines. Hence I have had an opportunity of meeting with a number of plants, either very rare or entirely new to me; to say nothing of animals, insects, and minerals. As to what I have already seen of the birds of this delightful region, they surpass every thing that travellers have any where noticed.

I now send you a description of the new genus of plants already mentioned, subjoining some observations on *Plumeria*, *Carica*, and what I formerly took for *Krameria* *. I shall proceed to describe some other very interesting genera hereafter.

Adieu!

Near Cacota of Surat, Oct. 3, 1767.

I fear you must have thought me altogether forgetful of my respect and regard for you. This I should never have apprehended, were it not for the accidents to which I perceive epistolary correspondence is liable. Your last long-expected letter, informing me of the receipt of the *Cinchona*, found me in the town of Bogotá, just before my departure for a very long journey, and I read it with the most lively pleasure. No succeeding ones have reached me. A letter which I wrote to you two years ago, containing a few observations and descriptions, as I feared to trouble you with too many, seems never to have come to hand. I take the present

* *Acæna elongata. Linn. Mant.* 2. 200. These descriptions have furnished matter for several articles in the second *Mantissa*, and the *Supplementum*, of Linnæus, where the name of Mutis often appears.

opportunity, therefore, of selecting from my papers whatever is most novel or curious in this way. I wish these few matters, in which I have been studious of brevity, may please you, the great arbiter of natural science. I anxiously desire your opinion upon them. Farewell!

Santa Fé de Bogotá, May 15, 1770.

Nothing could be more welcome than your letter of the 10th of April 1769, in which you acknowledge the receipt last autumn of mine of the 19th of May 1767, as well as of one I wrote you from Cacota of Surat, Oct. 3, 1767, which had then just reached you. I perceive, however, that you had not received what I wrote from the same place March 3, 1768. I have always lamented, as I still do, the vast distance between us, which delays our correspondence for years together, and, what is much worse, occasions the loss of many of our letters. This last I strongly suspect to have been the case with what you may have written to solve my doubts concerning the genera I communicated, and on which I requested your opinion. Your usual accuracy and promptitude induce this suspicion.

I should never have aspired to those most flattering acknowledgments you are pleased to make for my communications. They are rather due from me, who am so happy to fulfil any of your wishes, and value your commands so highly, that the obligation lies on my part. I am the more happy to declare this, as you inform me that my little packet of remarks proved not unworthy of your approbation, and even delighted you like a draught of *Nepenthes*.

Your questions, whether the young branches of *Cinchona* are milky, in what soil it grows, and under what degrees of heat or cold, I confess myself utterly unable to answer. I have never visited the province of Quito, in which the native places of this valuable plant, Caxaminia, Loxa, and Cuenca, are situated, because of its great distance from the districts of Carthagena, Bogotá, Pampilona, and Girona. As far as I am able to judge, I conceive it to grow upon mountains, whose height would scarcely seem credible to Europeans, were it not verified by the most accurate observations with the barometer. Their altitude indeed is about half that of the most elevated spots in the Torrid zone. At Quito the mercury was observed by the celebrated De la Condamine scarcely ever to rise above 20 inches and 1 line; whence he justly determined the elevation of that place to be 1462 Parisian fathoms of six feet. Such also is nearly that of the country of Bogotá above the level of the sea, according to the observations I have here, for the first time, made. The temperature of both these towns is nearly the same; for I find the spirit of wine in Reaumur's thermometer here scarcely exceeds 18 degrees above the freezing point. Nevertheless, it appears that the *Cinchona* *officinalis* grows no where, under this temperature, except in the province of Quito, from the line to the fifth degree of south latitude. It is indeed asserted, by Mr. Santistevan, to grow also in the second degree of north latitude, near Popayan. I have often heard this gentleman declare that he had there gathered the *Cinchona* in flower, and that it is known by the name of *Palo del requeson.* He gave me some of the leaves, which are twice as large as those of the *officinalis;* but he brought away none of the flowers, which he declared to be constantly furnished with six stamens. I shall accompany this letter with the description of another *Cinchona* *, called by me *gironensis.*

You wonder, not without reason, at my having met with a Cape tree in this country. Yet the characters detailed in my description appear to prove this tree a *Brabeium* rather than any thing else. I confess my surprise at the plants of various climates, which I have either seen growing wild here, or readily propagating themselves when once introduced. Numerous European plants are here to be seen, either wild or cultivated. Hence I have seen the splendid table of our amiable Viceroy furnished, at all times of the year, with the most delicious vegetables, such as Italy might envy us. The most excellent strawberries have there been constantly supplied for these last ten years. This plant, which

* This appears by the fruit, a *berry* of five cells, to be no *Cinchona.*

we now see propagating itself so extensively, was obtained by the Viceroy, at my suggestion, by means of seeds imported in the dried fruit.

I now understand, what I was not aware of, that there is already a genus established by the name of *Jacquinia*. As you ask me what genus I should wish to bear my name, I confess I should like to have the tree N° 3, belonging to the *Pentandria Monogynia*, of which I sent you a description in my letter of May 15, 1767, provided there be no doubt as to the genus. If this be not undoubtedly new, I would have N° 11, which I had intended to name *Jacquinia*. If both are new, I should prefer the former, leaving the latter for any distinguished botanist whom you may think deserves it best.—*Solanum*, *n*. 2, *Bejaria*, *n*. 8, *Quadria* *, *n*. 12, I perceive you declare to be new; while you still continue in doubt as to the genera of 1, 7, and 9. You lament that you cannot certainly make out the place of my abode. In fact, our Cacota of Surat, not to be found in the maps, is an Indian town, two days journey distant from the American Pampilona. I have made this the centre of my excursions for four years past; and as I am likely to be, for some time longer, resident here, your letters may safely reach me through the care of Mr. Bellmann. I wish I could soon obtain the sixth edition of the *Genera Plantarum*, and the new edition of the *Systema Naturæ* which you have promised me. These, and any thing else, may easily be sent by Mr. Bellmann.

* This proved an *Hypericum*.

The orders of the Viceroy have brought me to this place for the third time. I arrived but a few days since. Here I found your letter, which I have over and over again pressed to my lips, waiting for me. Henceforth I promise myself more frequent opportunities of writing to you. In my next letter, as I write this in haste, I will enclose flowers of the two genera abovementioned.

I have many things to communicate to you. Having, for nearly ten years past, been engaged in long and not unpleasant journeys over these extensive American regions, I found means to collect a surprising number of plants. Formerly, before I left Europe, I could scarcely give credit to the accounts of the vast fertility of the country about the river of the Amazons. But since I have been an eye-witness of such fertility, though I could not previously conceive an idea of it, I can readily give my testimony to the declaration of the learned M. de la Condamine — that many years would scarcely be sufficient for an able botanist and expert draughtsman to delineate, describe, and arrange systematically, the immense variety of plants found in that country. Though I made a great collection of plants, doubtless many have escaped me.

The *Begonia* * I first met with, long ago, in a very celebrated forest called Tequendama, bordering

* *B. ferruginea, Linn. Suppl.* 419, *Sm. Plant. Ic. t.* 46, must be the species here intended. The only species besides, sent by him, was *B. urticifolia, ibid. t.* 47 ; but this made a part of his second collection.

our Bogotá river; though it was not there very plentiful. Afterwards I found it abundantly near Pampilona.

I do indeed greatly lament that our botanist Quer, so tenacious of Tournefort's method, should have thought of attacking your sexual system. That his proceedings are not founded on any solid reasons, I am well aware. I am highly disgusted with his *Flora*; the perusal of which has but lately fallen in my way. It is in every respect a rustic performance. I should point out some things to you, did I not apprehend that you have already heard of what I should think most remarkable.

My collection of descriptions of Birds, drawn up in my own way, is very considerable. I should have sent you a few of the most curious, but I wished first to correct them, as well as to accommodate them to your principles. Your long desired work upon animals, known to me as yet by name only, I have not been able to procure in Spain at any price. I do not admire Klein's system. I cannot longer conceal from you how much I wish to be associated with the illustrious members of the Society of Upsal. Adieu!

Description of MUTIS's Cinchona gironensis, *subjoined to the foregoing letter of May* 15, 1770.

A middle-sized tree, with crowded, uniform, rather short, moderately spreading, round, smooth *branches*, crossing each other; the young ones compressed and downy. *Leaves* crossing each other in pairs, spreading, ovate or oval, sometimes slightly heart-shaped; some occasionally orbicular, others fiddle-shaped; all are quite entire, flat, acute, five inches long, clothed on both sides with soft down; furnished beneath with transverse parallel veins, alternately larger and smaller; the rib round, prominent, hairy. *Foot-stalks* round, flattened above, furrowed, hairy, scarcely an inch long. *Stipulas* opposite, one at each side, sessile, awl-shaped, concave, fringed, very short. *Flower-stalks* axillary, (from the bosoms of the upper leaves) erect, compressed, swelling upwards, solitary, rather shorter than the leaves. *Spikes* two-ranked. *Flowers* nearly sessile, one always at each fork of the common flower-stalk. *Bracteas* two or three to each flower, at the base of its germen, small, awl-shaped, hairy, on a level with the top of the calyx, withering, many of them deciduous. *Calyx* a perianth of one leaf, tubular, abrupt, notched, rigid, downy, superior, longer than the germen, permanent. *Corolla* of one petal, pale violet, salver-shaped; tube cylindrical, rather swelling upwards, long, softly downy, membranous; limb in five oblong, entire, obtuse, concave, spreading segments, one third the length of the tube. *Nectary* a minute, slightly five-sided border, in the bottom of the tube, surrounding the style, permanent. *Stamens* five, scarcely visible, inserted into the throat of the corolla. *Anthers* linear, incumbent, slightly cloven at the base, of two cells, bursting inward. *Germen* inferior, minute, roundish, downy. *Style* thickened and most downy at the base; slender, tubular, with ten furrows, slightly downy, in the upper part; erect, the length of the corolla. *Stigma* thickish, with five obsolete anglés. *Pericarp* an obovate *berry*, crowned with the calyx and nectary, juicy, softly downy, of five cells. *Seeds* numerous, minute.

The fruit never bursts. Its cells are five distinct cartilaginous tubes, scarcely one third of a line in diameter, combined lengthwise, and not easily separable, occupying the centre of the berry. This fruit is not ripe till after June. I have rarely seen it, having always met with the tree in full bloom. Children eat the ripe berries as Mulberries; hence this is called the *Morito*, or Mulberry-tree, because of an imaginary but mistaken idea of the fruit.

This plant grows plentifully in the province of Giron, near the town of the same name, towards Canta, in a very hot climate, where the thermometer of Farenheit is at 87. The flower not only loses occasionally a fifth in each part of fructification, but sometimes adds a sixth, as I have remarked in another species found near Pampilona.

—

[*Note.* Specimens of this plant do not occur among the communications of Mutis to Linnæus. If not an *Hamellia*, it constitutes perhaps a new genus in the ninth section of Jussieu's *Rubiaceæ*. The alleged resemblance to a Mulberry, though not confirmed by the scientific description, may induce a suspicion of its belonging to Jussieu's tenth section.

Descriptions of *Manettia*, Linn. Mant. 2. 553, 558, and a few other things of small importance conclude this letter.]

Santa Fé de Bogotá, June 6, 1773.

As there is so vast a distance between us, it would be some satisfaction to me if your letters were more frequent. I cannot but complain of my hard fate, and sincerely lament that several years have passed without my receiving a letter from you. I have so long known your diligence and exactness in epistolary correspondence, that I suspect those who are entrusted with our mutual communications have not been so faithful as they ought. However this may be, you may depend on my attachment remaining unimpaired. Our Viceroy has brought with him your most valuable present of several of your works, which I have just received from his hands, and have many a time pressed to my lips. I cannot express how ardently I have longed for these books, which no gold could purchase in Spain. I immediately perceived the honourable mention you have made of me, under the genus *Cinchona* *, which I gratefully acknowledge. I wish it were in my power to contribute to your gratification. I

* *Linn. Syst. Nat. ed.* 12. *v.* 2. 164.

have sent you a small tribute of respect, by my friend Don Ruiz-Pavon, which he takes with him to Upsal, and which, though not equal to my wishes, will show something of my good will.

Our illustrious Viceroy, just arrived in this town from Spain, is a most ardent promoter of science. He has become acquainted with our correspondence in consequence of your present of books, confided to his care; and he is much interested in what passes between us. He generally enters into conversation with me, after dinner, about you; and makes me read passages out of your letters, highly flattering to me, in which he takes great delight, though they put me to the blush. This benevolent man, a few days since, took me with him into the hilly country, where he went for the purpose of planting strawberries, now one of our luxuries, in order that they may become naturalized all over these mountains.

I will shortly send you a catalogue of what I last forwarded. At present I have not time to prepare it, my friend's departure being so near at hand. I entreat you to bestow your kind attention upon him. You will not, I am sure, withhold your accustomed favour from a stranger, to whom your protection will be of the first importance. The frequent conversations I have had with him about you have excited in him so strong an inclination to become acquainted with you, that he cannot controul his desire to visit Upsal, for the purpose of seeing and being known to you, of conversing with you, and profiting by your instructions. He hopes also, through your interest, to take lessons in the art of metallurgy from the learned Wallerius. Thus, through your means, Sweden, so famous for science, may boast the honour of giving masters in botany and mineralogy to the far distant regions of India. I cannot but envy the fortunate lot of my friend, while I admire his well-directed courage and zeal. He will present you with my most cordial regards.

Farewell!

From the Mines of Ybague, Feb. 8, 1777.

Your very kind acknowledgments, my dear friend, for the collection of plants which I formerly transmitted, induce me to hope that what I have sent you, by Mr. James Gahn, will not prove unacceptable. Our friend Ruiz, after a long journey, which has occupied him for three years, is now safely returned to America, and I have passed many delightful days with him in hearing all he could tell me of you and your concerns, as well as of your worthy son. The art of assaying, which he learned at *Celeferd*, he has practised, under my inspection, for two whole months, with the greatest success. We came together from Bogotá, a few days since, to these mines of Ybague, that he may put in practice every thing that he has further acquired in this science during his stay in the Upper Hartz. I beg leave to return you my best thanks for the very great kindness with which you were pleased to receive him. Ruiz himself, as well as Escallon, and all my pupils here, desire their compliments. Pray present mine to your son. Farewell!

Remarks on my first Collection.

N° 9 you have determined to be a *Callisia* *; but, with your leave, it is certainly a *Tradescantia*. The filaments are six, hairy, &c.

N° 13. I am glad you esteem this, as I do, a new genus. Pray give it a name †.

N° 14. *Alchemilla aphanoides* ‡. I have always been doubtful to which of these two genera I should refer this plant. It has really two pistils and three stamens. The seeds are two. I have another species, resembling an *Alchemilla*, which is likewise digynous, with two seeds; but the stamens are two only.

N° 16. *Rhexia bogotensis* of my manuscripts. I do not well recollect whether my drawing, tab. 6, was this plant, or another species now sent you under N° 111. The anthers of this latter are versatile, and it is destitute of an outer calyx; while the former has upright anthers, and an exterior calyx of two leaves §. I meet with no trifling difficulties in

* This lies in *Callisia* in the Linnæan herbarium, having never been described by Linnæus. It is certainly not his *C. repens*, Jacquin's *Hapalanthus*, *Stirp. Amer.* 11. *t.* 11.

† *Escallonia myrtilloides. Linn. Suppl.* 156. *Sm. Plant. Ic. t.* 30.

‡ Since published in *Linn. Suppl.* 129.

§ Rather bracteas.

the genus *Rhexia*. Every species that I have hitherto seen has anthers opening by terminal pores, except a very remarkable one, found here, whose stamens are placed in a double row, the innermost being shortest; and the anthers burst by a longitudinal fissure *.

Nº 17. Not determined by you for want of ripe fruit. This is Nº 1 of my drawings †.

Nº 20. Certainly *Datura arborea*, represented in my drawings, t. 2, 3, 4.

Nº 21. I thank you, as I ought, for the very remarkable genus which you have dedicated to me ‡. I have known and described this plant ever since the year 1762.

Nº 28, 29. This was my *Logia*, of which I have sent you three drawings §.

Nº 31. *Spiræa argentea (Linn. Suppl.* 261), mistaken by me for *Dalibarda*.

Nº 56. I cannot sufficiently admire your Lyncean eyes, for determining the genus of this plant ||,

* Nº 16, as well as Nº 115 of Mutis's second list, is *Rhexia glutinosa, Linn. Suppl.* 216. Nº 111, of the second list, remains in the Linnæan herbarium, as yet undescribed. The former is tab. 6 of Mutis's drawings. Of the species whose anthers open lengthwise, I find a few flowers but no figure.

† *Lisianthus glaber. Linn. Suppl.* 134. *Sm. Plant. Ic. t.* 29.

‡ *Mutisia Clematis. Linn. Suppl.* 373. *Cavan. Ic. t.* 492. The latter describes and figures ten more species of this noble genus.

§ *Calceolaria pinnata, perfoliata,* and *ovata;* see *Sm. Plant. Ic. t.* 3 and 4.

|| *Viola parviflora. Linn. Suppl.* 396.

as well as for making out the flower of *Gomezia* *, enclosed in a little bit of paper under N° 15. The present, 56, is certainly a *Viola;* and I have always so considered it since the year 1761.

N° 57, 58, both the same plant †, whose fructification, as far as I see, is exactly that of the new *Hypericum*, n. 10 (*mexicanum, Linn. Suppl.* 345), except that the latter has but three styles, while this has five. I confess I am not yet able to understand the real limits of *Hypericum*.

N° 59. Through haste in selecting my specimens, I sent for *Polygala æstuans* (*Linn. Suppl.* 315), a plant of a new genus. If it were what I suppose, it is remarkable, and indeed altogether singular, for the inconstancy of its parts of fructification. But of this I propose to say more another time. You will find a drawing of its fructification in tab. 13. f. 2 ‡.

N° 70—72. You call this an *Andromeda.* I have the living plant every day before my eyes, and its characters, especially the perianth, becoming pulpy, or changed into a berry, forces me to take it for a *Gaultheria* §.

* *Nerteria depressa. Gærtn.* 124. *Sm. Pl. Ic. t.* 28.

† *Brathys juniperina, Linn. Suppl.* 268. *Hypericum Brathys. Sm. Pl. Ic. t.* 41.

‡ The specimen in the Linn. herb. is certainly *P. æstuans*, a *Monnina* of *Fl. Peruv.* and very different from the drawing indicated.

§ This is *Andromeda anastomosans, Linn. Suppl.* 237. The ripe downy capsules of the specimen, already bursting, altogether superior, and distinct from the calyx, may perhaps justify

Nº 88. My *Quadria*, your *Hypericum laurinum* (*H. petiolatum, Linn. Suppl.* 345), though the fruit is a berry? Jacquin, without seeing the fruit, long ago suspected this might be a new genus *. I had described the plant, and had made a new genus of it, before I saw Jacquin's work. But let it remain an *Hypericum*.

Nº 89. Under this number I find you have written *Carica Papaya*; but in my list it is *Cinchona bogotensis*, the same with the *peruviana (officinalis)*, observed by me about the town of Bogotá, ever since the year 1772. I have had a most beautiful drawing made of it †.

Nº 93. *Hermesias* of Loefling, *Brownea* of Jacquin (and Linnæus). Our people call it Bloodwood, meaning that the wood is excellent for stopping hemorrhages; whence also the name given by Loefling ‡. I have constantly found eleven stamens in perhaps a thousand flowers that I have examined. The pericarp is a legume.

Linnæus in this instance; for though the calyx itself really appears to be succulent, it does not invest any part of the capsule.

* *Hypericum cayennense. Jacq. Amer.* 213. *Linn. Suppl.* 343. The Surinam specimens really seem not to differ from Mutis's, and therefore *H. petiolatum* is to be struck out.

† Mutis's specimen, with flowers and fruit, scarcely seems the same with the true officinal plant brought by M. de la Condamine.

‡ *Loefl. It.* 278. This author's appellation evidently alludes to the fine red flowers.

N° 110. I long to see the flowers of this. I have very frequently, in my last journey, met with the plant laden with legumes, but could never find a single flower (1).

N° 119. You have passed this over without notice. I think it a *Melastoma* (2). I send it again under n. 79 of the present, or second, collection. The fructification is drawn in tab. 13. f. 5.

N° 122. This also is passed over. I take it for one of the *Myrtoideæ* (3). Here is another specimen.

N° 123. Of this you ask my opinion. I make it a *Scutellaria* (4).

N° 126. You say this is a pentandrous plant, unknown to you. It is certainly decandrous, with an inferior germen and seems to belong to *Vaccinium*, if not a new genus (5).

N° 138, 139. A *Melongena* according to Feuillée. You doubt whether it be a *Solanum* or a *Capsicum*. I say the former (6).

N° 140. A floret of this plant is delineated in my tab. 13. f. 4 (7).

(1) Perhaps another *Brownea*.

(2) *Melastoma squamulosa*, Sm. in Rees's Cycl. Mutis here confounds two species. See n. 79 of the second list.

(3) It is a *Rubiacea*, probably an *Erithalis* hitherto nondescript.

(4) Most evidently so, and very near *S. minor*.

(5) *Arbutus ferruginea. Linn. Suppl.* 238. Rather a decandrous *Vaccinium*, as L'Heritier long ago made it.

(6) Very near *Melongena laurifolia*, &c. *Feuillée*, 735, *t.* 26, if not the same.

(7) *Cineraria americana. Linn. Suppl.* 373.

N° 143. This you mark as a probable species of *Sibthorpia*, but you could not see it sufficiently clearly to be certain. Pardon me therefore, my dear Sir, if I send you another specimen. I esteem the plant a new genus, with five stamens; a monopetalous wheel-shaped corolla, in five deep segments; two styles; and a tumid fruit of two cells. See tab. 13. f. 6. If I am right, I beg of you to name it *Escallonia*, in honour of a man, deeply versed in your system, who is the indefatigable companion of my excursions *. In the plant N° 30 and 63, which you also denominate *Sibthorpia*, and which is accidentally omitted in my drawings, I cannot now examine the characters, having none but dried specimens. I will take the first opportunity of investigating the living plant, which I had referred to *Pentandria* †.

Of the first Collection of Drawings.

Tab. 1, 2, 3, are species of my *Logia*, see n. 28, 29, so named in 1767. I am not as yet acquainted with the characters of your *Calceolaria*.

Tab. 14. A very singular plant, which is by no means a *Tradescantia*, but undoubtedly belongs to *Gynandria Diandria*. The flower has three, nearly equal, exterior petals, in the place of a calyx; as

* *Dichondra repens. Forst. Gen. t.* 20. *Sm. Pl. Ic. t.* 8.

† This is *Sibthorpia retusa, Humb. Nov. Gen. et Sp* v. 2. 391. *t.* 177, very near *europæa*, but with more numerous and unequal lobes, or notches, in the leaves.

well as three interior ones, of which the middle one is broader than the rest, and turned upwards. The nectary, occupying the centre of the flower, is very hairy *.

Tab. 19. A dioecious tree. Male flowers destitute of corolla and calyx, except a small scale. Stamens several. Filaments scarcely any. See tab. 13. f. 3. The female flowers have neither calyx nor corolla. There are two germens, in some degree combined, and two styles. If this should prove a new genus, as I suspect, please to give it a name, which I wish you likewise to do to all the new genera that want one, as I have more than once signified. I prefer the names of botanists to all others †.

Notes on the second Collection of Specimens.

N° 58. A large and most beautiful flower of an *Aristolochia*, which I have preserved ever since 1761, when I gathered it on the banks of the river Magdalena ‡.

N° 61. *Davilia*, a new genus of mine. *Octandria Monogynia.* Capsule of three lobes, three cells, and three valves. Seeds three §.

* *Tradescantia nervosa*, *Linn. Mant.* 223, but undoubtedly, as Mutis indicates, one of the *Orchideæ*.

† This seems to be *Trophis laurifolia*, *Willd. Sp. Pl. v.* 4. 733, except that the drawing represents four seeds.

‡ This magnificent flower exceeds in size every other species. It is *Aristolochia ringens* of Vahl, *Willd. Sp. Pl. v.* 4. 155. *Vahl. Symb. v.* 2. *t.* 47. *v.* 3. 99.

§ The late M. L'Heretier informed me that the Spaniards

N° 71. *Lythrum? dipetalum (Linn. Suppl. 250).* I have constantly, during 16 years' frequent observation, found but two petals. In my present journey, however, I have noticed rudiments, as it were, of two other petals.

N° 80. The same as N° 119 of my first list, on which you made no remark. I make it a *Melastoma* *.

N° 86. Of this † I sent a description in a letter, to which I have had no answer, and in which I described my *Krameria*.

N° 90. A large leaf of a *Passiflora* ‡.

N° 91. Is this different from the foregoing? §

N° 93. A *Gualtheria*, or very near it ||.

N° 116. *Calceolaria perfoliata*, of which you desired me to send you a specimen.

I send by this opportunity a number of syngenesious plants, which cause me much perplexity. I rely on your friendship to give me your sentiments upon them, and all the rest. I send you a *Fulica*.

call this plant *Rosario*, and that it belongs to *Monoecia Octandria*. It is akin to *Croton*, but the pubescence is not stellated.

* N° 80 is *M. ligustrina*, Sm. in Rees's Cycl. indubitably distinct from N° 119, *squamulosa*, which has three-ribbed leaves, hoary beneath, and an angular scaly calyx; while *ligustrina* has triple-ribbed, broader, quite smooth leaves, and a furrowed naked calyx.

† *Buchnera grandiflora, Linn. Suppl.* 287. *Escobedia scabrifolia. Humb. Nov. Gen. et Sp. v.* 2. 371. *t.* 174.

‡ *P. clypeata. Sm. in Rees's Cycl. n.* 20.

§ Most widely! being very near *rubra*.

|| *Vaccinium villosum*. Sm. in Rees's Cycl. n. 23.

Another species, with a slightly curved bill, I have removed to *Ardea*, perhaps not correctly.

In a few months I propose to send you a very large collection. Farewell.

[This was indeed a last farewell, as Linnæus died in the following January. The intended collection of dried plants came to the hands of his son. See the following letters.]

CHARLES VON LINNE JUNIOR TO MUTIS.

No date.

The 10th of January in the present year (1778), a most fatal day to me, has deprived me of my beloved father. His bodily frame had, for two years past, been debilitated by three paralytic strokes, and he died at last of a gouty suppression of urine, terminating in gangrene. Nothing could tend more to soothe the feelings of a son, thus deprived of such a parent, than the contemplation of that treasure of rare and beautiful plants which you had destined for him. I cannot describe to you the grateful sensations with which I have, this summer, performed the task of examining them. I forgot for a time the bitterness of my grief, and almost forgot myself. I want words to express my thanks. I wish I might ever be able to testify my gratitude by deeds. Happy shall I be if you allow me to inherit your friendship for my father.

I put every thing else aside till I had turned over all your specimens, and every day was devoted to their investigation and admiration. I now send you a catalogue of the whole, as my father was accustomed to do, which I wish may meet your approbation. Tell me if you would have me do it otherwise. I shall ever be devoted to your service.

It is much to be regretted we are so far distant from each other that it requires a year or more to receive any communication. In that space of time natural productions can scarcely fail to receive injury from insects, as has now happened to several of your specimens. Our friend Mr. Gahn, who was this summer at Upsal, has however promised to facilitate our intercourse.

This year I have prepared a *Supplementum* to the *Systema Vegetabilium*, in which you will find your name as the discoverer of many rare plants. I quote your figures as follows; for example, under *Datura arborea, Mutis Amer. in MSS. fasc. 2. tab.* I shall send you this little work.

I have raised a very wonderful plant this year in my garden. It is a new species of *Hedysarum* from Bengal. This plant has a spontaneous motion in its foliage, which seems almost voluntary. You are aware that various parts of the vegetable body, especially those subservient to impregnation, can be so stimulated by the touch, as to exhibit some kind of movement. You know also the motion of some kinds of *Mimosa* and *Oxalis*, as well as of the *Dionæa muscipula*, arising from the touch of any extraneous body, or from the agitation of the wind. But the plant in question is not affected by either of these causes. Whether in the open air or in a close room, it spontaneously moves its leaflets, now one way, now another, one, two, or more, at a time; not all at once, nor all in one direction; and this takes place whether the air be serene, cloudy, or rainy. It has not yet flowered, but I expect that event in the course of the autumn, and if it produces seeds you shall have some. The plant requires great heat.

Please to make my grateful acknowledgments to Mr. Escallon for the beautiful plants he has sent. Give my compliments also to your most skilful mineralogist, my worthy friend Mr. Ruiz. I often call to mind his agreeable conversations at Upsal, and I now profit by what he has taught me. I hope he will not forget me. I have a collection of minerals, but not much from your part of America. He can easily, if he pleases, enrich my stock.

MUTIS TO PROFESSOR CHARLES VON LINNE', JUN. THE WORTHY SON OF THE PRINCE OF NATURALISTS SIR CHARLES VON LINNE', KNIGHT OF THE POLAR STAR, &c.

From the Mines of Ybagua, Sept. 12, 1778.

This letter, which I have many a time, in the joy of my heart, had it in contemplation to write to you, my worthy friend, I find myself now scarcely able to begin, on account of the grief with which yours just received has overwhelmed me. As I opened this letter, enclosed in one from a beloved brother of mine who lives at Cadiz, I did not at once discover from whom it came, the superscription being in an unknown hand; but I feared it might bring me an account of the precious life of my valued friend the Chevalier von Linné being either in danger, or perhaps extinct. When I had read it, I perceived, but too certainly, the truth of what had been announced in the publick papers, that this great man, your illustrious father, was no more. To cultivate his faithful friendship has, for many years, been my chief ambition, in spite of the wide distance between your polar region and the equator. I wanted resolution to open, soon afterwards, a packet from Mr. Gahn, whose hand-writing I recognized in the direction, lest I might perhaps find a letter, the last, and now posthumous, pledge of his friendship, flattering me with hopes which I had already abandoned. Allow me therefore, my dear Sir, to recall to your mind those recollections which, however sad, we ought not to forget. If it were possible for you to overcome the feelings of nature, I cannot satisfy the claims of friendship without lamenting, with you, our common loss.

Let me inform you therefore, that, so long ago as the year 1761, when I ventured to introduce myself to this great man, by a trifling communication, as I had not enjoyed any intercourse with him before my departure from Europe, I was first favoured, in this my distant abode, with one of those letters, so highly valued by the most learned men in Europe. In this, according to his usual custom, your distinguished father endeavoured, in the most attractive style, to stimulate my youthful ardour more and more for the study of Nature. From that period I rejoiced to devote myself to his service, and our correspondence was kept up for 18 years, as regularly as the great distance between us, the negligence of those in whom we confided, and my occasional extensive journeys, would admit. By some unavoidable accidents, indeed, many of my letters never reached him; and I have also, too late, discovered that many of his had been lost. Meanwhile our communications were confidential and exclusive, not extended on my part to any other persons, whether my countrymen or not; for I devoted all my discoveries and all my labours to his immortal genius alone. A little while ago, when I still supposed him living (as I saw the illustrious name of Von Linné among the members of the Royal Academy of Paris, in a list at the end of the *Connoissance des tems)*, I was particularly happy to obtain the complete fructification of that most elegant tree which yields the Peruvian Balsam, in order that I might satisfy his curiosity, so often expressed, on the subject of the genus of this tree, either by describing it among my new genera, or by transmitting any observations for his use. But when I had just overcome the difficulties which had so long deprived me of this acquisition, and was anticipating the pleasure my excellent friend would receive from the communication, the world was deprived of him. You have lost an affectionate parent, and I a most highly esteemed patron. I trust that you, my honoured friend, will, with his blood, inherit his exalted genius, his ardent love of science, his kind liberality to his friends, and all the other valuable endowments of his mind. On my part, I shall show my gratitude to his memory by teaching and extolling the name of Linnæus, as the supreme prince of naturalists, even here under the equator, where the Sciences are already flourishing, and advancing by the most rapid steps; and where, I am disposed to believe, the Muses may, perhaps, in future ages, fix their seat. If my opinion be of any weight as a naturalist, I must declare that I can find no name, in the whole history of this department of knowledge, worthy to be compared with the illustrious Swede. Of this at least I am certain, that the merits of Newton in philosophy and mathematics are equalled in botany, and all the principles of natural history, by the immortal von Linné. These great men stand equal and unrivalled, in my judgment, as the most faithful interpreters of Nature's works. I trust, Sir, you will not take amiss this testimony of mine in favour of your distinguished parent; for, as you are closely allied to him in blood, I feel myself scarcely less intimately attached, by the particular friendship with which he was so good as to favour me. His memory will ever be cherished by me, as that of a beloved preceptor; and I shall value, as long as I live, every pledge of his regard.

At length, my dear Sir, I am favoured with your letter, so well calculated to mitigate and to repair my loss, as I already perceive in it traces of your father's character. You tender me your friendship in the place of his; I, in like manner, promise you every thing I had destined for his use, which I shall gladly transfer to you as, in every respect, his rightful heir. I see how much you are disposed to imitate his liberality, in your present of the last edition of the *Systema Naturæ*, with the two *Mantissæ*, which I had so long wanted, but which no Spanish gold could purchase. At my first departure from Spain I acquired, through the friendship of my friends Alstroemer and Logie, the edition of the *Systema* which was then current, the valuable Travels of Loefling, and the *Philosophia Botanica*. By a lucky chance I purchased, at the same time, the first and second volumes of the *Amœnitates*, the first edition of the *Species Plantarum*, and the fourth of the *Genera*. These were all my stock. They were to me an inestimable acquisition, for I perceived I could make no progress without them. I subsequently obtained the fifth edition of the *Genera Plantarum*; and at length, through the bounty of your father, I became possessed of the edition of his *Systema Naturæ* published in 1767, the *Species Plantarum* of 1762, and the *Fauna Suecica* of 1761. Whatever has appeared since these, as well as many earlier works of your father's, are entirely unknown to me, and I suspect many have recently been printed. I beg of you not to suffer me to remain longer in ignorance of what would be so extremely welcome, and so eminently serviceable. I long, beyond measure, to peruse these precious records, particularly the *Museum Ludov. Ulricæ*, and the four subsequent volumes of the *Amœnitates*, which are every where quoted. My anxiety to possess these books arises, not only from a wish to draw up my descriptions after the precepts and example of your father, but also that I may see what objects, within my reach, require further examination or illustration. Let me not however, at the beginning of our intimacy, intrude any further on you, my valued friend, than to intreat your kind attention in forwarding what I want. I shall most cheerfully, as I ought, request Mr. Gahn to pay all expenses for me, once more beseeching you to excuse the trouble I give you. You would the more readily pardon my indiscretion, if you, who are seated in the centre of science, could be aware of the extortion to which I am exposed in this remote corner of the world, on account of the scarcity of books in general. My library is, nevertheless, a very ample one, nor has any thing comparable to it been hitherto seen in America. I must throw myself upon you to forgive all my imperfections, relying on that friendship which I enjoyed with your father, and which, on that ground, I am anxious to cultivate with you. I am impatient to know the opinion of your father, relative to some of my earliest discoveries, among which many things were sent without names, that he might be at liberty to apply such as he thought most fit. I wrote to him from this place, where I have lived two years, in February 1777; but I perceive my letter did not arrive so soon as it ought. It was accompanied by a catalogue of what was sent by Mr. Davila, of the Royal Museum, as well as some remarks upon the former collection. Mr. Gahn informs me all these things are now in your possession. When I then wrote, I promised, as I intended, to send a large collection in the space of a few months. But a long illness, and the requisite attention to my laborious duties as superintendant of these mines, have prevented the performance of my promise. When I was just beginning to set about it, I suffered severely, for many days, from an accident, very frequent here, caused by an insect, which, like fleas, attacks men, cattle, and dogs, but which is different from the *Oestrus bovinus*, or Common Gad-fly. On the 24th of February, 1777, I clearly perceived that this troublesome inmate had lodged itself in my leg, where a tumour was apparent. By an imprudent application of the juice of tobacco, and afterwards of the milk of the fruit of *Musa guineensis*, according to the practice of the country people here, I brought on a most violent erysipelas, which quickly suppurated. My life was in the utmost danger, and it was long before I was restored to health. The frequency of these insects, added to the dread of serpents, made me almost resolve to quit my country residence; and I was preparing to do so, when my mind recovered its wonted resolution by the animating hope of new discoveries, while the recollection of past perils grew more and more feeble. Thus fortified, and rendered familiar with the danger, I thought it best to begin my enquiries with an examination of the very insect in question. For this I soon found a peculiarly favourable opportunity, and I have from time to time made very curious observations, recorded in my journal, relative to this insect, whose history has hitherto remained concealed from all mankind, being very imperfectly known even to the inhabitants of this neighbourhood. You shall hear more about it another time. As far as I can discover from the records of travellers, the species is entirely new. It may be named *Oestrus hominis*, very widely differing from all others of this genus, especially from the *Oestrus bovis*, likewise very frequent here, which is so fully described by Reaumur. I find no account in any scientific traveller, nor in any descriptive work, answering to our insect; which is about the size of a common house-fly, *Musca domestica*. The body of the female is covered with a number of little imbricated tubes, formed like a wasp's nest, lodging above fifty minute *larvæ*, or maggots. Her resemblance to a common fly causes her approach to be viewed without apprehension, by those who have never experienced the dreadful mischief she brings. Her pernicious progeny readily quit their retreat, and she confides them to our unsuspecting care for food, education, and even transformation, if our patience should so far endure, and with scarcely any compunction for what she has inflicted upon us, she takes her leave, speedily to perish herself. On the 24th of May last I was fortunate enough to dislodge one of these pernicious intruders from my arm, just in time to prevent the greater part of its mischief, and without much difficulty. But I must say no more of this most wonderful insect at present.

I am situated at a very great distance from the place where I met with the Peruvian Bark, *Cinchona officinalis*, growing wild, where also grows the *Mutisia*. I sent all my specimens of the *Cinchona*, before my departure, along with a most elegant drawing, to the Royal Museum, bringing with me a still better figure. Whenever I go to the town of Bogotá I will forward the latter, as well as a specimen, to you. I have not yet met with the *Caranna* tree. Many of the other things about which you enquire are, as yet, unknown to me. Of the Elastic Gum, called by our people *Caucho*, and its various kinds, I am furnished with pretty complete information. How many curious particulars will you find, on this subject, in my letters to your father! I have never seen the tree which yields this gum, and which grows in the province of Chocò. From analogy of what I have seen, I conclude it must be one of the American species of *Ficus*. These American Figs have about them insects of the same genus (whether they belong to *Ichneumon* or not I am ignorant) as the Figs of Europe, but of a different species. I have frequently observed them, with much satisfaction. I perceive by your letters that there exists some doubt respecting the tree which yields the Dragon's Blood. But I have never met with the trees described under this popular name, of *Sanguis Draconis*, by Loefling and Jacquin, and under the scientific one of *Pterocarpus*. Of this only I am certain, that the name of *Draco*, or Dragon, is universally given, throughout the most distant parts of this country, to a tree producing a very similar exudation, which, if I mistake not, is a species of *Croton*. You have a specimen of it under N° 1 * of my last parcel; and perhaps of another species at N° 41 †. — Concerning Jalap, I confess I have nothing worth communicating. Plants of the tribe to which it belongs are cultivated, for ornament, at Carthagena, and elsewhere. I have never stopped to examine them, supposing them well known to every body. The root, serving for medicinal use, is brought to us from a great distance, or even from Spain itself. I have been very anxious to get the *Ipecacuanha*, but as yet I have never seen the living plant. The roots are gathered about Simiti, and in other very hot countries, and brought to market at Mompoxia. I perceive you are still in doubt on this subject; which reminds me of a conjecture I formed, long ago, to the following effect. It is some time since I made a description of a

* A new species, near *Croton lacciferum* of Linnæus.

† This is very near *Croton flavens* of Linnæus; but differs in having ovate or elliptical (not heart-shaped nor taper-pointed) leaves, whose upper side is extremely soft and silky, scarcely stellated in its pubescence like most of the genus.

plant *, which several persons thought the *Raizilla besuquillo ipecacuanha;* that is, "the Little root, whose stem climbs up the surrounding trees;" by which appellation the *Ipecacuanha* is known among us. This plant was found by a kind of empiric, in the province of Girona, the climate of which differs little from the burning temperature of the known country of this drug. The person who gathered the plant asserted it to be precisely the same as what is brought from Simiti, with which he was well acquainted. Hence I made a point of obtaining a specimen, as fresh as possible, from the governor of Giron, for examination and description. But as I became satisfied that the genuine *Ipecacuanha* belonged to a very widely different genus, so I should have no scruple in asserting, before any body, that the new plant was either no *Ipecacuanha* at all, or, if it were the same with that of Simiti, as the abovementioned empiric declared, they had nothing to do with the Brasilian *Ipecacuanha.* The medicine, if I mistake not, was originally brought from Brasil, and is still universally known among the Indians by this its Brasilian appellation. Nor am I surprised that our plant should be very different, though endowed with the same virtues. For I know another plant, whose roots very nearly agree with *Ipecacuanha* in appearance, as well as qualities, which nevertheless is a species of *Viola.* You may find a specimen in my first collection, N° 56 (*V.*

* *Psychotria emetica. Linn. Suppl.* 144.

parviflora; Linn. Suppl. 396). The supposed *Ipecacuanha* of Giron, of which I made a description in 1768, sent in 1774, belongs to *Pentandria Monogynia.* You may depend on my making further inquiries about these matters, though my present abode is very distant from Simiti. Nor will I neglect any fit opportunity of obtaining information concerning the remainder of your questions. With regard to the remedies of which I make use in my own practice, I would remark, by the bye, that I believe the whole of practical medicine lies in a small compass; in which opinion the most eminent men concur. The more simple, and free from a jumble of numerous articles, it is, the better, as you and every intelligent practitioner must soon have discovered, in spite of the preconceived opinions of the vulgar herd of physicians. The *materia medica* to which I have recourse in my practice is therefore of the most simple kind. The reputation I have acquired among the Americans is such, that I am beset with a crowd of sick people, who flock after me, even into my rural retirement, having learned by experience that it is possible to be well cured of their diseases at a very moderate expense. An European could scarcely believe how little these people would spend upon physick, if the apothecaries' shops were, happily, to the great advantage of humanity, banished from the country. The American is accustomed to content himself with the simple produce of the fields. Indeed, to express my opinion freely, I do not think Flora's most liberal and delightful gifts are intended for the nauseous purposes of pharmacy.

You would scarcely, my friend, believe how generally useful I have found a simple decoction of *Scoparia dulcis,* in intermittent fevers, and, with the addition of the acid juice of oranges, in continued ones; occasional injections, of the most simple materials, being also used. I am much surprised that Cartheuser, if I remember right, should have wished to expel the *Sarsaparilla* from the shops, while I long to have it reinstated there. Scarcely any *lues venerea* resists my method of administering a drink of this medicine. Sometimes I add to my own method that of Van Swieten, used here perhaps, for other purposes, before it was known to that great man. I must inform you also, that I have, in two different instances, experienced the power of this despised medicine in curing a palsy preceded by cholick; a fact perhaps not uninteresting to our worthy friend Alstroemer.

You enquire concerning the Palm which affords us butter. This is one of the most common of its tribe, and attains a considerable height, greatly resembling in habit the *Areca oleracea* of Jacquin, who did not examine the fructification of that plant in a complete state. From his description, however, I suspect ours to be different. The character he gives of the *Spatha* by no means answers to the very peculiar structure of our Palm, not to mention great differences in the fruit. Hence I am induced to consider ours as a distinct species, belonging to the same genus *. The nuts, bruised and rudely ground, are thrown into water, where they undergo a slow maceration, without any artificial heat or pressure, and the kernels gradually dissolve. The butter floats on the surface, the heavier matter sinking to the bottom of the water. Three washings are sufficient to extract all the oily matter, which, under the twentieth degree of Reaumur's thermometer above the freezing point, has the consistence of butter, of a whitish colour; but at the twenty-third becomes liquid like oil. I do not doubt that a similar oil may be obtained from the nuts of all the Palms. Ours is in general use, and by no means unpleasant. The pulp of the fruit is juicy, very mucilaginous, and rather sweet; it serves to fatten hogs. The Palms which bear tallow and wax are known to me by description only. I have never seen them, which I much wish for, in order to complete my history of this family.

My letter has, unawares, extended to a greater length than I proposed. I gladly accept your offered friendship. Pray write frequently. Farewell, my dear Sir; and do not forget your far-distant friend.

P. S. The fructification of the Balsam of Peru tree is nearly the same as that of the *Toluifera*. I send you a specimen †, as the first pledge of my friendship. The seed vessel is a legume.

* Here follows the long botanical description, printed in *Linn. Suppl.* 454, where this Palm is named *Cocos butyracea*.

† This very complete specimen remains in the Linnæan herbarium. See the article *Myroxylon* in Rees's Cyclopædia.

MR. DRU DRURY, GOLDSMITH, Nº 1, LOVE-LANE, ALDERMANBURY, LONDON, TO LINNÆUS.

MOST EXCELLENT SIR, London, Aug. 30, 1770.

I cannot better express the strong inclination I have of testifying my respect to you, as the greatest master of natural history now existing, than by presenting you a copy of a work I have just published here *.

Believe me, Sir, it is not from vanity I take the liberty of making you this offering; nor, poor as it is (for I am truly sensible of its defects), would I make it to any person that is inferior to Linnæus in the study of nature. But to whom should I pay my acknowledgments of this sort but to the *father* of natural history? You, Sir, I look on as that *father*, and therefore I beseech your kind acceptance hereof; a circumstance that will do me great honour and favour, and at the same time countenance my weak endeavours to promote a study that I must confess I prefer to every other.

Permit me also to take this opportunity to congratulate you on the effects which your *Systema* has had among the followers of natural history here in London; the number of whom, although not equal to those found in many other countries, is yet every day increasing, to such a degree as could not have been expected, a little time ago, by the most sanguine well-wishers of this science.

* Illustrations of Natural History, wherein are exhibited upwards of 240 figures of Exotic Insects, on 50 copper plates. By D. Drury. London, 4to, 1770.

That it may still increase and flourish, and that you may, with health, live to see its study carried to the furthest ends of the earth, is the hearty prayer of, Sir, your sincere admirer and most humble servant,

D. DRURY.

JEAN JACQUES ROUSSEAU TO LINNÆUS.

A Paris, le 21 Septembre, 1771.

Recevez avec bonté, Monsieur, l'hommage d'un très ignare mais très zélé disciple de vos disciples, qui doit en grande partie à la méditation de vos écrits la tranquillité dont il jouit, au milieu d'une persécution d'autant plus cruelle qu'elle est plus cachée, et qu'elle couvre du masque de la bienveillance et de l'amitié la plus terrible haine que l'enfer excita jamais. Seul avec la nature et vous, je passe dans mes promenades champêtres des heures délicieuses, et je tire un profit plus réel de votre *Philosophia Botanica* que de tous les livres de morale. J'apprends avec joye que je ne vous suis pas tout à fait inconnu, et que vous voulez bien même me destiner quelques unes de vos productions. Soyez persuadé, Monsieur, qu'elles feront ma lecture chérie, et que ce plaisir deviendra plus vif encore par celui de les tenir de vous. J'amuse ma vieille enfance à faire une petite collection de fruits et de graines. Si parmi vos tresors en ce genre il se trouvoit quelques rebuts dont vous voulussiez faire un heureux, daignez songer à moi ; je les recevrois, Monsieur, avec une reconnoissance, seul retour que je puisse vous offrir, mais que le cœur dont elle part ne rend pas indigne de vous. Adieu, Monsieur, continuez d'ouvrir et d' interpréter aux hommes le livre de la nature. Pour moi, content d'en déchifrer quelques mots à votre suite dans le feüillet du regne végétal ; je vous lis, je vous étudie, je vous médite, je vous honore et vous aime de tout mon cœur. J. J. Rousseau.

PROFESSOR ROBERT RAMSAY TO LINNÆUS.

Sir, Edinburgh, June 29, 1773.

Although I have not the honour to be known to you, I have long been an admirer of your writings, and no less so of your private character; together they make you respected and esteemed all over Europe. Having heard that you intend to give a new edition of your *Systema Naturæ*, and being assured by my friend Dr. Gahn that you would receive favourably any information regarding natural history, I used the freedom some time ago to send you a box with two preserved birds, male and female, which I believe are peculiar to the British Islands, and which you never have yet mentioned in any edition of your *Systema Naturæ*. They belong to your genus *Tetrao*; are the Red Game, Gorcock, or Moorcock, of Willoughby; the Moorcock or Moorfowl of Sibbald; the Red Grous of Mr. Pennant. For a further account of this I refer you to the above authors. I shall only add that they are a perfectly distinct species from the *Tetrao Lagopus*, as I am persuaded you will be convinced when you see them. They inhabit moist, heathery plains, *Ericeta,* and the sides of hills, and never change their colour in winter. We never find the *Lagopus* but on the dry, rocky summits of our highest mountains, and they constantly change to white in winter. I wish they may arrive to you safe and in good condition, and that they may be acceptable.

It will now be necessary to inform you who your correspondent is; my name is not like yours, a name of fame, but is confined to a very small circle. I have the honour to be the King's Professor of Natural History here, and I have, for some years, endeavoured to explain your *Systema Animalium* in our University. I wish to show you my gratitude for the pleasure and instruction I have received from your writings; and I hope that if I can be in any way useful to you in this part of the world, you will be so good as to favour me with your commands, and give me opportunities of showing you the perfect regard and esteem, with which I have the honour to be, Sir, your most obedient and most humble servant, ROBERT RAMSAY.

RIGHT HON. JAMES BURNET, LORD MONBODDO, TO LINNÆUS. [Latin.]

SIR, No date.

I have often wondered, while you are so illustrious for your knowledge, that what Lucretius falsely says of his Epictetus may most truly be applied to you—"all Nature lying plain and open before you"—there should be any persons who could shut their eyes against so brilliant a light. Among such I am not ashamed to enumerate the French author Buffon, who, though he has written many excellent things on natural history, in one point has wandered materially from the truth. He not merely withholds his approbation of your classes and their subdivisions, but has even presumed to treat them with ridicule. Yet those who have merely made themselves acquainted with the first rudiments of philosophy, cannot possibly be ignorant, that a distribution into genera and species is the foundation of all human knowledge; and that to be acquainted with an individual, as they term it, or one single thing, is neither art nor science. Buffon therefore is not your adversary so much as he is at once the enemy of all philosophy, and of all human knowledge. But above all things I discommend this writer, to come to the subject upon which I wish to consult you, that, whereas, in the history of animals, there is nothing more important to us than what is related of such as most nearly approach our own genus, differing from us only in being less accomplished in arts and learning, I speak of the wild men of the woods, vulgarly called the *Ouran Outang;* yet on this subject Buffon is eminently culpable. He does not scruple to contemn and reject what Bontius, a physician and most credible author, relates, but also what you have collected from many witnesses, respecting such animals. He will not believe that such a creature, though, as he allows, endowed with a human shape, can either speak or think. But if what is reported in the sixth volume of your *Amœnitates Academicæ*, on the faith of a certain Swede named Köping, be true, his own words being given in an academical oration delivered by one Mr. Hopp in your Upsal Academy, in the year 1760, under your auspices, though by no means adorned with your genius or sagacity; there can be no sort of reason why we should doubt that this wild man, or *Ouran Outang*, is a real man, or at least a being most nearly allied to us in his rational faculty. Indeed that Swede, according to Mr. Hopp, asserts, that he had seen animals, so unlike mankind that they were even furnished with tails, like cats, and yet making use of boats, and exercising traffick.

Now I am anxious to know who this traveller was? where did he live? and how far is he worthy of credit? from what port did he sail? and in what country did he meet with such a prodigy? Upon all these points Hopp is silent. I should also wish to be informed whether he has published an account of his travels? If he be living, I would have him asked whether these men with tails used any language in their traffick? whether they walk upright, and are of the usual human stature? I should likewise be glad to know, if, Sir, I do not give you too much trouble, what writers you have relied on in your relation concerning the wild man, particularly *Koep. Itin.** *cap.* 86, and *Dalin. Orat.* 5, whose names are known to us here by your mention of them only, and we are altogether ignorant how far they are worthy of credit.

Sincerely wishing you may continue to be the glory of your country and of all Europe, and long enjoy your reputation and your health, to the great advantage of philosophy, I remain, &c.

CAROLINE LOUISA, MARGRAVINE OF BADEN, TO LINNÆUS. [Latin.]

ILLUSTRIOUS SIR, Carlsrhue, Aug. 4, 1775.

I lately received, by the post, a drawing of an exotic plant, said to be a native of Surinam and Mexico, to which I found, with great surprise, my own name affixed †. A systematic description of this plant, under your signature, accompanied the drawing. It was not without the highest pleasure that I perceived, by these tokens, the good opinion I had obtained of the Prince of Botanical Science,

* Nils Matson Köping, or Koeping, *En reesa*, or Travels, in Asia, Africa, &c.; a quarto volume in Swedish, published at Wisingsborgh, in 1667. This is the person abovementioned, quoted in the *Amœn. Acad.* On turning to the passage, the most charitable opinion we can form is, that the author, or editor, of this book had confounded the original minutes of the voyage, and jumbled together what belonged partly to savages and partly to monkies.

† *Carolinea princeps. Linn. Suppl.* 314.

in consequence of my own endeavours to promote this study; and, at the same time, that he did not altogether disapprove of the devotion of a woman to such pursuits; though I am afraid my claims to your good opinion may have been over-rated by Mr. Björnstahl, in his letters to you.

However this may be, I will venture to declare, that, having long been devoted to this study, and continually occupied with your works, which have justly obtained such high authority, my ardour for botany has so powerfully been excited by your approbation and encouragement, however partial, that I promise you I will not neglect to have my rare plants engraved, however far the execution of this project may fall short of your expectation or of my own.

Your favour to me, though far exceeding my deserts, will lead me to do all in my power to make your name, long consigned to immortality, more extensively known in this country. I shall leave you to judge whether I could show my gratitude in any manner more suitable to your feelings or mine.

May you, Sir, long remain at the head of botanical science! You may be assured that this wish is felt by no one more sincerely than by her who is happy to have an opportunity of writing this letter, and who esteems herself most highly honoured by the tribute of respect which she is thus enabled to pay to a man of such distinguished merit.

CAROLINE LOUISA, MARGRAVINE OF BADEN,
BORN LANDGRAVINE OF HESSE DARMSTADT.

MR. FRANCIS MASSON TO LINNÆUS.

HONOURABLE SIR, London, Dec. 26, 1775.

I hope your goodness will excuse the liberty I have taken in addressing myself to you, as it proceeds from a knowledge of your superior merit, and your exalted character in natural history. I have been employed some years past, by the King of Great Britain, in collecting of plants for the Royal gardens at Kew. My researches have been chiefly at the Cape of Good Hope, where I had the fortune to meet with the ingenious Dr. Thunberg, with whom I made two successful journeys into the interior parts of the country. My labours have been crowned with success, having added upwards of 400 new species to his Majesty's collection of living plants, and I believe many new genera.

I expect soon to go out on another expedition, to another part of the globe, to collect plants for his Majesty. If I should be so fortunate as to discover any thing new in any branch of natural history, I should be happy in having the honour of communicating it to you. I had the pleasure of seeing Mr. Sparrmann at the Cape, and received from him a parcel of seeds, which he collected in the Southern Islands, and which I now send you. I could not presume to send you any Cape plants, as I suppose Dr. Thunberg has sent you every kind that he collected, which are much the same as mine. I also intended sending you a collection of the St. Helena plants; but that I shall defer until your ports are open for shipping.

The inclosed specimen I think is a new genus, to which my worthy friend Mr. Thunberg had a great desire of giving the name of *Massonia,* honouring me with this mark of his friendship. But notwithstanding the good will of Dr. Thunberg and many other botanical friends, I have declined receiving that honour from any other authority than the great Linnæus, whom I look upon as the father of botany and natural history, in hopes that you will give it your sanction. I am sorry that the leaves are not more perfect; but it is the only specimen I have. I shall take the liberty to give a description of the root and leaves; the flowers being perfect, I refer them to your better judgment.

Radix bulbosa, tunicata, subglobosa, diametro sesquiunciali.

Folia duo, radicalia, ovata, subrotunda, acuminata, glabra, lævia, carnosa, nervosa, nervis immersis, palmaria, adspersa maculis numerosis purpurascentibus.

Scapus brevissimus, sinubus foliorum quasi immersus, teres, glaber.

Habitat in campis elevatis in regione *Rogefeld.*

I have seen another species, with narrow leaves, which flowered in Kew garden. This I shall send you by next opportunity. I shall add no more; but am, with great esteem, your most obedient, humble servant,

FRANCIS MASSON.

HONOURABLE SIR, Madeira, Aug. 6, 1776.

I had the pleasure of receiving yours on the 19th of May, on which day I left England. This prevented my writing you an answer; nevertheless, I left orders with Mr. Lee to send you a specimen of the *Massonia angustifolia*, with an engraving of both species; also several other dry plants, which I thought were new genera. I am glad of the intelligence concerning the *Massonia*. I was so weak a botanist as to think its fruit, being a capsule, would entirely separate it from the genus *Hœmanthus;* but I beg leave to offer an argument in favour of my former opinion; which is, that the aforesaid plant has *germen superum*, whereas *Hœmanthus* has *germen inferum*. Your good nature and generosity, in acquiescing to confer the honour of a genus on so young a botanist as I am, deserve my most grateful thanks.

I have observed many new plants here, especially trees, which I think are new genera, but have not seen their fructification perfect. I have noticed that most of the Syngenesious plants are of the first sub-division, *Corollis ligulatis*, which to me are a little difficult, having found several shrubby plants which I can refer to no genus but *Sonchus*, which is very uncommon. I have observed that all the rare plants grow either on the high cliffs near the sea, or in horrible deep chasms, that run towards the middle of the island. But towards the top, which is more than 5000 geometrical feet perpendicular, I have found nothing but a few European plants, especially *Spartium scoparium*. I have taken the freedom to send you a few plants, which I should be glad to know your opinion of, there being many which I could not determine.

I have sent you a plant under the name of *Aitonia rupestris* of Forster's *Genera Plantarum*, which he discovered at Madeira, and named it in honour of William Aiton, his Majesty's botanical gardener at Kew. But if the said gentleman deserved that honour, I think Forster did him great injustice, to give an ingenious gardener a plant which can never be introduced into Europe, and consequently its existence will still be doubted. I know Mr. Aiton has no ambition for that honour; but still, for the respect I have for so ingenious a gardener, I took the liberty to give N° 19 * that name, thinking it a new genus, and a plant which will soon be introduced into the European gardens; but this matter I must submit to your better judgment. I shall now conclude with imploring the Divine Being to grant you still a longer existence on earth, to patronize the great study of the works of Nature; which is the earnest wish of, Sir, your most obliged, humble servant, FRANCIS MASSON.

* This was no other than *Campanula aurea*. *Linn. Suppl.* 141.

FROM THE SAME TO THE YOUNGER LINNÆUS.

SIR, Madeira, Dec. 12, 1778.

In June last I was honoured with your long-desired epistle, dated December 1777. Being at that time in the Canaries, I had no opportunity of acknowledging the favour; but now being again arrived in the island of Madeira, I have taken the earliest opportunity of acknowledging how much I am indebted to you for your civility.

I condole with you on the death of your great father; a loss which every lover of arts and sciences must ever feel.

I am sorry that the small parcel of specimens which I sent was damaged in the passage; but am happy to know a better way of conveyance than by London. The hurry and confusion, peculiar to great cities, will account for the difficulty of obtaining such objects. I humbly thank you for the names of those that were perfect, but cannot help offering my opinion concerning the *Echium (candicans)* and *Digitalis (Sceptrum)*. The first is a very large shrub, having spikes of blue flowers near two feet long at the extremities of the branches, which make a glorious appearance when in bloom. I think the *Echium argenteum*, from the Cape of Good Hope, is far from being so fine a plant. I think also the *Digitalis* differs from the *canariense*, in size of the shrub, mode of flowering, and figure of the corolla. I am sorry it is not in my power to renew the specimens of those which were lost, having sent all the Madeira collection to England. But as the Canary collection still remains in my possession, I have sent a few seeds and specimens, which I hope you will receive in better order.

Respecting the Physicians and Botanists in Portugal, I have not the honour of being acquainted with any of them. But since my residence here I have contracted a correspondence with Mr. De Visme, a merchant in Lisbon, who has a fine botanic garden, and is an enthusiastic lover of plants. By his last letter he informed me that the Queen of Portugal had sent several young botanists (students of Vandelli) to their settlements of Angola and Brazil to explore these rich countries. But I have a poor opinion of the genius of the Portuguese for such researches. I have now finished my peregrinations among the Fortunate Islands, and have found great pleasure, as well in the novelty of their productions as the singularity of the country. Madeira consists of tremendous broken precipices, covered with the most luxuriant evergreens. The woods are shady and moist, abounding with the most curious species of *Filices* and *Musci.* The Azores are remarkable for abundance of hot springs and other vestiges of volcanos ; but their natural productions come nearer to those of Europe. The Canaries, remarkable for the enormous height of the land, abound with rare plants, which nearly approach the productions of Africa. I have made no collection of animals in these islands. The birds are but few, and those well known in Europe. Of insects and shells I have seen none worthy of observation. I now wait for a passage to the West Indies; but the present war, in which my country is involved, will, I fear, render my voyage less extensive than I at first expected. I shall always esteem your correspondence the greatest honour, and cheerfully contribute my mite for the advancement of natural knowledge.

I long to hear the fate of Mr. Thunberg, with whom I had the honour to be acquainted at the Cape, we having together made two long journeys into the interior parts of that country, in which time I profited much by his extensive knowledge in the history of Nature. Inclosed is a list of the specimens, with some local observations. I have the honour to be, with all possible respect, your most obedient, humble servant, FRANCIS MASSON.

DR. PAUL DIETERICK GISEKE TO LINNÆUS. [Latin.]

Hamburgh, Feb. 16, 1776.

MY EVER-HONOURED PRECEPTOR,

Your very kind letter, received in December last, afforded me and my friend Gruno the most lively satisfaction. I should have answered it sooner, had I not been so severely afflicted this winter with chilblains, and with consequent ulcerations of my hands, as to be unable to write. What chiefly delights me is your truly paternal regard towards me, which made my time pass so happily while with you, and which I earnestly wish and hope you may never withdraw. I rejoice no less in that active vigour of body and mind, at your advanced age, which has enabled you, in so short a time, to complete another *Mantissa*, as well as a new edition of your *Systema*. In these we shall, doubtless, find all the communications of the King of Sweden, announced in your last letter but one. How happy should I be to cast my eager eyes over these treasures, and to hear them explained by yourself in person! I cannot help expressing a wish that you would subjoin that new arrangement of your Natural Orders, which you mentioned to me, where you divide the *Dicotyledones* into two families *; and moreover that you would insert your new genera among them. I am the more desirous of this, as the new edition of your *Genera Plantarum*, to which you spoke of annexing this new arrangement, seems to be put aside by the publication of this intended *Mantissa*. Forgive my importunity, and, if possible, grant my request!

As to my friend Gruno, I return you, in his name, my most grateful thanks for your kind consent to receive him under your tuition. He is desirous, if it please God, to pass two summers with you; but he wishes to know what time of the year your lectures commence? If your Italian pupils are likely to arrive time enough for you to begin about Easter, he must very soon take his departure from hence. But if it will not be before Midsummer, he will wait for a better season for travelling. Neither he nor I can tell whether Mr. Murray intends to

* Now published in the editor's Grammar of Botany, p. 209.

bring his Italian friends by Hamburgh, or any other way; nor whether he is anxious not to miss the beginning of your lectures.

Mr. Gruno would not object to a residence in the country, and, if we may take so great a liberty, we would beg the favour of one of your servants to provide a lodging for him. We are very unwilling to give you so much trouble, but we neither of us know any person to whom we can apply on this occasion. If he should determine upon a lodging in town, he might have that which Mr. Adolphus Murray occupied, in the Upsal hospital, which is the best that I am acquainted with, provided it be at liberty; nor would he much complain of the price, as it is usual to give six or seven Louis d'or a year for a room in the German Universities. If this cannot be had, he might take the lodging which some Danish students occupied in my time, situated, if I mistake not, in the suburbs. We do not however limit you, either with regard to expense or situation; and if we are too troublesome, be so good as to tell us of any body to whom we may write on this subject. I fear the Italian and other students, said to amount to more than a thousand, will already have engaged all the best lodgings.

Farewell, my dear Sir; please to present my best respects to your highly distinguished family, and grant me a continuance of your favour.

P. D. GISEKE.

LINNÆUS TO HIS EXCELLENCY GOVERNOR RYK TULBAGH*, AT THE CAPE OF GOOD HOPE. [Latin.]

SIR, No date.

I received last summer a second most noble and valuable present from you, of more than 200 specimens of plants, with several birds well preserved, a very numerous collection of bulbs, and 50 sorts of seeds.

In the month of August last I made my best acknowledgments to you for these favours, sending you the names of the plants, according to their numbers, and a drawing of the *Tulbaghia*, which plant will, I trust, remain a lasting monument to your honour, among botanists, as long as the vegetable tribe shall endure. My letter was sent to the house of the East India Company. at Amsterdam, as usual, and has doubtless reached you long ago.

I sent you, some time since, the *Museum Reginæ Ludovicæ Ulricæ*, but know not whether you have received it, and therefore send, by the present occasion, another copy of the same work, that you may see how faithfully I have acknowledged the cu-

* This gentleman was appointed Governor of the Cape of Good Hope in 1750, and held that important office more than 20 years. He was no less attentive to the duties of his station, and the interests of the colony, than to every branch of science. Astronomers, geographers, and naturalists alike experienced his assistance and protection. He enriched the cabinets of Holland with rare animals and fossils, and the gardens of all Europe with new or curious plants, one of which, allied to *Narcissus* and *Pancratium*, was named *Tulbaghia* by Linnæus.

rious insects with which you have, long ago, so generously and kindly furnished me.

Be assured, Sir, that I never received a more welcome communication than your last, nor one that gave me more satisfaction, as containing so great a number of rare Cape plants, which I had never before seen. I had thought that the greater part of the vegetable productions of that country, if not all of them, had come under my inspection already; but your collection convinces me that I was previously acquainted with but a small portion, there being in that collection as many quite unknown to botanists as there are of plants already known. I have sown all your seeds, but none of them have yet come up, though several of the bulbs are putting forth leaves. I have carefully dissected and examined the flowers of the dried specimens, and have referred them to their proper genera, with suitable specific distinctions. These, if it please God, may one day be made known to the publick, accompanied with those full descriptions of your own which you have communicated to me, and which cannot but redound to your scientific reputation.

I wish, my honoured friend, that you would favour me with a bulb or two of your *Tulbaghia*, that it may be propagated and dispersed throughout the gardens of Europe, so as to render your name familiar to all the lovers of rare and beautiful plants.

The captain who is so good as to take charge of this letter is requested by me to ask you for living plants, in pots, of two Cape plants which I particularly wish to procure, the *Liparia globosa* and *Xeranthemum canescens,* a favour which I hope you will not refuse me. I have shown the captain dried specimens of both, that he may be sure of them.

May you be fully aware of your own fortunate lot, not only in being permitted by the Supreme Disposer of events to inhabit, but also to enjoy the sovereign controul of, that paradise upon earth, the Cape of Good Hope, which the Beneficent Creator has enriched with his choicest wonders. Certainly, if I were at liberty to change my fortune for that of Alexander the Great, or of Solomon, Croesus, or Tulbagh, I should without hesitation prefer the latter.

May you long enjoy your enviable situation, and allow me to remain your sincere and most devoted servant, &c.

LINNÆUS TO PETER CUSSON, M. D. OF MONTPELLIER.

[Latin.]

Without date or conclusion.

The accelerated approaches of old age have long oppressed me, and I have been still more disabled by the severest winter I ever knew. A slight paralytic stroke attacked me on one side; and though, in due time, I recovered from this attack, it has left me feeble and timid; nor have I ventured, for the last half year, to enter my museum, or to mount up to it. Even if I live, I do not propose to visit it till a milder season has rendered the air more salubrious.

The whole of my collection of dried plants is glued upon paper with isinglass, so that the specimens cannot be taken off. Many of them were collected, as usual, in a flowering state, and are without the seeds. I wish I could send the whole class * for your inspection; but I must be sure of receiving it again, lest there should remain an important chasm in my herbarium.

I rejoice to hear news of Monsieur Commercon, respecting whom I had, till now, but little information. I wish I had the advantage of his acquaintance. So acute an observer must doubtless have made many interesting and valuable acquisitions. How is it possible to conceive that his collection can amount to 2500 species? I wish, as a certain blind man said, I could see it. I lament that my age is so advanced as to afford me no hope of ever seeing his discoveries published.

Ever since the second edition of my *Species Plantarum* came out, I have still been labouring, with the help of my friends. My pupil Solander, who has just been round the world with an Englishman of the name of Banks, thought he had collected a thousand new plants, chiefly from the new-discovered countries; but they will never amount to 500. He has promised me specimens of the whole, before he goes out on another voyage. However this may

* Meaning the Umbelliferous tribe.

be, I have no doubt that Commerçon's botanical collection is still more extensive; and if he has but half the number abovementioned, he will exceed all the botanists who ever existed. But when may we look for the appearance of this great work? When will his numerous Fishes be published? Natural History is more defective in this department than in any other. Pray give me all the intelligence you can of this excellent man.

What is your new *Angelica semisterilis?* Have you another half-barren umbelliferous plant, besides that which comes from the Cape?

M. DE CONDORCET TO THE YOUNGER LINNÆUS.

A Paris, ce 15 Aout, 1778.

Recevez, Monsieur, l' assurance de la part que je prens avec toute l' Europe savante à la perte qu' elle a faite par la mort de votre illustre pere. Je tacherai de rendre à ses talens et à ses travaux la justice qu'ils meritent, dans le tribut que j'ai à rendre à sa mémoire. Nous connaissons ici ses ouvrages; mais nous n' avons connu sa personne que pendant un voyage fort court. S' il y a dans sa vie quelques anecdotes dignes d' entrer dans l' éloge d' un homme célébre, j' ose vous supplier de les remettre, écrites en Latin, à M. Vargentin, qui aura la bonté de me les faire passer, par M. le Comte de Creutz, Ambassadeur de Suede en France.

L'Académie m'a chargé de vous presenter ses remerciments et ses regrets.

Agréez, je vous prie Monsieur, l'assurance de mon sincere et respectueux devouement.

LE MARQUIS DE CONDORCET.

A Monsieur Monsieur Von-Linné.

LINNÆUS THE YOUNGER TO DR. DU ROI. [Latin.]

No date.

Being uncertain of the address of Mr. Ehrhart, I have thought it safest to communicate with you, Sir, on the subject of the *Supplementum Plantarum*, he having entrusted to you the superintendance of the printing of that work in his absence.

What principally displease me are pages 69—74, containing the following genera of Mosses — *Hedwigia*, *Pottia*, *Georgia*, *Grimmia*, *Webera*, *Catharinea*, *Weissia*, and *Andreœa* — with which I have no sort of concern. Every body will think me mad, if these should come forth under my authority; especially as I have, this very year, in an academical dissertation already published, reformed the genera of Mosses, according to principles of whose solidity I am convinced; and have also, in the same work, given names to several of them. I would therefore have the above pages, containing these genera, cancelled; or the whole sheet may be reprinted. An interval in the paging would be of no consequence. If Mr. Ehrhart is desirous that the genera in question should be published in this work, they may form an appendix; but in his name, not in mine. Nothing could be more unexpected on my part. No explanation can ever convince me of the propriety of allowing these genera to remain as they are, if the work is in any respect to be attributed to me *.

RIGHT HON. SIR JOSEPH BANKS, BART. K. B. P. R. S. TO THE YOUNGER LINNÆUS.

SIR, Soho Square, Dec. 5, 1778.

With pleasure I received your favours, and the first edition of your learned father's *Systema*, for which I return you my thanks. I always had the highest respect for that valuable man, and shall pay every duty to his memory which gratitude can dictate. I have invariably studied by the rules of his System, under your learned friend Dr. Solander; so that the plants in my intended publication will be arranged according to his strictest rules. Such as are of genera described by him will have his names. The new ones, which I think will almost outnumber them, will be named, either in honour of distinguished botanists, or, according to rules in the *Philosophia Botanica*, by names derived from the Greek.

* This letter is without date, and seems to end abruptly. The sheet alluded to was cancelled; but the editor was favoured, by Mr. Ehrhart, with an impression. The discarded genera, with their characters, are published in this author's *Beiträge*, vol. I. 174.

29.

Altho I have not the honour to be known to you, I have long been an admirer of your writings, & no less so of your private character together, they make you respected & esteemed all over Europe

Robert Ramsay

30.

L'académie m'a chargé de vous presenter ses remerciemens et ses regrets.

Agreez, je vous prie, Monsieur l'assurance de mon sincere et respectueux devouement le Marquis de Condorcet

31.

when you have not a duplicate a small branch or part broken from your specimen may serve without damaging it so much I shall be able & willing to make

Elsewhere

yours with all due attention
your affectionate
& Faithfull Servant
Jos: Banks P.R.S

R.M, sc

The material originally positioned here is too large for reproduction in this reissue. A PDF can be downloaded from the web address given on page iv of this book, by clicking on 'Resources Available'.

Uninterruptedly, however, as I have applied to the work of engraving for near five years, I have not yet advanced above half of my intended progress. About 550 plates are engraved, and I think, if circumstances as yet unexpected do not oblige me to cut it short, it will extend to double that number. Understand by this how impossible it will be for you to quote it in a work intended for publication in the course of this year.

The plants which you so kindly sent me by Mr. Troille I fear are lost. The ship has not been heard of, though more than a year is elapsed since she sailed.

In one thing it will be in your power to oblige me much, and I shall not want for gratitude; if you will kindly undertake to supply me with as good a collection of Mutis's plants as you can spare, without damaging your own collection *. A small bit, you know, is of great use to a botanist. When you have not a duplicate, a small branch, or part, broken from your specimen, may serve, without damaging it too much. I shall be able and willing to make returns, in things which you cannot easily obtain elsewhere.

I am, with all due attention, your affectionate and faithful servant, JOS. BANKS, P. R. S.

* These wishes the editor had the pleasure of fulfilling, after he acquired the Linnæan herbarium, in 1784.

DEAR SIR, Soho Square, Dec. 30, 1782.

I should have answered your favour long ago had I been certain how to have directed it. But as I despaired of hitting you in the course of your flight through Germany, I resolved to wait till the time when I had reason to expect that you had returned to Upsala.

Thank you for your information concerning *Calligonum*. It has rectified a mistake, which might have led me, at some time, into a real scrape. For *Pimento* I have many specimens, both cultivated and wild, every one of which has *folia opposita*, and they are quite like the figure in Hughes's History of Barbadoes, drawn by Ehret.

The figure in Browne's Jamaica (tab. 25. f. 2.) is not referred to from the letter-press. It has clearly been constructed from a bad specimen; but as that very specimen is, in all likelihood, to be met with in your herbarium *, you will do well to enquire whether, as I suspect, it is not taken from a different plant. Solander, who constituted a new genus under the name of *Myrcia*, very soon indeed after he came to England, meant that with opposite leaves, which flowers, you know, in our gardens, and is highly aromatic.

I have been always inclined to think that the blunder which originated in Browne, from a bad specimen, has been followed since by every writer.

* It does not exist there. The leaves in the figure are alternate, which is the blunder adverted to in the subsequent part of this letter.

I should like therefore to hear Mr. Ortega upon the subject, though, I confess, I esteem him but as a nettle * in the garden of botanic science. Be so good as to send me the title of his essay.

Van Royen sent me a grass long ago as *Bobartia,* which proved quite unlike it, and so trifling that I really forget what it was. Pray send me some information about the Danish *Bobartia.* The generation of Eels may be seen at large in Vallisneri. He discovered it in 1707, and published it in 1710, with a figure. I have not the book, but it is copied in his *Opera omnia,* and in *Valentini Amphitheatrum Zootomicum,* where you may see it.

The compliments of Christmas attend you from all our family. We shall think of you when we cut the twelfth cake on Monday. Indeed we should sincerely rejoice if future fortune would allow you once more to partake of it. Yours faithfully,

Jos. Banks.

FROM THE SAME TO THE EDITOR.

My dear Sir James, Soho Square, Dec. 25, 1817.

* * * * * *

My chief reason for troubling you with this is to tell you that I have paid obedience to your mandate, by reading your article on Botany, in the Scotch *Encyclopædia,* which, conceiving it to be an elementary performance, I had neglected till now to peruse.

* Alluding to the meaning of his name in Spanish.

I was highly gratified by the distinguished situation in which you have placed me, more so, I fear, than I ought to have been. We are all too fond of hearing ourselves well spoken of, by persons whom we hold in high regard. But, my dear Sir James, do not you think it probable, that the reader, who takes the book in hand for the purpose of seeking botanical knowledge, will skip all that is said of me, as not at all tending to enlarge his ideas on the subject?

I admire your defence of Linnæus's natural classes. It is ingenious and entertaining, and it evinces a deep skill in the mysteries of classification; which must, I fear, continue to wear a mysterious shape, till a larger portion of the vegetables of the whole earth shall have been discovered and described.

I fear you will differ from me in opinion, when I fancy Jussieu's natural orders to be superior to those of Linnæus. I do not however mean to allege that he has even an equal degree of merit in having compiled them. He has taken all Linnæus had done as his own; and having thus possessed himself of an elegant and substantial fabrick, has done much towards increasing its beauty, but far less towards any improvement in its stability.

How immense has been the improvement of botany since I attached myself to the study, and what immense facilities are now offered to students, that had not an existence till lately! Your descriptions, and Sowerby's drawings, of British plants, would have saved me years of labour, had they then existed. I well remember the publication of Hudson *, which was the first effort at well-directed science, and the eagerness with which I adopted its use.

Believe me, &c.

The last letter, coming from a man of such distinguished talents and experience, is so valuable a commentary on several leading subjects of the present volume, that the editor could not withhold it from the publick. He must rely on the favour of his readers, not to attribute to a foolish vanity this exposure of what gives an important sanction to his own sentiments, while it displays at once the knowledge, the indulgence, and the unassuming candour, of the writer. The hand that traced these lines is no longer held out to welcome and encourage every lover of science; and the homage of the motley crowd, of which science formed but the livery, has passed away. The lasting monument of botanical fame, of whose judicious and classical plan so interesting a memorial is left us, in the first of Sir Joseph Banks's letters to the younger Linnæus, has been sacrificed to the duties incumbent, for almost half a century, on the active and truly efficient President of the Royal Society. Its loss would ill have been supplied by ever so stately a mausoleum of marble; and even this mausoleum has been suffered to crumble, in embryo, into dust! The

* In 1762.

names of Banks and of Newton are, indeed, alike independent of an abortive or a mutilated monument; and inscriptions on brass or on marble now resign their importance and their authority to the more faithful records of history and science, perpetuated for ever, if they deserve it, by the phœnix-like immortality of the press.

END OF THE CORRESPONDENCE.

INDEX.

A.

INDEX.

INDEX.

B.

C.

D.

E.

F.

G.

H.

INDEX.

I and J.

K.

L.

M.

N.

O.

P.

Q.

INDEX.

R.

S.

T.

V.

U.

W.

Y.

Z.

ERRATA.

Vol. I. p. 15, l. 15, read *Bæck.*
113, note, read *Tabernæmontana.*
121, l. 7 (from the bottom); read *Warneria.*
174, l. 18, read *Iguana.*
218, l. 9, read *Tænia.*
247, l. 5 (from the bottom), read *seed-lobe.*
373, l. 19, read *Lithophyta.*

Vol. II. p. 122, note, read *Gypsophila.*
123, note, read *Isopyrum.*
and *fumarioides.*
163, l. 3 from the bottom, read *or.*
257, l. 1, for *nails,* read *tails.*

London: Printed by John Nichols and Son, 25, Parliament Street.

Printed in the United States
By Bookmasters